Energy, Environment, and Sustainability

This books series publishes cutting edge monographs and professional books focused on all aspects of energy and environmental sustainability, especially as it relates to energy concerns. The Series is published in partnership with the International Society for Energy, Environment, and Sustainability. The books in these series are editor or authored by top researchers and professional across the globe. The series aims at publishing state-of-the-art research and development in areas including, but not limited to:

- Renewable Energy
- Alternative Fuels
- Engines and Locomotives
- Combustion and Propulsion
- Fossil Fuels
- Carbon Capture
- Control and Automation for Energy
- Environmental Pollution
- Waste Management
- Transportation Sustainability

More information about this series at http://www.springer.com/series/15901

Shantanu Bhattacharya · Avinash Kumar Agarwal
Om Prakash · Shailendra Singh
Editors

Sensors for Automotive and Aerospace Applications

Springer

Editors
Shantanu Bhattacharya
Department of Mechanical Engineering
Indian Institute of Technology Kanpur
Kanpur, Uttar Pradesh, India

Om Prakash
Design Program
Indian Institute of Technology Kanpur
Kanpur, Uttar Pradesh, India

Avinash Kumar Agarwal
Department of Mechanical Engineering
Indian Institute of Technology Kanpur
Kanpur, Uttar Pradesh, India

Shailendra Singh
Maruti Suzuki India Limited
Gurugram, Haryana, India

ISSN 2522-8366 ISSN 2522-8374 (electronic)
Energy, Environment, and Sustainability
ISBN 978-981-13-3289-0 ISBN 978-981-13-3290-6 (eBook)
https://doi.org/10.1007/978-981-13-3290-6

Library of Congress Control Number: 2018961717

This Springer imprint is published by the registered company Springer Nature Singapore Pte Ltd.
The registered company address is: 152 Beach Road, #21-01/04 Gateway East, Singapore 189721, Singapore

Preface

Energy demand has been rising remarkably due to increasing population and urbanization. Global economy and society are significantly dependent on the energy availability because it touches every facet of human life and its activities. Transportation and power generation are two major examples. Without the transportation by millions of personalized and mass transport vehicles and availability of 24 × 7 power, human civilization would not have reached contemporary living standards.

The International Society for Energy, Environment and Sustainability (ISEES) was founded at Indian Institute of Technology Kanpur (IIT Kanpur), India, in January 2014 with the aim of spreading knowledge/awareness and catalysing research activities in the fields of energy, environment, sustainability and combustion. The society's goal is to contribute to the development of clean, affordable and secure energy resources and a sustainable environment for the society and to spread knowledge in the above-mentioned areas and create awareness about the environmental challenges, which the world is facing today. The unique way adopted by the society was to break the conventional silos of specialications (engineering, science, environment, agriculture, biotechnology, materials, fuels, etc.) to tackle the problems related to energy, environment and sustainability in a holistic manner. This is quite evident by the participation of experts from all fields to resolve these issues. ISEES is involved in various activities such as conducting workshops, seminars and conferences in the domains of its interest. The society also recognizes the outstanding works done by the young scientists and engineers for their contributions in these fields by conferring them awards under various categories.

The second international conference on "Sustainable Energy and Environmental Challenges" (SEEC-2018) was organized under the auspices of ISEES from 31 December 2017 to 3 January 2018 at J N Tata Auditorium, Indian Institute of Science Bangalore. This conference provided a platform for discussions between eminent scientists and engineers from various countries including India, USA, South Korea, Norway, Finland, Malaysia, Austria, Saudi Arabia and Australia. In this conference, eminent speakers from all over the world presented their views related to different aspects of energy, combustion, emissions and alternative energy resources for sustainable development and a cleaner environment. The conference presented five high-voltage plenary talks from globally renowned experts on topical themes, namely “Is It Really the End of Combustion Engines and Petroleum?” by Prof. Gautam Kalghatgi, Saudi Aramco; “Energy Sustainability in India: Challenges and Opportunities” by Prof. Baldev Raj, NIAS Bangalore; “Methanol Economy: An Option for Sustainable Energy and Environmental Challenges” by Dr. Vijay Kumar Saraswat, Hon. Member (S&T), NITI Aayog, Government of India; “Supercritical Carbon Dioxide Brayton Cycle for Power Generation” by Prof. Pradip Dutta, IISc Bangalore; and “Role of Nuclear Fusion for Environmental Sustainability of Energy in Future” by Prof. J. S. Rao, Altair Engineering.

The conference included 27 technical sessions on topics related to energy and environmental sustainability including 5 plenary talks, 40 keynote talks and 18 invited talks from prominent scientists, in addition to 142 contributed talks, and 74 poster presentations by students and researchers. The technical sessions in the conference included Advances in IC Engines: SI Engines, Solar Energy: Storage, Fundamentals of Combustion, Environmental Protection and Sustainability, Environmental Biotechnology, Coal and Biomass Combustion/Gasification, Air Pollution and Control, Biomass to Fuels/Chemicals: Clean Fuels, Advances in IC Engines: CI Engines, Solar Energy: Performance, Biomass to Fuels/Chemicals: Production, Advances in IC Engines: Fuels, Energy Sustainability, Environmental Biotechnology, Atomization and Sprays, Combustion/Gas Turbines/Fluid Flow/Sprays, Biomass to Fuels/Chemicals, Advances in IC Engines: New Concepts, Energy Sustainability, Waste to Wealth, Conventional and Alternate Fuels, Solar Energy, Wastewater Remediation and Air Pollution. One of the highlights of the conference was the Rapid Fire Poster Sessions in (i) Energy Engineering, (ii) Environment and Sustainability and (iii) Biotechnology, where more than 75 students participated with great enthusiasm and won many prizes in a fiercely competitive environment. More than 200 participants and speakers attended this four-day conference, which also hosted Dr. Vijay Kumar Saraswat, Hon. Member (S&T), NITI Aayog, Government of India, as the chief guest for the book release ceremony, where 16 ISEES books published by Springer, under a special dedicated series “Energy, Environment, and Sustainability” were released. This is the first time that such significant and high-quality outcome has been achieved by any society in India. The conference concluded with a panel discussion on “Challenges, Opportunities & Directions for Future Transportation Systems”, where the panellists were Prof. Gautam Kalghatgi, Saudi Aramco; Dr. Ravi Prashanth, Caterpillar Inc.; Dr. Shankar Venugopal, Mahindra and Mahindra; Dr. Bharat Bhargava, DG, ONGC Energy Center; and Dr. Umamaheshwar, GE Transportation, Bangalore. The panel discussion was moderated by Prof. Ashok Pandey, Chairman, ISEES. This conference laid out the road map for technology development, opportunities and challenges in energy, environment and sustainability domains. All these topics are very relevant for the country and the world in the present context. We acknowledge the support received from various funding agencies and organizations for the successful conduct of the second ISEES conference SEEC-2018, where these books germinated. We would therefore like to acknowledge SERB, Government of India (special thanks to Dr. Rajeev Sharma, Secretary); ONGC Energy Center (special thanks to Dr. Bharat Bhargava); TAFE (special thanks to Sh. Anadrao Patil); Caterpillar (special thanks to Dr. Ravi Prashanth); Progress Rail, TSI, India (special thanks to Dr. Deepak Sharma); Tesscorn, India (special thanks to Sh. Satyanarayana); GAIL, Volvo; and our publishing partner Springer (special thanks to Swati Meherishi).

The editors would like to express their sincere gratitude to a large number of authors from all over the world for submitting their high-quality work in a timely manner and revising it appropriately at short notice. We would like to express our special thanks to Dr. Nripen Chandra, Dr. Rishi Kant, Dr. Vinay Kumar Patel, Dr. Saibal Banerjee Dr. Ankur Gupta, Mr. Aviru Kumar Basu, Ms. Geeta Bhatt, Ms. Anubhuti Saha, Mr. Pankaj Singh Chauhan, Mr. Kapil Manoharan, Mr. Mohit Pandey and Ms. Poonam Sundriyal, who reviewed various chapters of the book and provided very valuable suggestions to the authors to improve their manuscript.

The book covers different aspects of Sensors for Aerospace and Automotive Applications. The content is divided into two streams: one is related to sensors for aerospace, and the other is related to sensors for automotive applications. The sensors for aerospace section deals with different aspects of functional films for gas sensing, related corrosion monitoring and control aspects, energy harvesting techniques for powering wireless sensor networks, sensors for health monitoring, sensors for flying and leakage monitoring in space inflatables. The other domain deals with electric vehicles and related sensors, different fabrication processes for automotive sensors, wireless sensing systems, sensors for automotive assembly shop, automotive joining, welding and paint process and MEMS sensors for automotive.

Kanpur, India — Shantanu Bhattacharya
Kanpur, India — Avinash Kumar Agarwal
Kanpur, India — Om Prakash
Gurugram, India — Shailendra Singh

Contents

Editors and Contributors

About the Editors

Shantanu Bhattacharya, Ph.D. is Professor of mechanical engineering and Head of Design Program at Indian Institute of Technology Kanpur. Prior to this, he completed his M.S. in mechanical engineering from Texas Tech University, Lubbock, Texas, and Ph.D. in bio-engineering from the University of Missouri at Columbia, USA. He also completed a postdoctoral training at the Birck Nanotechnology Center at the Purdue University. His main research interests are design and development of micro- and nano-sensors and actuation platforms, nano-energetic materials, micro- and nano-fabrication technologies, water remediation using visible light photocatalysis and product design and development. He has many awards and accolades to his credit which includes the Institution of Engineers Young Engineer Award, the Institute for Smart Structures and Systems' Young Scientist Award, the Best Mechanical engineering design award (National Design Research Forum, IEI), fellowship from the high energetic materials institute at Australia, fellowship of the Institution of Engineers of India. He has guided many Ph.D. and master's students and has many international journal publications, patents, books and conference proceedings.

Avinash Kumar Agarwal is Professor in the Department of Mechanical Engineering at Indian Institute of Technology Kanpur. His areas of interest are IC engines, combustion, alternative fuels, conventional fuels, optical diagnostics, laser ignition, HCCI, emission and particulate control, and large bore engines. He has published 24 books and more than 230 international journal and conference papers. He is Fellow of SAE (2012), ASME (2013), ISEES (2015) and INAE (2015). He received several awards such as the prestigious Shanti Swarup Bhatnagar Award-2016 in engineering sciences; Rajib Goyal Prize-2015; NASI-Reliance Industries Platinum Jubilee Award-2012; INAE Silver Jubilee Young Engineer Award-2012; SAE International's Ralph R. Teetor Educational Award-2008; INSA Young Scientist Award-2007; UICT Young Scientist Award-2007; INAE Young Engineer Award-2005.

Om Prakash is Director of Materials & Manufacturing Research in Boeing's Research and Technology Division in Bangalore, India. He has done his M.E. in mechanical engineering and Ph.D. in engineering materials from IISc Bangalore and the University of Cambridge, respectively. He has since worked in McMaster University (Canada), IIT Kanpur (India) and GE Global Research (India). He has published more than 20 technical reports, 30 journal articles and filed 18 patents.

Shailendra Singh is Vice President for warehouse-parts and accessories-marketing and sales, and materials management in Maruti Suzuki India Limited. After completing his B.Tech. from HBTI Kanpur, India, he has done his MBA from ICS Hitotsubashi (JAPAN) and is a EFPM candidate at IIM Lucknow.

Contributors

Adit School of Mechanical Sciences, Indian Institute of Technology Bhubaneswar, Bhubaneswar, Odisha, India

Avinash Kumar Agarwal Department of Mechanical Engineering, Indian Institute of Technology Kanpur, Kanpur, UP, India

Aviru Kumar Basu Design Programme, Indian Institute of Technology Kanpur, Kanpur, UP, India; Microsystems Fabrication Laboratory, Department of Mechanical Engineering, Indian Institute of Technology Kanpur, Kanpur, UP, India

Geeta Bhatt Microsystems Fabrication Laboratory, Department of Mechanical Engineering, Indian Institute of Technology Kanpur, Kanpur, UP, India; Design Programme, Indian Institute of Technology Kanpur, Kanpur, India

Shantanu Bhattacharya Microsystems Fabrication Laboratory, Department of Mechanical Engineering, Indian Institute of Technology Kanpur, Kanpur, UP, India; Design Programme, Indian Institute of Technology Kanpur, Kanpur, UP, India

Pankaj Singh Chauhan Microsystems Fabrication Laboratory, Department of Mechanical Engineering, Indian Institute of Technology Kanpur, Kanpur, UP, India

Ankur Gupta School of Mechanical Sciences, Indian Institute of Technology Bhubaneswar, Bhubaneswar, Odisha, India

Sudarsana Jena School of Mechanical Sciences, Indian Institute of Technology Bhubaneswar, Bhubaneswar, Odisha, India

Rishi Kant Microsystems Fabrication Laboratory, Department of Mechanical Engineering, Indian Institute of Technology Kanpur, Kanpur, UP, India

Kapil Manoharan Microsystems Fabrication Laboratory, Department of Mechanical Engineering, Indian Institute of Technology Kanpur, Kanpur, UP, India

Pramod Pal School of Mechanical Sciences, Indian Institute of Technology Bhubaneswar, Bhubaneswar, Odisha, India

Mohit Pandey Microsystems Fabrication Laboratory, Department of Mechanical Engineering, Indian Institute of Technology Kanpur, Kanpur, UP, India; Design Programme, Indian Institute of Technology Kanpur, Kanpur, UP, India

Pramod Kumar Parida School of Mechanical Sciences, Indian Institute of Technology Bhubaneswar, Bhubaneswar, Odisha, India

Rohit Kumar Pippara School of Mechanical Sciences, Indian Institute of Technology Bhubaneswar, Bhubaneswar, Odisha, India

Om Prakash Materials and Manufacturing Research, Boeing Research & Technology, India Center, Bengaluru, Karnataka, India

Shailendra Singh Design Programme, Indian Institute of Technology Kanpur, Kanpur, UP, India; Maruti Suzuki India Limited, Gurugram, Haryana, India

Poonam Sundriyal Department of Mechanical Engineering, Indian Institute of Technology Kanpur, Kanpur, UP, India; Microsystems Fabrication Laboratory, Indian Institute of Technology Kanpur, Kanpur, UP, India

Shreyansh Tatiya Microsystems Fabrication Laboratory, Department of Mechanical Engineering, Indian Institute of Technology Kanpur, Kanpur, UP, India; Design Programme, Indian Institute of Technology Kanpur, Kanpur, UP, India

Chapter 1
Introduction to Sensors for Aerospace and Automotive Applications

Shantanu Bhattacharya, Avinash Kumar Agarwal, Om Prakash, Shailendra Singh, Mohit Pandey and Rishi Kant

Abstract The sensors in aerospace and automotive industry play a vital role in navigation, detection, monitoring, and control of various systems related to ground and air vehicles. The sensors may be of analog or digital type which can be utilized at various places as per engineering requirements of the systems. The advancement in fabrication technologies related to such sensors has offered a wide variety of sensors ranging from nano to macroscopic length scales. Nowadays, sensors have not only become an important component to automate the production process and to perform record-keeping at different stages of manufacturing of ground and air vehicles but are also used to ensure the smooth functioning of systems by continuous monitoring of the prognostic behavior of the vehicles while in use. The sensors utilize different types of signal transduction mechanisms and signal conversion schemes for their working. A sensing device comprises of an analyte (external to device) detection and signal transduction to machine read the signals based on which closed-loop control can be carried out intelligently without human intervention. This book discusses about various types of sensors used in different sectors of aerospace and automotive industries. The book includes fabrication

S. Bhattacharya (✉) · M. Pandey · R. Kant
Microsystems Fabrication Laboratory, Department of Mechanical Engineering,
Indian Institute of Technology Kanpur, Kanpur 208016, UP, India
e-mail: bhattacs@iitk.ac.in

R. Kant
e-mail: dsrishikant@gmail.com

S. Bhattacharya · S. Singh · M. Pandey
Design Programme, Indian Institute of Technology Kanpur, Kanpur 208016, UP, India

S. Singh
Maruti Suzuki India Limited, Gurugram 122015, Haryana, India

A. K. Agarwal
Department of Mechanical Engineering, Indian Institute of Technology Kanpur,
Kanpur 208016, UP, India

O. Prakash
Materials and Manufacturing Research, Boeing Research & Technology, India Center,
Bengaluru 560001, Karnataka, India

S. Bhattacharya et al. (eds.), *Sensors for Automotive and Aerospace Applications*, Energy, Environment, and Sustainability,
https://doi.org/10.1007/978-981-13-3290-6_1

aspects of sensors based on MEMS technologies, talks about aircraft corrosion monitoring, wireless sensing networks, embedded sensors, and discusses about various sensors used in flying of air vehicles. The book also covers sensor especially used in different section of automotive industry like in assembly shop, during joining and welding, paint processes, etc.

Keywords Sensors · Detection · Monitoring · MEMS · Aircraft Automotive

We are living in the modern world where the cutting edge technology related to manufacturing processes and techniques help in producing robust and smart products that contribute to human life modernization. Sensors are one of the products which play a crucial role in the manufacturing of automated machinery for various fields like chemical, mechanical, biological, automobile, aerospace products, etc. Ways and means of faster and comfortable and safe transportation have always challenged the human civilization by putting demands to maximize automation and control in dynamic systems which can be provided by real-time monitoring of functionality on a subsystem level of such machinery. The only approach to do this rapidly is through the use of advanced sensors and robust feedback mechanisms for self-correction. The automation of a vehicle would minimize the risk associated with accidental and rollover events which generally occur during mobility. Apart from these, there is a huge requirement of sensors in manufacturing processes to monitor and control defects in the product at every stage of the production process and especially when it comes to the production of ground and air vehicles they become all the more important. In aerospace applications, the sensors are almost used in every stage from taking off—to the landing stage of air vehicles. There are various sensors utilized in the aircrafts to monitor different components such as nose wheel, steering, leading edge flap, trailing edge flap, lower paddle, spoilers, ailerons, tail elevators all of which help in flying. The fabrication of these sensors requires high-end fabrication protocols at different length scales.

There is an enormous demand for robust, highly sensitive, selective and precise sensors. To fabricate such sensors there are challenges involved at different stages like in (1) the fabrication protocol (2) the material used (3) the cost of manufacturing (4) the selectivity and sensitivity (5) the robustness, etc. Finding the right parameters during specific application is very important for the selection of manufacturing tools for a sensor. Many sensors having the same functioning may have different base materials depending on the usage environment. Also, depending on the scale many sensors may have different packaging architecture. Many industries rely on the functioning of the sensors and have a proper set of protocols which they follow for the selection of different types of sensors like medical diagnostic industries emphasize on the micro-/nano-level detection with high specificity and selectivity while production industries emphasize on the high working life, robustness and reliability of the sensor without service challenges or fewer replacements.

In this monograph, sensors for automotive and aerospace industries in various sectors are covered as the products of these industries have very different working conditions due to which for the same application, different parameters may be tuned on same or multiple sensors to get the right results. Sensors find their application in various sectors in these industries and thus, advance technological development is always focused on decreasing the overall cost and improving the working lifetime while maintaining the quality of the sensors.

Gas sensing is essential to quantify leakages of different gases in trace levels while the automotives or air vehicles are in operation. Advance gas sensing elements with the modern approach for the fabrication replaces the old gas sensing devices. Functionalized molecules are used to detect the analyte at a very small scale and are applied in aerospace and automotive industries for detection of any leakage of gases, monitoring the exhaust gases, etc. Advance thin film technology creates a very economical and robust solution of the functional material in the gas sensing applications. The machines and equipment used in these industries are very expensive, and thus, health monitoring becomes one of the important aspects of the industry. Many sensors are used to monitor the health of the vehicles and manufacturing machines and equipment. The world has witnessed many aerospace failures especially due to the material loss and fatigue cracking which is primarily due to corrosion. The types and causes of corrosion are studied to fabricate the different types of sensors and different schemes of corrosion control in air vehicles. In aerospace industries, most of these sensors are mounted on board which are powered by the conventional battery system and which provides the data to the user wirelessly. These sensors are mostly used in health monitoring as well as in flight control and navigation. Due to the heavy weight, large size, and short cycle life, conventional batteries and power systems are replaced by the advance energy harvesting systems and devices. These systems harvest energy from the local environment such as high pressure, vibrations, wind kinetic energy, thermal gradients, and solar source by using various principles like piezoelectric effect, electromagnetic effect, thermoelectric effects, etc. The main concern for the industries are the manufacturing of these sensors at a relatively low cost and to overcome the challenges of their integration with other sensor networks or a central system wirelessly.

Similarly, many other sensors are embedded in the aircraft for maintaining the complex operational aspects of an air vehicle during flight on a 24×7 basis. Various smart sensors are used to detect any failure or need of maintenance so that necessary preventive actions can be taken immediately before it fails. The breakdown maintenance is very costly for the industries in terms of time, money, and workforce. In aerospace vehicles, passenger safety is of paramount importance, and thus, various parameters of flight are monitored and controlled by using various sensors for flying. As so many lives are onboard therefore sensor systems have to be ultra-precise and must inform in advance so that disasters do not occur. The usage of traditional flight sensors is discussed along with the changes occurring due to the use of MEMS-based sensors. Different MEMS fabrication techniques fulfill the application-based requirements of different type of air vehicle systems during flight.

These sensors are coupled with the central control system to control and simulate the flight of the air vehicle and are very useful for unmanned aerial vehicles (UAV).

In the space vehicles through which the modular inflatable space structures are sent into space for space exploration with very less folded volume, helium and nitrogen are generally used to inflate the space structures to deploy them. These polymer-based structures are very light weighted and having very low volume are inflated through internal pressurization to attain the structural integrity. In such a harsh environment, these structures are prone to lose their structural integrity due to gas leakage and thus, monitoring the compressed gases and the detection of any leakage becomes very important. The main concern in these type of applications is the high sensitivity and selectivity for the gases at the micro- and nanoscale due to which semiconductor metal oxides like vanadium pentaoxide (V_2O_5) is used which can detect the very small concentration of He gas through resistive changes. The miniaturization of sensors is possible due to advancement in MEMS technology and is used for various control, aviation, health, and monitoring purposes.

In the next part of this monograph, the need and application of sensors in the automotive industries are discussed. One of the vast applications of the sensors is in the electric vehicles. Considering the environmental issues associated with the use of fossil fuels in petrol/diesel vehicle, the electric vehicle came up as a good alternative. The advance miniaturized components having integrated sensors and the use of MEMS technology reduces the load on the environment and also makes a significant impact on the power system. Different types of electric vehicles, their design, and applications in daily life change the sensing devices and their fabrication strategy and the material use accordingly. The MEMS technology is advancing day by day with improvement in micromachining methods and silicon-microelectronics offers very adaptable and flexible solutions due to its low power consumption and compactness.

With the improvement and invention of various materials and fabrication techniques, MEMS have covered a wide range of sensing applications in embedded and wireless sensing. Various wireless sensors are applied in the automobile sector and are remotely operated with RFID technique, Radar systems, optical systems, ZigBee wireless modules and many more. The data from the remotely operated sensor is sent to the receiver with acoustic waves, optical, and electromagnetic waves, etc., and is analyzed at the receiver end. The swift progress in microelectronics has helped us to integrate the wireless module with the sensing elements, to get the real-time monitoring of the analyte. The main challenge with these sensors is their integration with the embedded nanoscale detection element, and the high cost of usage and thus low-cost solid-state sensors are focused in research to solve the existing problems.

Moving from automobile to its manufacturing plant, many applications of macro-, micro- and nano-length scale sensors are very important in the manufacturing processes. With the expansion of the market, there has always been a possibility to explore more and to advance in the manufacturing technology while keeping the motive of risk-free production. While enhancing the quality and performance of the manufacturing processes, the need to reduce the size and increase the sensitivity of the sensor always remains a challenge. The integration of MEMS has helped in increasing the quality and to refine the performance of the vehicle and its manufacturing unit. The manufacturing unit consists of many subsections through which the vehicle is actually produced.

Starting with the joining process which is the key process for any automobile industries, welding is used to connect various parts to form a basic structure. Different type of welding and joining process is used to form and connect different types of parts. From monitoring the current and voltage values for proper welding to ensuring the exact position of the weld, automation has drastically increased the accuracy and thus the quality of the vehicle. Various sensors are applied in every subunit to sense the analyte and thus provide the data to the central system to further proceed the automation process. Safety sensors are implemented to minimize the human risk in the production unit due to use of the high current and voltage in the welding processes. They are generally integrated into the control unit to stop the process immediately when some safety breach is detected, to ensure that there will not be any material or human loss.

Similarly, for ensuring that the vehicle paint job which is considered as the heart of the manufacturing process, is executed properly, many different sensors are implemented in the various sub-processes. A huge amount of money is invested in this job to ensure that the vehicle comes out in its best look. This is a rigorous and time-consuming process which includes pre-treatment, electrodeposition, sealing, priming, coating, and ending up with an inspection of the whole vehicle's paint job. Following the paint job, all the subassemblies are connected to the vehicle in the assembly shop, to get the vehicle on the road after the inspection and testing.

The whole automobile manufacturing unit consists of various sensors and robotic arms interconnected to each other to work flawlessly. Some sensors are present to ensure the working of the robotic arm while others are present to prevent any miss-happening due to malfunction of any machine. Thus, advance embedded and wireless sensors are not just good for the automobile and aerospace industries but are an essential part of the industries to ensure that good quality check for the manufacturing products and human working conditions.

In this monograph, we have presented the fundamental as well as applied innovative strategies on several emerging fields of advance sensing technology involving micro-/nano-level mechanical, micro-electro mechanical systems (MEMS), optical, electrochemical, solid-state semiconductors, etc. We have covered the following topics:-

- Functional thin films for the gas sensing applications
- Corrosion monitoring and control in aircraft
- Wireless sensing application and energy harvesting techniques for them in aircraft applications
- Embedded sensor for health monitoring in aircraft
- Sensors for flying for various air vehicles
- Leakage monitoring for inflatable space structures
- Electric vehicles and sensor application in them

- Fabrication processes for sensors for automotive applications
- MEMS sensor for automobile applications
- Sensor for joining and welding process
- Sensor for paint processes in automobile manufacturing
- Sensors for automobile assembly shop

The topics are categorized into two sections: (a) Aerospace industries and (b) Automotive industries.

This monograph is intended for all the graduate-level researchers, product developers and industry professional working in sensors domain for the automobile and aerospace industries. It will not just provide you with the systematic information of the sensors used in the automobile and aerospace, but it will also provide you with the synergy between fabrication protocols, challenges, types, and integration according to the application of the sensors.

Chapter 2
Functional Films for Gas Sensing Applications: A Review

Ankur Gupta, Pramod Kumar Parida and Pramod Pal

Abstract There are enormous functional materials which have been explored for gas sensing applications due to the fact that when a particular gas interacts with their surfaces, some alteration in their properties (optical, electrical, etc.) is observed. These functional materials are also termed as sensing films, to be utilized as one of the essential components in a gas sensor. Other components include electrodes connected with sensing film and data acquisition system coupled with it. During the past several decades, various metals, semiconductors, ceramics, and hybrid materials have been extensively explored for gas sensing applications. Selection of functional film is dependent on the gas to be detected. A gas sensor should have some characteristics, viz., higher sensitivity, selectivity for target gas, least response and recovery time, higher reproducibility, and stability. Therefore, selection of appropriate sensing film is highly required for a well-efficient gas sensor development. This article reviews the various kinds of sensing films, their fundamental aspects along with the sensing mechanisms. Morphological changes in the materials and doping of other functional materials also affect the performance of a gas sensor. Hence, issues related to the efficient gas sensing are also covered in this article.

Keywords Sensing films · Functionalized films · Gas sensing
Nanomaterials

2.1 Introduction

Gas sensor is a device which is capable to detect trace amount of gas or any other specific type of gaseous emission in a particular area. A gas detector can employ a variety of means to realize the presence of gas in particular surrounding. It may be

A. Gupta (✉) · P. K. Parida · P. Pal
School of Mechanical Sciences, Indian Institute of Technology Bhubaneswar, Bhubaneswar 752050, Odisha, India
e-mail: ankurgupta@iitbbs.ac.in; ankurgupta.k10@gmail.com

S. Bhattacharya et al. (eds.), *Sensors for Automotive and Aerospace Applications*, Energy, Environment, and Sustainability,
https://doi.org/10.1007/978-981-13-3290-6_2

provided with sound alarm in the definite area of the gas leakage, or it may be quantified on the basis of display. There are many gases that are very harmful to organic life such as humans and animals. Gas sensor can be used to detect combustible, flammable, toxic, and nontoxic gases. These types of sensors are generally used in industry, oil rigs, to monitor manufacturing process, emerging technology, firefighting, etc. Some of the common gases detected by gas sensors are ammonia, VOCs, CO, CO_2, H_2S, aerosol, hydrogen, oxygen, NOx, etc. When gas sensors interact with a gas of different concentrations, the behavior of the sensing films changes due to electron or ion interaction between the desired gas and sensing films used in the sensor. The sensing films are most important part of the gas sensor. The films are prepared by chemical processing for a particular gas that it will detect this particular gas even if at low concentration. The different types of gas sensors (Semiconductor gas sensors, catalytic gas sensors, electrochemical gas sensors, optical gas sensors, acoustic gas sensors, and thermal conductivity gas sensors) are used for different gas detections. The performance characteristic of each gas sensor depends on sensitivity, selectivity, detection limit, and response and recovery time limit. For making gas sensor consumer friendly, sensor should be wireless, room temperature based, low power consumption, and small size. In short, performance of any gas sensor depends on 3S (sensitivity, selectivity, and stability) and 3R (response, reversibility, and reliability).

Catalytic sensors are simple to fabricate and low-cost technology. It detects flammable gases. The disadvantage of this type of sensors is that it requires oxygen or air to work and can be poisoned by lead, chlorine, and silicones. Thermal sensors are robust and simple in construction. This works in the absence of oxygen, and the measuring range of thermal sensors is wide. Thermal sensor reaction occurs due to heating of wire. Electrochemical sensors can be used for wide range of gases detection. It measures toxic gas at relatively low concentration. The advantages and disadvantages of optical, infrared, semiconductor, and surface acoustic wave sensors are briefly described in Awang (2014).

Sensing films may be prepared by chemical processing of different chemical compounds in particular environments. There are various types of sensing films such as oxide thin films, semiconductor thin films, polymer thin films, and composite thin films which are summarized in Table 2.1. Sensing film acts as an active medium between electrode and desired gas to be sensed. The physical principle of thin film sensitivity toward probing gas strongly depends on absorption, adsorption, chemisorptions, and coordination chemistry (Emmer 1992). The sensing film is an active layer which is generally deposited on interdigitated electrode by various methodologies. The methods of deposition of sensing thin films on electrode are drop casting, CVD, PVD, sputtering, ion beam, lithography technique, inkjet printing technology, spin coating, etc. Films are generally used for converting the chemical information to electronic signals such as frequency change, current change, voltage change, or resistance change. Some useful information such as functional characteristic of thin films, target gas to sense, sensitivity of thin films toward particular gases, and advantages of each film are tabulated here. Some sensing films are very reactive to exposed gas at high temperature and some are at room temperature.

2.2 Types of Sensing Films

There are broadly three kinds of sensing films, viz., metallic, semiconductor-based, and polymer-based films. Some of most explored films can be found as semiconductor-based sensing film, polymer-based, MOSFET structures, MIM impedance structure, resistive structure, vibrating piezoelectric sorption structure, etc. (Nafarizal 2016; Kelly and Arnell 2000; Adzhri et al. 2015; Ranade et al. 1993).

2.3 Fabrication Methodologies of Functional Films

Various metal oxides are used in the form of thin films, depending on types of gases and operating temperature. These metal-oxide-based thin films are easily prepared and fabricated by various methods such as reactive magnetron sputtering, ion etching, spray pyrolysis, vacuum evaporation, pulse laser deposition, sol–gel, screen printing technology, and so forth. When probing gas gets absorbed and desorbed on the surface of the thin films, the gas and thin film interaction takes place and can be detected by investigating the change in properties, viz., change in the electrical resistance, optical properties, and reaction energy released due to reaction in between them (exothermic or endothermic). The electrical and optical properties are dependent on fabrication technique. Hence, it requires proper analysis and investigation before choosing correct fabrication technique for metal oxide thin films. For example, fabrication of metal oxide thin films by magnetron sputtering technique provides easier control of repeatable thin films and good adhesion of metal oxide thin films on various substrates.

2.3.1 Fabrication Techniques

There are various fabrication techniques explored for the fabrication of metal-based thin films. Description of some of them is illustrated as follows:

(a) Magnetron Sputtering

In magnetron sputtering technique, plasma diagnostics require the understanding mechanism and formation behavior of thin films. The copper metal is converted into oxide mode thin films (Cu_2O or CuO) by reactive magnetron sputtering

Table 2.1 Description of the various sensing films and their sensing characteristics

S. no.	Name of sensing film	Target gas	Sensitivity values in %	Other sensing/functional characteristics	References
1	Polyaniline flexible films (PANI)	NH_3	S = 30% at 200 ppm and 130% at 1000 ppm in room temperature	It also senses HCL, CO_x, CH_x, ethanol, and hexane Advantage: It is room temperature-based gas sensor. It has good reproducibility and long-term stability. Mechanically robust and bent sharply without damage. These films have strong adherent properties to substrate and also transparent. In humidity atmosphere, its sensitivity toward ammonia decreases, while in dry air condition, it shows better sensitivity	Kumar et al. (2017)
2	PCDTBT flexible films	NO	S = 80.6% for 100 ppm	It can also sense CO, NH_3, H_2S, CL_2, ethanol, NO_2 gases Stable material in ambient environmental condition, chemically durability, constant conductive, change in humidity may increase or decrease in electrical conductivity of polymer, under bending sensing current decreases and sensing recovery time increases, long-term environmental stability. It shows high sensitivity in high relative humidity condition	Gusain et al. (2017)
3	TiO_2 thin films	H_2	S = 99% for 500 ppm in room temperature (zero humidity condition)	These films are suitable for room temperature-based gas sensor. Humidity negatively affects the sensor's performance. It has very good adhesion to polyamide foil. There is no change in sensitivity in bending the films Used in various fields like photocatalysis, antibacterial and protective coatings, anti-reflecting coatings, dielectric films	Krško et al. (2017)
4	Amorphous indium-gallium-zinc oxide (IGZO) thin films	Methanol–	45%	Also sensitive for acetone, toluene, chlorobenzene Advantage: Room temperature gas sensing, good response and reproducibility, long-term stability, high gas sensitivity at UV intensity	Jaisutti et al. (2016)
		Ethanol–	40%		
		Isopropanol	35%		
5	ITO thin films (In_2O_3 + 17%SnO_2)	Methanol	ITO film of 100 nm thickness shows maximum sensitivity S = 325% and S = 15% at 450 ppm	CO_2 (S = 250% at 1000 ppm, 575 K) CCl_4 (S = 120% at 1250 ppm, 448 K) NO_2 (S = 750% at 250 ppm, 350 °C) NO (S = 80% at 155 ppm, Room temperature) Chlorine gas (S = 300% at 10 ppm, 300 °C) Advantage: Room temperature-based gas sensor, ITO film conductivity increases with increase in methanol concentration, weight %age of SnO_2 up to 17% in In_2O_3 increases sensitivity, addition of catalytic layer on ITO films improves sensitivity, selectivity with stimulating layer, stability, short response time, reliability, reproducibility	Patel et al. (1994, 1995, 2003), Sberveglieri et al. (1988, 1990), Galkidas et al. (1992), Zhang et al. (2002)
	ITO/Cu-		30%		
	ITO/Au-		23%		
	ITO-		18%		
	ITO/Pt-		8%		
6	SnO_2/Ca, Pt			Methyl alcohol, ethyl alcohol, toluene, benzene, acetone	Lee et al. (2001)

(continued)

Table 2.1 (continued)

S. no.	Name of sensing film	Target gas	Sensitivity values in %	Other sensing/functional characteristics	References
	Additives used with this film: Pt La_2O_3 CuO Pd Sc_2O_3 TiO_2 WO_3 ZnO Pd + V_2O_5	Ethyl alcohol, Toluene, Methyl alcohol, Acetone, Benzene	S = 60% at 500 ppm for ethanol and toluene S = 70% at 500 ppm for methanol and acetone S = 50% at 500 ppm for benzene (400 °C)	Advantage: This gas sensor shows high and selective sensitivities to low gas concentration at high temperature of 400 °C Sensitivity increases by adding dopants (Pt, La_2O_3, CuO, Pd, Sc_2O_3, TiO_2, WO_3, ZnO, Pd + V_2O_5) High and broad sensitivity, reproductively	
7	ZnO thin films	Acetone	S = 100% for 500 ppm at 400 °C	It is also sensitive toward isopropanol, ethanol Advantage: high repeatability, stable, reversible, films are highly transparent	Al-Hardan et al. (2010)
8	Thin film carbon black–polymer composite	Toluene Methanol 2-propanol Hexane Ethyl acetate Acetone Chloroform Benzene Acetone		Advantage: It can be modified and also customized for a specific chemical. Another advantage is that conductivity and sensing properties can be controlled through change in composition of composite polymers	Lonergan et al. (1996)
9	Pd thin films	H_2	S = 1% for 600 ppm at 25 °C and 15% for 2000 ppm	The sensitivity of this film decreases with increase in temperature. The structural deformation occurs when thickness of films will increase. 6 nm Pd film shows reversible and highly sensitive. Sensitivity is proportional to hydrogen partial pressure. It can be deposited on both flexible and hard substrates	Öztürk and Kılınç (2016)
10	MoS_2 thin films	NO_2	S = 7% for 1.2 ppm in 2 nm film thickness	The sensitivity of this film decreases with increase in thickness of film. It has high reproducibility and flexibility. Functional characteristics of MoS_2 films with Pt nanoparticles further increases the sensitivity several times	He et al. (2012)
18	PANI/α-Fe_2O_3 hybrid composite thin flexible films	NH_3	S = 72% for 100 ppm	NO_2, LPG, methanol, ethanol It is a room temperature-based gas sensor	Bandgar et al. (2017)

(continued)

Table 2.1 (continued)

S. no.	Name of sensing film	Target gas	Sensitivity values in %	Other sensing/functional characteristics	References
19	PANI/Carbon nanotube nanocomposite thin film	NH_3	S = 25% for 50 ppm	Methanol, ethanol, acetone, dichloromethane, ethyl glycol, IPA, pyridine It shows fast response/recovery time, room temperature operable, reliable flexibility	Xue et al. (2017)
20	Carbon aerogel/Polymer thin film composite	Hexane	S = 10 ppm^{-1}	Good recovery in nitrogen atmosphere, best sensitive to polar and nonpolar VOCs	Thubsuang et al. (2015)
		Acetone	S = 1 ppm^{-1}		
21	ITO thin film and ITO/Au/ITO thin film	Methanol	S = 2.5% for ITO films and 4.5% for ITO/Au/ITO films in 750 ppm	Operable at room temperature	Jeong et al. (2010a)
22	GO-H_2SO_4	CO	S = 10% for NO gas and 5% for CO gas in GO-S films at 50 ppm S = 22% for CO and 26% for NO in GO-P films at 50 ppm	It has high response, rapid and low temperature operable. Sensitivity increases with respect to temperature up to 350 K. Sensitivity increases as concentration of target gases increases	Karaduman et al. (2015)
	GO-H_3PO_4	NO			
23	SnO_2/rGO	NO_2	3.31% for 5 ppm at 50 °C	Rapid response, good selectivity, and reproducibility. Low temperature operable. Composite SnO_2-rGO film exhibits high response (3.31%) than rGO film response (1.13%)	Chatterjee et al. (2015) Zhang et al. (2014a)
	SnO_2/rGO	NO_2	11% for 100 ppm, at RT	Room temperature based. SnO_2 is n-type and rGO p-type semiconductor. So the composite makes p–n junction and depletion zone. Sensor response depends on ratio of SnO_2/rGO. Increase in SnO_2 content in rGO enhances NO_2 response. The penetration of SnO_2 nanoparticle in rGO leads to open more active layer as vacancy defect, oxygen functional group, and sp^2-bonded carbon. Indium (In) and ruthenium (Ru) dopants in SnO_2 increase the sensitivity and lower the operating temperature. Slow recovery of sensor, poor adherence of SnO_2 with electrode. This films also good sensitivity toward ammonia. For ammonia, it has high response magnitude, fast response, and recovery, good reversibility and repeatability, low temperature operable. SnO_2 only exhibits low conductivity and no response to analyte. Go play a crucial role in gas sensing with SnO_2	Cui et al. (2013)

(continued)

Table 2.1 (continued)

S. no.	Name of sensing film	Target gas	Sensitivity values in %	Other sensing/functional characteristics	References
	SnO_2/CVD graphene	NO_2	11% for 0.1 ppm at 150 °C	Reversible response, fast response, and recovery (less than 50 s). Low temperature operable. Long-term stability in environmental condition	Van Quang et al. (2014)
	SnO_2/rGO/CNT	NO_2	2.53% for 5 ppm at RT	It has high response, fast recovery rate, good stability, high selectivity, room temperature operable. This sensor is also sensitivity to CO, ammonia, chlorine	Liu et al. (2015a)
	SnO_2/rGO nanosheet	NO_2	100% for 5 ppm at 200 °C	This sensor also senses oxygen, SO_2, hydrogen, CO, C_6H_6, C_7H_8. The sensor response depends on amount of rGO present and operating temperature	Lee et al. (2015)
	SnO_2/rGO	Ethanol	22% for 1.12 ppm at 300 °C	This gas sensor shows better sensing behavior, short responding time and recovery time to ethanol, and H_2S at low concentration. It is of low cost and has high sensitivity used for environmental monitoring area	Chang et al. (2013)
	SnO_2 + rGO Polyaniline	NH_3	1.6% for 20 ppm at RT		Ye et al. (2015)
	SnO_2 + CVD graphene sheet	H_2S	2.1% for 1 ppm at 260 °C		Zhang et al. (2011)
	SnO_2 + rGO-Au	NO_2	1.6% for 5 ppm at 50 °C	Low temperature operable. This sensor exhibits much better response and recovery time. By the addition of Au, it can sense 1 ppm level NO_2. This sensor is also selective toward H_2, Cl_2, NO gases	Zhang et al. (2014b)
	SnO_2 + exfoliated graphene	Ethanol	965% for 100 ppm at 350 °C	This film also senses H_2S, NO_2, H_2, and methane gases at 350 °C. It shows short response time 2 s for 1000 ppm ethanol at optimal temperature of 300 °C. In-doped in this film increase the specific surface area, high porous reactive site of composite so it shows superior sensitivity. Fast response and recovery. In-doped increase and graphene loading, and SnO_2 nanoparticle exhibit high response to ethanol	Inyawilert et al. (2015)
	SnO_2 + rGo	Ethanol	4% for 100 ppm at RT	It has low detection limit of 1 ppm at room temperature, fast response and recovery time, and good repeatability. It can also sense methanol, methane, CH_4, H_2, CO_2, SO_2	Zhang et al. (2015a)
	SnO_2 + GN	NH_3	15.9% for 50 ppm at RT		Esfandiar et al. (2014)
	SnO_2 + rGO	NO_2	300% for 8 ppm at 50 °C		He et al. (2013)
	SnO_2 + rGO	NH_3	2.87% for 1% in air at RT		Mao et al. (2012)
		NO_2	1.12% for 1 ppm at RT	Low detection limit and repeatable performance to target gas at room temperature. Thermal treatment is provided for sensor performance	Mao et al. (2012)

(continued)

Table 2.1 (continued)

S. no.	Name of sensing film	Target gas	Sensitivity values in %	Other sensing/functional characteristics	References
				and stability. Thermal treatment removes the residual solvent from rGO	
	SnO_2 + rGO	Acetone	2.19–9.72 for 10–2000 ppm at RT	Fast response–recovery and good repeatability. Room temperature sensing	Zhang et al. (2015b)
24	ZnO/Ag/rGo	C_2H_2	19.2% for 100 pp, at 150 °C	The reported film showed good repeatability, fast response, and very good sensitivity. The sensing temperature range is 25–300 °C. But at 200 °C and 5% wt, Ag in films shows high sensor response time of 57 s at 3 ppm, good linearity and selectivity to acetone. The fabricated film has less humidity effect	Uddin et al. (2015a)
	ZnO/Ag/Graphene	C_2H_2	21.2% for 100 ppm at 150 °C	The optimum condition for sensitivity of this film is 150 °C and 3% wt of Ag in ZnO/Graphene. The film is selective to acetylene, showing excellent repeatability, fast response/recovery of 25/80, detection range of 1–1000 ppm, stable performance at 31% RH. It has also sensitivity to gases like H_2, CO, CO_2, NO_2, O_2 at 100 ppm in 150 °C	Uddin et al. (2015b)
	ZnO/Functionalize Graphene	Ethanol	93.5% for 100 ppm at 340 °C	The functionalized graphene and ZnO nanohybrid film improves the charge transport, enlarge adsorptivity, and narrow bandgap. The gas sensor based on functional graphene has fast response, higher sensitivity, and good selectivity to ethanol	Song et al. (2015)
	ZnO/rGO	Ethanol	96.4% for 50 ppm at 260 °C	It is selective to ethanol and higher photocatalytic activity in UV irradiation. Due to photocatalytic decontamination properties, this can be used in various electronic and photonic systems. It also senses CH_3COCH_3, HCHO, H_2S, NO_2, NH_3, H_2, CO	Zou et al. (2013)
	PEDOT/rGO	Ammonia	25% at 1 ppm in RT	Due to excellent porous structure of PEDOT results to provide excellent synergetic effect between PEDOT and rGO during gas sensing. In contrast, the composite film provides high sensitive to ammonia and other reductive analyte gases (H_2S, CH_2Cl_2, methanol, CH_3COCH_3, SO_2, $CH_3(CH_2)_3OH$)	Yang et al. (2014a)
	ZnO/rGO	NO_2	25.6% for 5 ppm at room temperature	This sensor exhibits short response/recovery time, high sensitivity at room temperature. This also senses gases like Cl_2, CO, and NO at room temperature. The response of sensor increases at temperature range of 0–400 °C, and after that, response decreases exponentially	Liu et al. (2014a)
	ZnO/graphene aerogel	NO_2	3.6–8.9% for 50–200 ppm at room temperature	It exhibits moderate sensitivity, fast response/recovery behavior at room temperature. Other gases could be sensed including CO, H_2,	Liu et al. (2015b)

(continued)

Table 2.1 (continued)

S. no.	Name of sensing film	Target gas	Sensitivity values in %	Other sensing/functional characteristics	References
				NH_3, C_3H_6O, methanol, CH_3 $(CH_3)_3OH$ at room temperature. Resistance of film is reduced with respect to increase in temperature	
	ZnO/rGo	CO	24.3% for 22 pp, in dry N_2 at RT		Singh et al. (2012)
		NH_3	24% for 1 ppm in dry N_2 at RT		Singh et al. (2012)
		NO	3.5% for 5 ppm in dry N_2 at RT	This film also senses ammonia, CO, and NO for concentration as low as 1 ppm at RT. This film is sensitive toward electron donor gases. It has large response and quick recovery time at room temperature. For improvising the selectivity of composite, tailoring should be done of graphene sheet edges with gas specification binding energy or different nanostructures of ZnO (nanorods, nanoplate, etc.), used	Singh et al. (2012)
	ZnO/rGO	H_2	3.5% for 200 ppm at 150 °C	Here, it is revealed that 1.2 wt% of graphene/ZnO shows best response in 200 ppm hydrogen at 150 °C as optimum temperature. Here, it is observed that the inclusion of graphene in ZnO reduced the operable temperature and increases the sensitivity. This can also sense the ammonia, ethanol, and LPG. Sensing response significantly increases with hydrogen concentration from 50 to 600 ppm but the response reduces at concentration range 0–50 ppm. If the wt% of GO increases in ZnO, then response increases till 1.2 wt% and after that falls suddenly. Resistance of the film decreases in the temperature range from 0 to 200 °C	Anand et al. (2014)
	ZnO/rGO	NO_2	35% for 50 ppm at 50 °C	Resistance change depends slightly on humidity. Increase in temperature reduces response/recovery time. The response of sensor drastically increases with temperature till 50 °C, and after that it started falling significantly. The concentration of gas also increases the response of sensor. The composite ZnO/rGO shows higher sensitivity than rGO and GO films	Kumar et al. (2015)
	ZnO/Chemically converted graphene	H_2S	0% for 2 ppm at RT	It is able to operate in RT. H_2S gas is tested in the environment of N_2 and O_2 gases mixture	Cuong et al. (2010)
25	WO_3 + rGO	NO_2	133% for 5 ppm at 250 °C	Nanocomposite with 5% wt of graphene shows best response and lowest response/recovery time at 250 °C. The sensing studies are	Srivastava et al. (2012)

(continued)

Table 2.1 (continued)

S. no.	Name of sensing film	Target gas	Sensitivity values in %	Other sensing/functional characteristics	References
				carried out in the temperature of 200–300 °C for 5 ppm concentration	
	WO_3 + rGO	NO_2	61% for 1 ppm at 300 °C	This paper presents photocatalytic behavior of WO_3/graphene composite and its effect toward adsorption chemical species, enhance light absorption, efficient charge separation, and transfer. It gives super sensitivity and selective to NO_2 gas	An et al. (2012)
	WO_3 + Graphene	Alcohol	5.8% for 100 ppm at 317 °C		Qin et al. (2011)
	WO_3 + rGO	NO_2	769% for 5 ppm at RT		Su and Peng (2015)
	WO_3 + Pd-Partially rGO	H_2	38% for 100 ppm at RT		Esfandiar et al. (2014)
26	Cu_2O + rGO	NO_2	67.8% for 2 ppm at RT		Deng et al. (2012)
	Cu_2O + Functionalized graphene sheet	H_2S	11% for 0.005 ppm at RT		Zhou et al. (2013)
	Cu_2O + rGo	NH_3	2.04% for 200 ppm at RT		Meng et al. (2015)
27	Co_3O_4 + rGO	NO_2	80% for 60 ppm at RT		Chen et al. (2013a)
	Co_3O_4 + rGO	Acetone	1.12% for 0.12 ppm at RT		Choi et al. (2014a)
28	NiO + rGO	NO_2	4.7% for 5 ppm at 200 °C		Tien et al. (2013)
29	In_2O_3 + rGO	NO_2	8.25% for 30 ppm at RT		Gu et al. (2015)
30	MoO_2 + rGO	H_2S	60% for 40 ppm at 110 °C		Bai et al. (2015)
31	PEDOT-PSS N-GQDs N-GQDs/PEDOT-PSS nanocomposite	Methanol Ethanol Acetone	120%, 30% and 10% at 50 ppm for methanol, ethanol, and acetone, respectively	It can also sense toluene, water, chlorobenzene, and propanol. It exhibits high sensitive, selective, rapid, and reversible response. The gas sensor has fast response and recovery behavior, excellent room temperature selective, and stability	Gavgani et al. (2015)
32	PG + N-doped	CO, NO RhB	No report on sensing parameters		Varghese et al. (2015), Ma et al. (2014)
33	PG + Li-doped	H_2S NH_3			Giannozzi et al. (2009)
34	PG + N and Silica Co-doped	NO_2	−26% for 21 ppm		Niu et al. (2013)

(continued)

Table 2.1 (continued)

S. no.	Name of sensing film	Target gas	Sensitivity values in %	Other sensing/functional characteristics	References
35	PG + P-doped	NH_3	5.4% for 100 ppm		Niu et al. (2014)
36	PG + p-PACA	H_2	8% for 10,000 ppm		Chen et al. (2013b)
37	PG + Ozone	NO_2	19.7% for 200 ppm		Chung et al. (2012a)
38	GO + Oley amine	Methanol	116%		Kim et al. (2014)
		$(CH_3)_2CO$	45.5%		
39	rGO + PPD	DMMP	8% for 20 ppm		Hu et al. (2012)
40	rGO + Cysteamine	NO_2	11.5% for 5 ppm		Su and Shieh (2014)
41	rGO + Tannic acid	NH_3	−87 5% for 2620 ppm		Yoo et al. (2014)
42	PG + Pd	H_2	5% for 1000 ppm		Pak et al. (2014)
		H_2	33% for 1000 ppm		Chung et al. (2012b)
	PG + Pt	H_2	3% for 3000 ppm		Zhu et al. (2014)
	PG + SnO_2	NO_2	−6% for 50 ppm		Liu et al. (2014b)
		CO_2	60% for 50 ppm		Nemade and Waghuley (2014a)
	PG + Al_2O_3	CO_2	1084% for 200 ppm		Nemade and Waghuley (2014b)
	PG + WO_3	NO_2	96% for 1 ppm		Srivastava et al. (2012)
	PG + Cu(x)O	NO_x	950 % for 97 ppm		Yang et al. (2014b)
	PG + Fe_2O_3	H_2S	450 for 15 ppm		Jiang et al. (2014)
	PG + Sb_2O_3	CO_2	60% for 50 ppm		Nemade and Waghuley (2014a)
	PG + Bi_2O_3	LPG	400% for 30 ppm		Nemade and Waghuley (2013)
	PG + TiO_2	O_2	20% for 100,000 ppm		Zhang et al. (2013)
43	GO + CNTs	H_2O	9.8%		Li et al. (2013)
	GO + ZnO	CO	243% for 22 ppm		Singh et al. (2012)
		NH_3	24% for 1 ppm		Singh et al. (2012)
		NO	−5% for 5 ppm		Singh et al. (2012)
44	rGO + CNTs	NO_2	20% for 10 ppm		Jeong et al. (2010b)
	rGO + NiO	NO_2	200% for 1 ppm		Tien et al. (2013)
	rGO + ZnO	NO_2	25.6% for 5 ppm		Liu et al. (2014a)

(continued)

Table 2.1 (continued)

S. no.	Name of sensing film	Target gas	Sensitivity values in %	Other sensing/functional characteristics	References
		H_2	250% for 200 ppm		Anand et al. (2014)
	rGO + Co_3O_4	Methanol	8% for 800 ppm		Chen et al. (2013b)
		NO_2	80% for 60 ppm		Chen et al. (2013b)
	rGO + $Co(OH)_2$	NO_x	70% for 97 ppm		Liu et al. (2014c)
	rGO + $C_{32}H_{16}CuN_8$	NH_3	15.4% for 3200 ppm		Zhou et al. (2014a)
	rGO + MnO_2	LPG	50% for 25 ppm		Nemade and Waghuley (2014c)
	rGO + Pd-WO_3	H_2	7200% for 100 ppm		Esfandiar et al. (2014)
	rGO + SnO_2	H_2S	2400% for 5 ppm		Choi et al. (2014b)
		$(CH_3)_2$ CO	900% for 5 ppm		Choi et al. (2014b)
		NO_2	231% for 5 ppm		Zhang et al. (2014a)
		NH_3	15.9% for 50 ppm		Latil et al. (2004)
	rGO + WO_3	NO_2	102% for 20 ppm		An et al. (2012)
	rGO + $ZnFe_2O_4$	$(CH_3)_2CO$	810% for 1000 ppm		Liu et al. (2013)
	rGO + Pt	H_2	22 mV for 10,000 ppm		Shafiei et al. (2009)
44	FGS + Cu_2O	H_2S	11% for 5 ppm	The preferred film is room temperature based. The optimum working temperature is 150 °C. Having worked in elevated temperature sensing properties improved, however, it triggers flammable or explosive risk. This film shows stability against oxidation	Zhou et al. (2013)
45	rGO- PANI hybrid	NH_3	59.2% for 50 ppm		Zhou et al. (2014a)
	Bare PANI nanofibers		13.4% for 50 ppm		
	Bare rGO		5.2% for 50 ppm		
46	Graphene foam	NH_3	6% for 50 ppm	It is room temperature-based low-cost production sensor, also used for NO_2 detection Mechanically, the foam is robust and flexible. The large porosity structures of foam make it feasible for gas to penetrate evenly into the entire structure of film ensuing high sensitivity and stable operation of the device	Yavari et al. (2011)
47	TiO_2-PANI hybrids	NH_3	35% for 50 ppm	The reported film work at room temperature. It can also be used to sense methanol, ethanol, NO_2, H_2S. TiO_2–PANI composite film reveals higher sensitivity value, and faster response and recovery rate	Pawar et al. (2011)

(continued)

Table 2.1 (continued)

S. no.	Name of sensing film	Target gas	Sensitivity values in %	Other sensing/functional characteristics	References
48	PG + Polyethyleneimine	CO_2	2.1% for 3667 ppm	It is room temperature-based gas sensor. In accordance with it, it has long-term stability, low concentration detection, performance stability, response, and recovery time of about 10 min	Zhou et al. (2014b)
49	PG + polypyrrole	H_2O	138%	The response and recovery time of this sensor are 15 s and 20 s, respectively	Lin et al. (2013)
50	PG + Polyaniline	NH_3	3.65% for 20 ppm	This sensor provides simple structure, low cost, easy fabrication, able to operate in RT. The addition of graphene with PANI increased the surface area of film, and ammonia absorption in this composite film is more than the single-layer PANI. The response time is 50 s, recovery time is 23 s, and the detection range is 1–6400 ppm	Wu et al. (2013), Cai et al. (2012)
51	PG + PMMA membrane-coated Pd NP	H_2	66.67% for 2 ppm	It is selective to hydrogen gas. In this journal, it has observed that Pd NP/SLG hybrid shows cross-sensitivity toward CO, NO_2, CH_4 at the absence of PMMA layer. On the contrary, there is no cross-sensitivity toward CO, NO_2, CH_4 at the presence of PMMA layer in film	Hong et al. (2015)
52	Pt/f-MWNT,	H_2	8% for response time 7 min	It is stable over repeated cycle of hydrogenation and dehydrogenation. The sensitivity of sensor increases in room temperature	Kaniyoor et al. (2009)
	Pt/f-G	H_2	16% for response time 9 min		
	rGO + Nylon-6				
53	rGO + Polymethyl methacrylate	HCHO	13.7% for 2 ppm	This is RT-based gas sensor. It is extremely selective to formaldehyde with comparatively broad range of concentration. This composite film could detect vapors as formaldehyde (CH_2O), methanol(MeOH), ethanol The sensing mechanism depends on the ratio of graphene/polymer (~ 5) in order to obtain higher sensitivity and selectivity	Alizadeh and Soltani (2013)
54	rGO + Poly (3-hexylthiophene)	NH_3	7.15% for 10 ppm	This film has superior sensing properties, linear response, excellent recovery characteristic, good selectivity, and restorability compared to single rGO film. It can also sense toxic gases like SO_2, CO, CO_2, NO_2	Ye et al. (2014)
55	RGO-A	NH_3	31.7% for 50 ppm	RGO-A is the aniline reduced graphene oxide. The device shows extraordinary repeatability and selectivity. RGO-A films exhibit 9.2 and 3.5 times better response than RGO-H (GO reduced by	Huang et al. (2013)

(continued)

Table 2.1 (continued)

S. no.	Name of sensing film	Target gas	Sensitivity values in %	Other sensing/functional characteristics	References
				hydrazine hydrate) and PANI. This can sense DMMF, methanol, dichloromethane, cyclohexane, chloroform	
	rGO + Polyaniline	NH_3	59.2% for 50 ppm	Excellent sensitivity and selectivity, repeatability to ammonia. The composite films show 3.4 and 10 times better response than bare PANI and graphene films. The hybrid composite film also senses DMMF, methanol, dichloromethane, cyclohexane, chloroform	Huang et al. (2012)
		$C_6H_5CH_3$	35.5% for 100 ppm	The sensor response increase with increasing temperature. By taking into account response time and recovery time, this film shows better response at 50 °C. This composite film exhibits better sensing behavior in terms of repeatability, response, and recovery time	Parmar et al. (2013)
		H_2	16.57% for 10,000 ppm	It is observed that the reported composite films have better response than only graphene (0.83% sensitivity) and PANI films (9.38% sensitivity). H_2O environments decorate the sensitivity of PANI films	Al-Mashat et al. (2010)
56	rGO + SnO_2	Ethanol	3.5–2212 for 5–500 ppm	This film can also sense methanol, acetone, benzene, toluene, m-xylene Sensitivity increases if RH increases	Zito et al. (2017)

plasma. Cu_2O thin films are produced from metal copper by introducing argon and reactive oxygen into vacuum chamber by mass flow controller. It has observed that when critical O_2 concentration is 7%, then Cu_2O thin film is produced. If O_2 concentration is greater than critical value (7%), then ($Cu_2O + 0.5O_2 = CuO$) CuO thin films are prepared. This film is sensitive toward ethanol. Several types of plasma diagnostics are tabulated below (Nafarizal 2016). There are several fields of magnetron sputtering including close field unbalanced magnetron sputtering, pulsed magnetron sputtering, variable field strength magnetrons, and duplex surface engineering technique (Kelly and Arnell 2000).

(b) Ion Etching

Etching processes are generally used to remove the thin films previously deposited on substrate for preparing desired pattern (isotropic/anisotropic). By this process, microfludic pattern, interdigitated electrode, and field-effect transistor are prepared. There are two types of etching process: (1) dry etching (reactive ion etching, sputter etching, and vapor phase etching) and (2) wet etching. Dry etching is easy to start and stop, less sensitive to change in temperature, and may have anisotropic and more repeatable. Wet etching is cheaper and high selectivity. In dry etching, etchant comes in the form of gases exposed in plasma, and as a result, etchant by-product steps into chemical reaction with previously deposited thin film substrate. For example, when CF_4 etchant exposed to plasma, it gives CF_3 and F and electron. If this by-product is exposed to silicon wafer, then SiF_4 volatile product is eradicated from the substrate surface (Fig. 2.1).

Wet etching requires a simple bath where liquid etchant is stored which is reactive toward thin films on surface of substrates. One should apply mask on required pattern manner such that the covered portion by mask will not react with etchant solution. Uncovered portion gets to wipe out due to chemical reaction and we will get proper pattern. For example, when copper reacts, ferric chloride $FeCl_2$ comes out as product. In below figure, copper thin film PCB substrate is exposed to ferric chloride solution and sellotape is used as a masked pattern of interdigitated electrode on thin-filmed substrate.

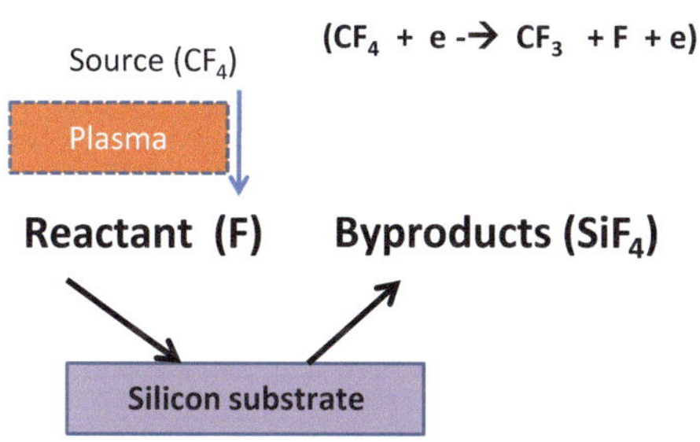

Fig. 2.1 Schematic showing dry etching over silicon substrate

$$FeCl_3 + Cu \rightarrow FeCl_2 + CuCl$$

The reactive ion etching of TiO_2 and gold thin film could be produced for gas sensing application (Adzhri et al. 2015; Ranade et al. 1993).

(c) Aqueous Chemical Synthesis

Fabrication of sensing films with aqueous chemical synthesis approach is the simplest and cost-effective way. It just requires some equipment such as arrangement to assist in the chemical reaction for essential bond formulations, some equipment to mix several chemical agents, some energy source to bring the overall molecular size down to sub-micron/nanorange, and heating arrangement to provide thermal treatment. Additional tool may be required for thin deposition to select substrates such as silicon, ITO substrate, etc. Some of the work based on sensing films fabricated with the help of aqueous chemical synthesis can be found in references (Gupta et al. 2013, 2014a, b, 2016). Gupta et al. fabricated ZnO-based sensing films simply by first mixing the chemical reagents followed by providing the ultrasonic treatment to bring down the size to nm scale. Solution obtained was then spin-coated to interdigitated electrodes made over silicon substrate to obtain thin sensing film over it (Gupta et al. 2013, 2014a, b, 2016).

2.3.2 Review of Some of the Fabrication Methodology of Sensing Film

(a) PANI Thin Film Fabrication and Characterization (Awang 2014)

PANI is a class of polymer which is derived from a polymer. The generalized composition of PANI is shown in Fig. 2.2.

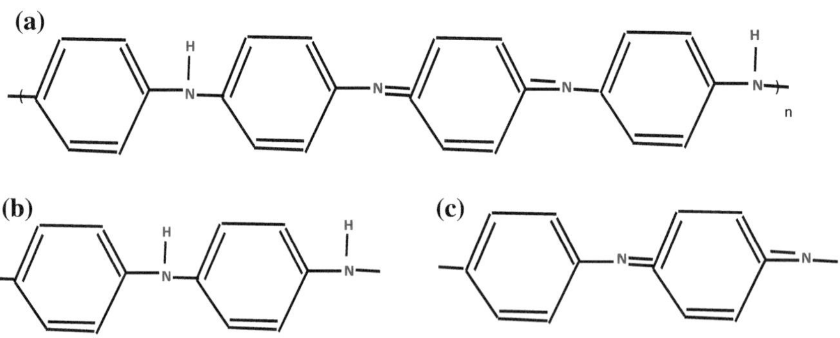

Fig. 2.2 **a** The structure of PANI unit, **b** reduced unit, and **c** oxidized unit

It consists of alternating reduced and oxidized repeat units. PANI has three different forms: leucoemeraldine base (LEB, fully reduced form), emeraldine base (EB, partially oxidized form), and base pernigraniline (PNB, fully oxidized form). Among the different oxidation states of PANI, the emeraldine salt (ES) is the only one that shows electrical conductivity. EB can be converted to ES and vice versa by protonation or deprotonation reaction. Polyaniline (PANI) sensing film is a very active layer for ammonia gas sensing. PANI films are deposited on PET flexible substrate by in situ polymerization technique (Awang 2014). PANI film is deposited on PET substrate when its PET substrate is dipped into a solution of aniline (0.1M) dissolved in HCL (M) and maintained at 0–5 °C. The solution is stirred after 30 min APS (0.1M) is added dropwise into the solution. After stirring for 30 min, different PET substrates had drawn out of solution at different time intervals of 5, 15, 40, 60 min, etc. Then, the substrates are washed properly after withdrawn from solution and dried in ambient air (Awang 2014). The SEM image was taken by FESEM of dried PET substrate, and the thickness of deposited PANI films was measured by Dektak Profilometer.

(b) IGZO-Based Sensing Films Fabrication and Characterization

IGOZ thin films are the composition of indium nitrate hydrate, gallium nitrate hydrate, and zinc acetate dihydrate. The amorphous IGZO films are deposited on ITO electrode for sensing VOCs gases. IGZO solution is prepared by dissolving powder of 0.085M indium nitrate hydrate, 0.0125M gallium nitrate hydrate, and 0.0275M zinc acetate hydrate in 2-methoxyethanol. The solution is stirred for 2 h at 75 °C. After spin coating on electrode, deep UV photochemical activation in N_2 atmosphere followed for 2 h (Kelly and Arnell 2000).

2.4 Sensing Mechanisms of Functional Films

Functional films may be composed of metal, semiconductors, or polymers. Each entity has their own functional properties and sensing mechanism. Description of sensing mechanism based on the findings reported in the various research papers is as follows.

2.4.1 Sensing Mechanism of Metal/Semiconductor-Based Films

Generally, conductance and resistance of sensor depend on sensing films. It is obligatory to expose the sensing mechanism of metal-oxide-based thin films

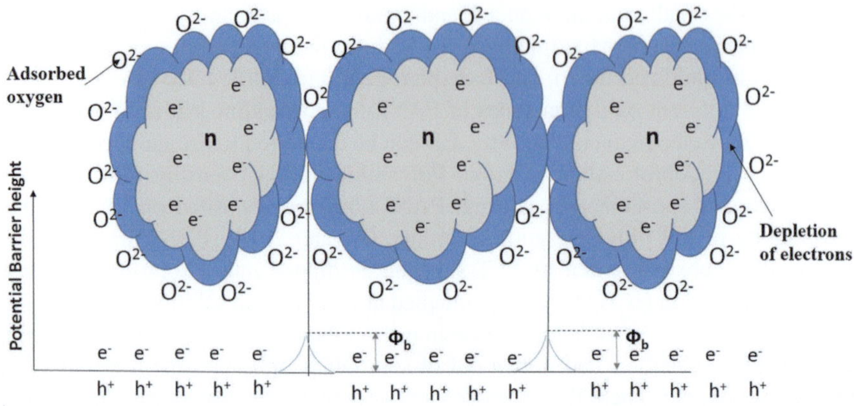

Fig. 2.3 The sensing mechanism scheme of semiconductor-based films (Reprinted with permission from Gupta et al. (2014a) copyright (2014) RSC advances)

because it gives a basic idea to design and fabricate of novel gas sensing films with outstanding recital. Typically, oxygen gets trapped and adsorbed into metal oxide at particular environment condition (temperature/pressure). The adsorbed oxygen molecules can confine electron from the interior of the oxide films. Thus, the negative charges trapped in the oxygen groups are the foundations of depletion layer ($O_2^-/O^-/O^{2-}$ layer) on oxide film surface, resulting in reducing conductivity of films. So when the sensor is exposed to reducing gases, the electrons trapped by oxygen species do return to metal oxide films resulting in a decrease in potential barrier height and also increase in conductivity. There are different molecular (O_2^-) ions and atomic (O^-/O^{2-}) ions present on the surface depending on functioning temperature. It has observed that in oxide films, below 150 °C, molecular species of oxygen ions stay alive in oxide films while above this temperature atomic species of ions are found. Oxygen vacancy performs as donor, increasing the surface conductivity, while adsorbed oxygen ions act as surface acceptor, binding electron, and reduce the surface conductivity of metal oxide films. The sensing properties of metal oxide gas sensor can be improved by controlling the morphology and the constitution of sensing material. The "small size effect" of metal oxide grain also affects the sensing properties (Gupta et al. 2014a, b, 2016) (Fig. 2.3).

On the surface of metal oxide film grains, adsorbed oxygen molecules take out electrons from conduction band and trap the electrons onto the surface of films in the form of ions, which produces a space charge layer on the surface of films. If the particle size (D) of sensing films is close to or less than double the thickness of space charge layer (L), the sensitivity of sensor increases surprisingly (Min 2003). In case if $D \gg 2L$, the conductivity of structure depends on inner mobility of charge carriers and the electrical conductivity varies exponentially on barrier height.

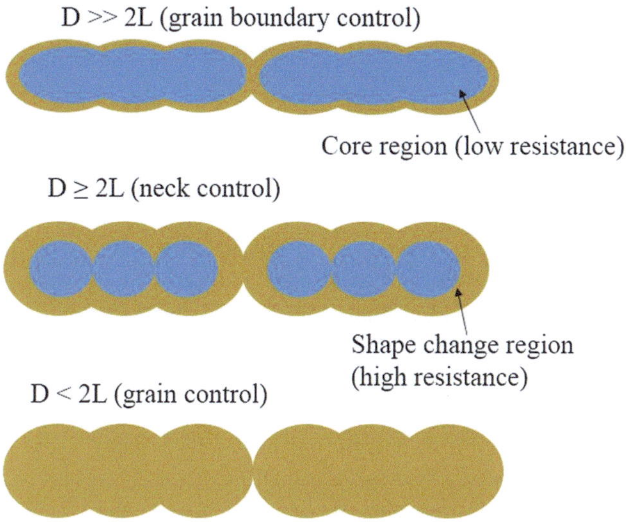

Fig. 2.4 Schematic model of space charge layer and grain size of films (Min 2003) (Reprinted with permission from Min et al. (2009). Copyright (2009) Elsevier.)

This type of films ($D \gg L$) is not so sensitive to the charges squared from surface reaction. For $D < 2L$, it means space charge layer dominates the whole particles and the crystallites are almost depleted of mobile charge carrier. Hence, energy bandgap is flat for whole structure of interconnected grains in films. Also for $D < 2L$, there is no barrier exists for intercrystallite charge transport and the conductivity depends on intercrystallite conductivity (Fig. 2.4).

$$O_2(\text{gas}) \rightarrow O_2(\text{adsorbed})$$
$$O_2(\text{ads}) + e^- \rightarrow O_2^-(T < 100\,^\circ\text{C})$$
$$O_2^-(\text{ads}) + e^- \rightarrow 2O^-(T = 100\text{–}300\,^\circ\text{C})$$

Metal oxide sensing films have divided into denser and porous medium. In denser oxide films, exposed gas does not adsorb into the surface of film but in case of porous film the analyte can easily penetrate into the grains films. A well-defined and uniform porous structure films is desired for the improvement of sensitivity because in porous layer the active grain is more than the denser layer (Barsan and Weimar 2001).

2.4.2 Sensing Mechanism-Based Polymer-Based Film

Conducting polymer such as polyaniline, polythiophene, polypyrrole, PPV, PEDOTS, polyacetylene, and their derivatives have been used as thin layer on electrode for gas sensing purpose. Polymer-based films are very popular across gas sensing area because of their sensitivity, short response time, room temperature based, and synthesis mechanism (Chemically/Electrochemically). By looking at chemically bond structure of polymer, typically all polymers have long-chain structure. And their chemical structure could be modified by copolymerization or structural derivation. Various types of mechanisms are there to prepare conducting polymer films in order to adapt different sensing materials and structural configurations of sensors. Some popular methods to deposit polymer films are electrochemical deposition, dip coating, spin coating, Langmuir–Blodgett technique, layer-by-layer self-assembly technique, thermal evaporation, vapor deposition polymerization, drop coating, and inkjet printing method. The concept of doping is the central theme which distinguishes the conducting polymer from other polymers. Conducting polymer could be doped by redox reaction or protonation, and oxidation reaction.

The physical properties of conducting polymer depend on doping levels. While most of the conducting polymer films are exposed to the analyte, the doping level of polymer could be altered by the transfer of electron from or to the analyte. This electron transfer causes the change in resistance of film and working function of sensor. Some gases are electron acceptor and some are electron donor; so when conducting polymer film opens to such gas, the electrical properties of films are changed. While electron acceptor gases (NO_2, I_2) come in contact with P-type conducting polymer, they do remove electrons from the aromatic rings of conducting polymers and enhance the conductance of polymer. On the other hand, while ammonia is electron donor gas, when it reacts with polypyrrole polymer, it is resistance down sharply. But after washing with dry nitrogen or air, that polymer film recovers its resistance totally or partially.

$$PPy^{+} + N{:}H_3 \rightarrow PPy^{0} + NH_3^{+} \quad \text{Adsorption}$$
$$PPy^{0} + N{:}H_3^{+} \rightarrow PPy^{+} + N{:}H_3 \quad \text{Desorption}$$

VOCs gases (benzene, toluene, and others) are not reactive to conductive polymer films at room temperature. It is also difficult to detect them by chemical reaction with conducting polymers. The physical interaction involving absorption or swelling between VOCs and conducting polymers influence change in properties of polymer films and make it detectable. Absorption is the first step in gas sensing technique. Analyte VOCs adsorption also enhances the potential barrier of the grain in films and changes the electrical properties of films. Interaction of VOCs analyte with the conducting polymer increases the interchain distance of polymer which affects the electron-hopping process between different polymer chains. In experiment, it has been observed that the change in electrical conductivity depends on relative permittivity of polymer and analyte. The conductivity will increase if electrical permittivity of polymer is less than analyte and decreases if electrical permittivity of polymer is greater than analyte. Hydrogen bonding and dipole–dipole interaction also play an essential role in VOC's gas sensing by conductive polymer. When polypyrroline film is exposed to acetone, H-bonds are formed in between C=O group of acetone and N-H group of pyrrole units (Ruangchuay et al. 2004). H-bonds are formed as a bridge in PANI chain when PANI molecules came in interaction with methanol. Some analyte gases especially alcohols and ketones interaction with polymer can change the crystallinity of polymer. In experiment, it was found that diffusion of methanol and ethanol was more efficient in polymer matrix than higher molecular weight alcohol did, because the high dielectric constant of small alcohol does strong interaction with nitrogen atoms in polymer. And also higher molecular weight VOCs analyte cannot diffuse into polymer matrix properly as light molecular weight molecules do because high molecular weight VOCs have long-chain length and nonpolar nature (Athawale and Kulkarni 2000).

As another widely used film, PANI is being utilized for sensing applications. Sensing mechanism/principle of PANI sensing films is as follows. PANI and ammonia produce reversible reaction. When PANI reacts with HCL, protonation reaction occurs. PANI doping in HCL gives conductivity of 0.143 s cm^{-1} (Gomes and Oliveira 2012). Due to protonation reaction, PANI molecule gains protons, forming energetically favorable N^{+}–H chemical bond. This results in the protonation or deprotonation of PANI (Gomes and Oliveira 2012). This leads to the formation of positively charge local centers located at nitrogen atom, which facilitates the valence electrons to move one such center to another, giving rise to the hopping of electron. PANI is the most special type of polymer that its conductivity could be increased by two completely different processes: one is oxidation doping and other is photonic acid doping. But conductive PANI film has to be protonic acid doping. In the fabrication part, 1M of HCL is added which results in protonation reaction given in figure (MacDiarmid 1996) (Figs. 2.5 and 2.6).

It is observed from the reaction that the protonated polymer is actually poly-semiquinone radial cations; one resonance form consists of two separated polarons. The oxide unit rings of PANI are equivalent and oxide state nitrogen become equivalent. The detailed reactions of PANI and HCl are given in MacDiarmid (1996). When PANI interacts with ammonia molecules, it takes proton from PANI and forms ammonium ions (NH_4^{+}) ions. However, with no ammonia replenishment, ammonium decomposes into ammonia and protons.

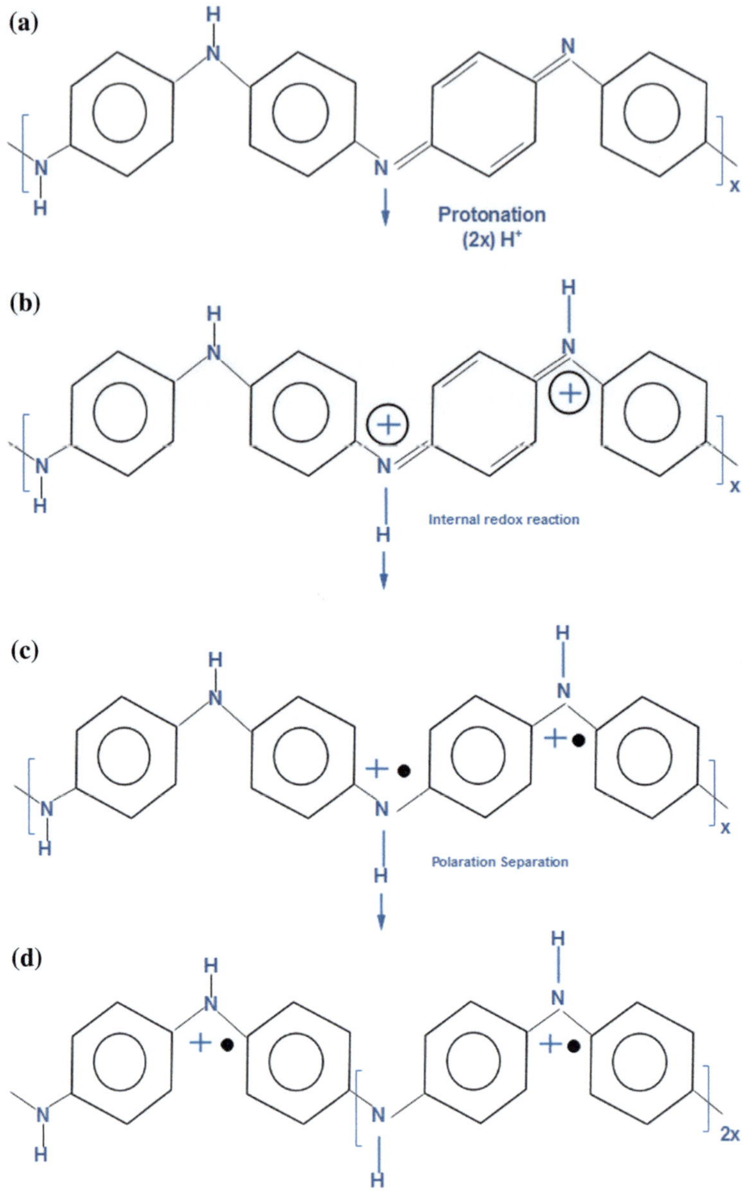

Fig. 2.5 Doping and de-doping process of PANI and mechanism of the formation of polaron structure of PANI (Stafström et al. 1987)

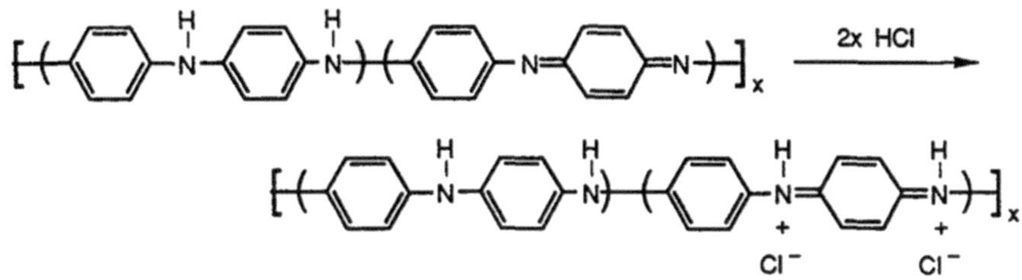

Fig. 2.6 Reaction of PANI with HCl (MacDiarmid 1996)

2.5 Other Important Aspects of Sensing Films

Apart from sensitivity of the functional film, there are other important aspects of sensing film which need to be carefully examined while selecting the film for a particular gas sensor. These are selectivity, stability, reproducibility, and least response and recovery time. Selectivity of any gas sensor is undeniably a very significant parameter in the selection of sensing films and is related to the discrimination capacity of a gas sensing element to specifically select a single gas from a mixture. The selectivity of metal oxides is extensive and they normally act in response likewise to all reducing gases which react with their surface oxygen. Catalytic additive leads to an overall enhancement of the sensor activity by providing a selective means to uphold reaction with target gas. It is the general place to modify the selectivity by using dopants, grain size controllers, catalysts, operating temperature, etc. Enhanced sensitivity when a metallic layer is added to the semiconducting material can be understood on the basis of band bending phenomenon which results in a decrease in the bandgap of semiconducting material on the metallic incorporation as shown in Fig. 2.7 (Gupta et al. 2013, 2014a, b, 2016).

Portability issue of gas sensing device is also substantial factor while designing the same. Portability issue and functionality wise up gradation of the sensing device can be brought by miniaturization of each component of the device. Performance of sensing film can be enhanced by bringing down the size of film or in other words, fabricating high specific surface area films. As we are aware of the fact that nanoscale of any material has distinct properties than the bulk form. That is why researchers are continuously exploring the high surface area based films, viz., nanowire, nanotubes, etc. so that within the minimum space occupied by the films, greater interaction of target gas molecules with the film could take place and as a result, enhanced performance of gas sensor could be achieved.

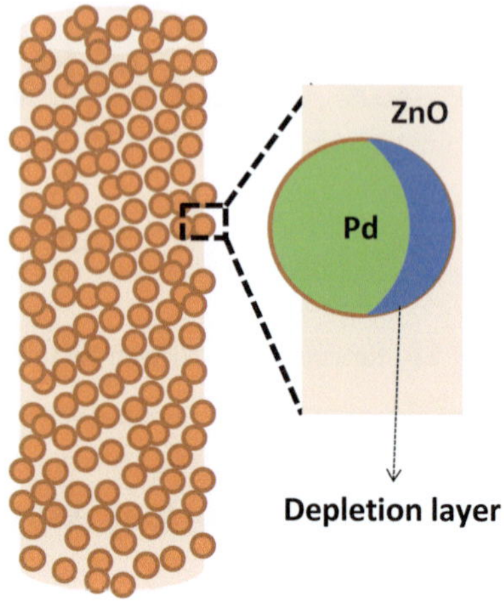

Fig. 2.7 The formation of metal–semiconductor interface (Reprinted with permission from Gupta et al. (2013). Copyright (2013) RSC advances

2.6 Application of Sensing Films in Aerospace and Automotive Field

In the miniaturized domain, thin sensing film has been developed which have already found suitable place in the aerospace and automotive applications. These fields require the material to be sustained in harsh environment such as high temperature, high pressure, etc. In aerospace, there is a requirement of compact, light-weighted sensors with wide temperature range detection, long-term stability. Chemical sensors are one of the major sensors in use and demand as well. It may include leakage detection, careful monitoring of emission of various gases, fire detection, etc. In automotive field, sensors from tens to few hundred in numbers are being placed in automotive vehicles. These vehicles are being operated in various environmental conditions. Nanostructured materials have enormous scope in the aerospace and automobile sector. Out of explored nanomaterials, carbon nanomaterial-based films have found superior place among other films. CNT-based polymer composite is now being explored in radiation detections, as heat dissipation coatings, high-strength components, and many more. Other films include indium oxide, tin oxide, wide bandgap nitrides, etc. Hydrogen sensors based on CNTs with Pd nanoparticles are finding immense utility in fuel cell car. Use of nanostructured coating made up of magnetic material is soon going to resolve problem of power dissipation. In short, nanostructured sensing films are enabling technologies to have positive impact on aerospace as well as automotive sector.

Table 2.2 Summarizes various sputter deposited thin films

Plasma diagnostics method	Sputtering deposition material thin film	Measurable parameter
Optical emission spectroscopy (OES)	ZnO	Excited state metal density
	WO_3	Excited state oxygen density
	TiO_2	Excited state argon density
ROAS and TD-LAS	TiO_2	Metal atom density
	ZnO and Al–ZnO	Metal ion density and oxygen density
	CuO and Cu_2O	Argon metastable density
Langmuir probe	ZnO	Electron density
	TiO_2	Ion density
	CuO	Electron temperature

2.7 Conclusion

In this article, review of sensing films used for gas sensing purpose has been performed. Table 2.2 summarizes various sensing films with target gas and reported sensitivity values. Some of the widely explored functional films in the research papers are ZnO-based films, PANI-based, SnO_2, TiO_2, etc. Carbon-based films (its various modified forms) are nowadays being utilized for developing cheaper functional films for various target gases. In short, this article provides insight into various polymeric, metallic, and semiconductor-based functional films widely used for gas sensing applications.

References

Adzhri R, Arshad MM, Fathil MFM, Hashim U, Ruslinda AR, Ayub RM, Azman AH (2015) Reactive ion etching of TiO_2 thin film: the impact of different gaseous. In: 2015 IEEE regional symposium on micro and nanoelectronics (RSM). IEEE, pp 1–4

Al-Hardan NH, Abdullah MJ, Aziz AA, Ahmad H, Low LY (2010) ZnO thin films for VOC sensing applications. Vacuum 85(1):101–106

Alizadeh T, Soltani LH (2013) Graphene/poly (methyl methacrylate) chemiresistor sensor for formaldehyde odor sensing. J Hazard Mater 248:401–406

Al-Mashat L, Shin K, Kalantar-zadeh K, Plessis JD, Han SH, Kojima RW, Wlodarski W (2010) Graphene/polyaniline nanocomposite for hydrogen sensing. J Phys Chem C 114(39):16168–16173

An X, Jimmy CY, Wang Y, Hu Y, Yu X, Zhang G (2012) WO_3 nanorods/graphene nanocomposites for high-efficiency visible-light-driven photocatalysis and NO_2 gas sensing. J Mater Chem 22(17):8525–8531

Anand K, Singh O, Singh MP, Kaur J, Singh RC (2014) Hydrogen sensor based on graphene/ZnO nanocomposite. Sens Actuators B Chem 195:409–415
Athawale AA, Kulkarni MV (2000) Polyaniline and its substituted derivatives as sensor for aliphatic alcohols. Sens Actuators B Chem 67(1–2):173–177
Awang Z (2014) Gas sensors: a review. Sens Transducers 168:61–75
Bai S, Chen C, Luo R, Chen A, Li D (2015) Synthesis of MoO_3/reduced graphene oxide hybrids and mechanism of enhancing H_2S sensing performances. Sens Actuators B Chem 216:113–120
Bandgar DK, Navale ST, Navale YH, Ingole SM, Stadler FJ, Ramgir N, Patil VB (2017) Flexible camphor sulfonic acid-doped PAni/α-Fe_2O_3 nanocomposite films and their room temperature ammonia sensing activity. Mater Chem Phys 189:191–197
Barsan N, Weimar U (2001) Conduction model of metal oxide gas sensors. J Electroceram 7 (3):143–167
Cai YM, Qin ZY, Zhou Z (2012) Nanocoating of polyaniline layer on the surface of graphene sheets for ammonia gas detection. In: Advanced materials research, vol 557. Trans Tech Publications, pp 1803–1806
Chang Y, Yao Y, Wang B, Luo H, Li T, Zhi L (2013) Reduced graphene oxide mediated SnO_2 nanocrystals for enhanced gas-sensing properties. J Mater Sci Technol 29(2):157–160
Chatterjee SG, Chatterjee S, Ray AK, Chakraborty AK (2015) Graphene–metal oxide nanohybrids for toxic gas sensor: a review. Sens Actuators B Chem 221:1170–1181
Chen N, Li X, Wang X, Yu J, Wang J, Tang Z, Akbar SA (2013a) Enhanced room temperature sensing of Co_3O_4-intercalated reduced graphene oxide based gas sensors. Sens Actuators B Chem 188:902–908
Chen X, Yasin FM, Eggers PK, Boulos RA, Duan X, Lamb RN, Raston CL (2013b) Non-covalently modified graphene supported ultrafine nanoparticles of palladium for hydrogen gas sensing. RSC Adv 3(10):3213–3217
Choi SJ, Ryu WH, Kim SJ, Cho HJ, Kim ID (2014a) Bi-functional co-sensitization of graphene oxide sheets and Ir nanoparticles on p-type Co_3O_4 nanofibers for selective acetone detection. J Mater Chem B 2(41):7160–7167
Choi SJ, Jang BH, Lee SJ, Min BK, Rothschild A, Kim ID (2014b) Selective detection of acetone and hydrogen sulfide for the diagnosis of diabetes and halitosis using SnO_2 nanofibers functionalized with reduced graphene oxide nanosheets. ACS Appl Mater Interfaces 6 (4):2588–2597
Chung MG, Kim DH, Lee HM, Kim T, Choi JH, Kyun Seo D, Kim YH (2012a) Highly sensitive NO_2 gas sensor based on ozone treated graphene. Sens Actuators B Chem 166:172–176
Chung MG, Kim DH, Seo DK, Kim T, Im HU, Lee HM, Kim YH (2012b) Flexible hydrogen sensors using graphene with palladium nanoparticle decoration. Sens Actuators B Chem 169:387–392
Cui S, Wen Z, Mattson EC, Mao S, Chang J, Weinert M, Hirschmugl CJ, Gajdardziska-Josifovskab M, Chen J (2013) Indium-doped SnO_2 nanoparticle–graphene nanohybrids: simple one-pot synthesis and their selective detection of NO_2. J Mater Chem A 1:4462–4467
Cuong TV, Pham VH, Chung JS, Shin EW, Yoo DH, Hahn SH, Kohl PA (2010) Solution-processed ZnO-chemically converted graphene gas sensor. Mater Lett 64(22):2479–2482
Deng S, Tjoa V, Fan HM, Tan HR, Sayle DC, Olivo M, Sow CH (2012) Reduced graphene oxide conjugated Cu_2O nanowire mesocrystals for high-performance NO_2 gas sensor. J Am Chem Soc 134(10):4905–4917
Emmer I (1992) Thin film gas sensors. Int J Electron 1;73(5):875–879
Esfandiar A, Irajizad A, Akhavan O, Ghasemi S, Gholami MR (2014) Pd–WO_3/reduced graphene oxide hierarchical nanostructures as efficient hydrogen gas sensors. Int J Hydrogen Energy 39 (15):8169–8179

Galkidas A, Marthunas Z, Setkus A (1992) Tin–indium oxide-based chlorine gas sensor. Sens Actuators B 7(1–3):633–636

Gavgani JN, Dehsari HS, Hasani A, Mahyari M, Shalamzari EK, Salehi A, Taromi FA (2015) A room temperature volatile organic compound sensor with enhanced performance, fast response and recovery based on N-doped graphene quantum dots and poly (3, 4-ethylenedioxythiophene)–poly (styrenesulfonate) nanocomposite. RSC Adv 5(71):57559–57567

Giannozzi P, Baroni S, Bonini N, Calandra M, Car R, Cavazzoni C, Dal Corso A (2009) Quantum Espresso: a modular and open-source software project for quantum simulations of materials. J Phys Condens Matter 21(39):395502

Gomes EC, Oliveira MA (2012) Chemical polymerization of aniline in hydrochloric acid (HCl) and formic acid (HCOOH) media. Differences between the two synthesized polyanilines. Am J Polym Sci 2(2):5–13

Gu F, Nie R, Han D, Wang Z (2015) In_2O_3–graphene nanocomposite based gas sensor for selective detection of NO_2 at room temperature. Sens Actuators B Chem 219:94–99

Gupta A, Pandey SS, Bhattacharya S (2013) High aspect ZnO nanostructures based hydrogen sensing. In: AIP conference proceedings, vol 1536, no 1. AIP, pp 291–292

Gupta A, Pandey SS, Nayak M, Maity A, Majumder SB, Bhattacharya S (2014a) Hydrogen sensing based on nanoporous silica-embedded ultra dense ZnO nanobundles. RSC Adv 4 (15):7476–7482

Gupta A, Srivastava A, Mathai CJ, Gangopadhyay K, Gangopadhyay S, Bhattacharya S (2014b) Nano porous palladium sensor for sensitive and rapid detection of hydrogen. Sensor Lett 12 (8):1279–1285

Gupta A, Gangopadhyay S, Gangopadhyay K, Bhattacharya S (2016) Palladium-functionalized nanostructured platforms for enhanced hydrogen sensing. Nanomater Nanotechnol 6:40

Gusain A, Joshi NJ, Varde PV, Aswal DK (2017) Flexible NO gas sensor based on conducting polymer poly [N-9′-heptadecanyl-2, 7-carbazole-alt-5, 5-(4′, 7′-di-2-thienyl-2′, 1′, 3′-benzothiadiazole)](PCDTBT). Sens Actuators B Chem 239:734–745

He Q, Zeng Z, Yin Z, Li H, Wu S, Huang X, Zhang H (2012) Fabrication of flexible MoS_2 thin-film transistor arrays for practical gas-sensing applications. Small 8(19):2994–2999

He JQ, Yin J, Liu D, Zhang LX, Cai FS, Bie LJ (2013) Enhanced acetone gas-sensing performance of La_2O_3-doped flowerlike ZnO structure composed of nanorods. Sens Actuators B Chem 182:170–175

Hong J, Lee S, Seo J, Pyo S, Kim J, Lee T (2015) A highly sensitive hydrogen sensor with gas selectivity using a PMMA membrane-coated Pd nanoparticle/single-layer graphene hybrid. ACS Appl Mater Interfaces 7(6):3554–3561

Hu N, Wang Y, Chai J, Gao R, Yang Z, Kong ESW, Zhang Y (2012) Gas sensor based on p-phenylenediamine reduced graphene oxide. Sens Actuators B Chem 163(1):107–114

Huang X, Hu N, Gao R, Yu Y, Wang Y, Yang Z, Zhang Y (2012) Reduced graphene oxide–polyaniline hybrid: preparation, characterization and its applications for ammonia gas sensing. J Mater Chem 22(42):22488–22495

Huang XL, Hu NT, Wang YY, Zhang YF (2013) Ammonia gas sensor based on aniline reduced graphene oxide. In: Advanced materials research, vol 669. Trans Tech Publications, pp 79–84

Inyawilert K, Wisitsoraat A, Sriprachaubwong C, Tuantranont A, Phanichphant S, Liewhiran C (2015) Rapid ethanol sensor based on electrolytically-exfoliated graphene-loaded flame-made In-doped SnO_2 composite film. Sens Actuators B Chem 209:40–55

Jaisutti R, Kim J, Park SK, Kim YH (2016) Low-temperature photochemically activated amorphous indium-gallium-zinc oxide for highly stable room-temperature gas sensors. ACS Appl Mater Interfaces 8(31):20192–20199

Jeong CW, Shin CH, Kim DI, Chae JH, Kim YS (2010a) An ITO/au/ITO thin film gas sensor for methanol detection at room temperature. Trans Electr Electron Mater 11(2):77–80

Jeong HY, Lee DS, Choi HK, Lee DH, Kim JE, Lee JY, Choi SY (2010b) Flexible room-temperature NO_2 gas sensors based on carbon nanotubes/reduced graphene hybrid films. Appl Phys Lett 96(21):213105

Jiang Z, Li J, Aslan H, Li Q, Li Y, Chen M, Besenbacher F (2014) A high efficiency H_2S gas sensor material: paper like Fe_2O_3/graphene nanosheets and structural alignment dependency of device efficiency. J Mater Chem A 2(19):6714–6717

Kaniyoor A, Jafri RI, Arockiadoss T, Ramaprabhu S (2009) Nanostructured Pt decorated graphene and multi walled carbon nanotube based room temperature hydrogen gas sensor. Nanoscale 1 (3):382–386

Karaduman I, Er E, Çelikkan H, Acar S (2015) A new generation gas sensing material based on high-quality graphene. Sens Actuators B Chem 221:1188–1194

Kelly PJ, Arnell RD (2000) Magnetron sputtering: a review of recent developments and applications. Vacuum 56(3):159–172

Kim Y, An TK, Kim J, Hwang J, Park S, Nam S, Park CE (2014) A composite of a graphene oxide derivative as a novel sensing layer in an organic field-effect transistor. J Mater Chem C 2 (23):4539–4544

Krško O, Plecenik T, Roch T, Grančič B, Satrapinskyy L, Truchlý M, Plecenik A (2017) Flexible highly sensitive hydrogen gas sensor based on a TiO_2 thin film on polyimide foil. Sens Actuators B Chem 240:1058–1065

Kumar N, Srivastava AK, Patel HS, Gupta BK, Das Varma G (2015) Facile synthesis of ZnO-reduced graphene oxide, nanocomposites for NO_2 gas sensing applications. Eur J Inorg Chem 2015:1912–1923

Kumar L, Rawal I, Kaur A, Annapoorni S (2017) Flexible room temperature ammonia sensor based on polyaniline. Sens Actuators B Chem. 1(240):408–416

Latil S, Roche S, Mayou D, Charlier JC (2004) Mesoscopic transport in chemically doped carbon nanotubes. Phys Rev Lett 92(25):256805

Lee DS, Jung JK, Lim JW, Huh JS, Lee DD (2001) Recognition of volatile organic compounds using SnO_2 sensor array and pattern recognition analysis. Sens Actuators B Chem 77(1–2):228–236

Lee J-H, Katoch A, Choi S-W, Kim J-H, Kim HW, Kim SS (2015) Extraordinary improvement of gas-sensing performances in SnO_2 nanofibers due to creation of local p–n heterojunctions by loading reduced graphene oxide nanosheets. ACS Appl Mater Interface 7:3101–3109

Li X, Chen X, Yao Y, Li N, Chen X, Bi X (2013) Multi-walled carbon nanotubes/graphene oxide composites for humidity sensing. IEEE Sens J 13(12):4749–4756

Lin WD, Chang HM, Wu RJ (2013) Applied novel sensing material graphene/polypyrrole for humidity sensor. Sens Actuators B Chem 181:326–331

Liu F, Chu X, Dong Y, Zhang W, Sun W, Shen L (2013) Acetone gas sensors based on graphene-$ZnFe_2O_4$ composite prepared by solvothermal method. Sens Actuators B Chem 188:469–474

Liu S, Yu B, Zhang H, Fei T, Zhang T (2014a) Enhancing NO_2 gas sensing performances at room temperature based on reduced graphene oxide-ZnO nanoparticles hybrids. Sens Actuators B Chem 202:272–278

Liu X, Cui J, Sun J, Zhang X (2014b) 3D graphene aerogel-supported SnO_2 nanoparticles for efficient detection of NO_2. RSC Adv 4(43):22601–22605

Liu S, Zhou L, Yao L, Chai L, Li L, Zhang G, Shi K (2014c) One-pot reflux method synthesis of cobalt hydroxide nanoflake-reduced graphene oxide hybrid and their NO_x gas sensors at room temperature. J Alloy Compd 612:126–133

Liu S, Wang Z, Zhang Y, Zhang C, Zhang T (2015a) High performance room temperature NO_2 sensors based on reduced graphene oxide-multiwalled carbon nanotubes-tin oxide nanoparticles hybrids. Sens Actuators B Chem 211:318–324

Liu X, Sun J, Zhang X (2015b) Novel 3D graphene aerogel–ZnO composites as efficient detection for NO_2 at room temperature. Sens Actuators B Chem 211:220–226

Lonergan MC, Severin EJ, Doleman BJ, Beaber SA, Grubbs RH, Lewis NS (1996) Array-based vapor sensing using chemically sensitive, carbon black–polymer resistors. Chem Mater 12;8 (9):2298–2312

Ma C, Shao X, Cao D (2014) Nitrogen-doped graphene as an excellent candidate for selective gas sensing. Sci China Chem 57(6):911–917

MacDiarmid AG (1996) The polyanilines: a novel class of conducting polymers. University of Pennsylvania

Mao S, Cui S, Lu G, Yu K, Wen Z, Chen J (2012) Tuning gas-sensing properties of reduced graphene oxide using tin oxide nanocrystals. J Mater Chem 22(22):11009–11013

Meng H, Yang W, Ding K, Feng L, Guan Y (2015) Cu_2O nanorods modified by reduced graphene oxide for NH_3 sensing at room temperature. J Mater Chem A 3(3):1174–1181

Min Y (2003) Properties and sensor performance of zinc oxide thin films. Doctoral dissertation, Massachusetts Institute of Technology

Nafarizal N (2016) Precise control of metal oxide thin films deposition in magnetron sputtering plasmas for high performance sensing devices fabrication. Procedia Chem 20:93–97

Nemade KR, Waghuley SA (2013) LPG sensing application of graphene/Bi_2O_3 quantum dots composites. Solid State Sci 22:27–32

Nemade KR, Waghuley SA (2014a) Role of defects concentration on optical and carbon dioxide gas sensing properties of Sb_2O_3/graphene composites. Opt Mater 36(3):712–716

Nemade KR, Waghuley SA (2014b) Highly responsive carbon dioxide sensing by graphene/Al_2O_3 quantum dots composites at low operable temperature. Indian J Phys 88(6):577–583

Nemade KR, Waghuley SA (2014c) Preparation of MnO_2 immobilized graphene nanocomposite by solid state diffusion route for LPG sensing. J Lumin 153:194–197

Niu F, Liu JM, Tao LM, Wang W, Song WG (2013) Nitrogen and silica co-doped graphene nanosheets for NO_2 gas sensing. J Mater Chem A 1(20):6130–6133

Niu F, Tao LM, Deng YC, Wang QH, Song WG (2014) Phosphorus doped graphene nanosheets for room temperature NH_3 sensing. New J Chem 38(6):2269–2272

Öztürk S, Kılınç N (2016) Pd thin films on flexible substrate for hydrogen sensor. J Alloy Compd 674:179–184

Pak Y, Kim SM, Jeong H, Kang CG, Park JS, Song H, Kim JT (2014) Palladium-decorated hydrogen-gas sensors using periodically aligned graphene nanoribbons. ACS Appl Mater Interfaces 6(15):13293–13298

Parmar M, Balamurugan C, Lee DW (2013) PANI and graphene/PANI nanocomposite films—comparative toluene gas sensing behavior. Sensors 13(12):16611–16624

Patel NG, Makhija KK, Panchal CJ (1994) Fabrication of carbon dioxide gas sensor and its alarm system using indium tin oxide (ITO) thin films. Sens Actuators B Chem 21(3):193–197

Patel NG, Makhija KK, Panchal CJ, Dave DB, Vaishnav VS (1995) Fabrication of carbon tetrachloride gas sensors using indium tin oxide thin films. Sens Actuators B Chem 23(1):49–53

Patel NG, Patel PD, Vaishnav VS (2003) Indium tin oxide (ITO) thin film gas sensor for detection of methanol at room temperature. Sens Actuators B Chem 96(1–2):180–189

Pawar SG, Chougule MA, Patil SL, Raut BT, Godse PR, Sen S, Patil VB (2011) Room temperature ammonia gas sensor based on polyaniline-TiO_2 nanocomposite. IEEE Sens J 11 (12):3417–3423

Qin J, Cao M, Li N, Hu C (2011) Graphene-wrapped WO_3 nanoparticles with improved performances in electrical conductivity and gas sensing properties. J Mater Chem 21 (43):17167–17174

Ranade RM, Ang SS, Brown WD (1993) Reactive ion etching of thin gold films. J Electrochem Soc 140(12):3676–3678

Ruangchuay L, Sirivat A, Schwank J (2004) Electrical conductivity response of polypyrrole to acetone vapor: effect of dopant anions and interaction mechanisms. Synth Met 140(1):15–21

Sberveglieri G, Groppelli S, Coccoli G (1988) Radio frequency magnetron sputtering growth and characterization of indium-tin oxide (ITO) thin films for NO_2 gas sensors. Sens Actuators 15 (3):235–242

Sberveglieri G, Benussi P, Coccoli G, Groppelli S, Nelli P (1990) Reactively sputtered indium tin oxide polycrystalline thin films as NO and NO_2 gas sensors. Thin Solid Films 186(2):349–360

Shafiei M, Arsat R, Yu J, Kalantar-Zadeh K, Wlodarski W, Dubin S, Kaner RB (2009) Pt/graphene nano-sheet based hydrogen gas sensor. In: Sensors, 2009 IEEE. IEEE, pp 295–298

Singh G, Choudhary A, Haranath D, Joshi AG, Singh N, Singh S, Pasricha R (2012) ZnO decorated luminescent graphene as a potential gas sensor at room temperature. Carbon 50 (2):385–394

Song N, Fan H, Tian H (2015) PVP assisted in situ synthesis of functionalized graphene/ZnO (FGZnO) nanohybrids with enhanced gas-sensing property. J Mater Sci 50(5):2229–2238

Srivastava S, Jain K, Singh VN, Singh S, Vijayan N, Dilawar N, Senguttuvan TD (2012) Faster response of NO_2 sensing in graphene–WO_3 nanocomposites. Nanotechnology 23(20):205501

Stafström S, Bredas JL, Epstein AJ, Woo HS, Tanner DB, Huang WS, MacDiarmid AG (1987) Polaron lattice in highly conducting polyaniline: theoretical and optical studies. Phys Rev Lett 59(13):1464

Su PG, Peng SL (2015) Fabrication and NO_2 gas-sensing properties of reduced graphene oxide/WO_3 nanocomposite films. Talanta 132:398–405

Su PG, Shieh HC (2014) Flexible NO_2 sensors fabricated by layer-by-layer covalent anchoring and in situ reduction of graphene oxide. Sens Actuators B Chem 190:865–872

Thubsuang U, Sukanan D, Sahasithiwat S, Wongkasemjit S, Chaisuwan T (2015) Highly sensitive room temperature organic vapor sensor based on polybenzoxazine-derived carbon aerogel thin film composite. Mater Sci Eng B 200:67–77

Tien HN, Chung JS, Hur SH (2013) Fabrication of a novel 2D-graphene/2D-NiO nanosheet-based hybrid nanostructure and its use in highly sensitive NO_2 sensors. Sens Actuators B Chem 185:701–705

Uddin AI, Lee KW, Chung GS (2015a) Acetylene gas sensing properties of an Ag-loaded hierarchical ZnO nanostructure-decorated reduced graphene oxide hybrid. Sens Actuators B Chem 216:33–40

Uddin AI, Phan DT, Chung GS (2015b) Low temperature acetylene gas sensor based on Ag nanoparticles-loaded ZnO-reduced graphene oxide hybrid. Sens Actuators B Chem 207:362–369

Van Quang V, Van Dung N, Sy Trong N, Duc Hoa N, Van Duy N, Van Hieu N (2014) Outstanding gas-sensing performance of graphene/SnO_2 nanowire Schottky junctions. Appl Phys Lett 105(1):013107

Varghese SS, Lonkar S, Singh KK, Swaminathan S, Abdala A (2015) Recent advances in graphene based gas sensors. Sens Actuators B Chem 218:160–183

Wu Z, Chen X, Zhu S, Zhou Z, Yao Y, Quan W, Liu B (2013) Enhanced sensitivity of ammonia sensor using graphene/polyaniline nanocomposite. Sens Actuators B Chem 178:485–493

Xue L, Wang W, Guo Y, Liu G, Wan P (2017) Flexible polyaniline/carbon nanotube nanocomposite film-based electronic gas sensors. Sens Actuators B Chem 244:47–53

Yang YJ, Yang XJ, Yang WY, Li SB, Xu JH, Jiang YD (2014a) Porous conducting polymer and reduced graphene oxide nanocomposites for room temperature gas detection. RSC Adv 4:42546–42553

Yang Y, Tian C, Wang J, Sun L, Shi K, Zhou W, Fu H (2014b) Facile synthesis of novel 3D nanoflower-like Cu_xO/multilayer graphene composites for room temperature NO_x gas sensor application. Nanoscale 6(13):7369–7378

Yavari F, Chen Z, Thomas AV, Ren W, Cheng HM, Koratkar N (2011) High sensitivity gas detection using a macroscopic three-dimensional graphene foam network. Sci Rep 1:166

Ye Z, Jiang Y, Tai H, Yuan Z (2014) The investigation of reduced graphene oxide/P3HT composite films for ammonia detection. Integr Ferroelectr 154(1):73–81

Ye Z, Jiang Y, Tai H, Guo N, Xie G, Yuan Z (2015) The investigation of reduced graphene oxide@ SnO_2-polyaniline composite thin films for ammonia detection at room temperature. J Mater Sci Mater Electron 26(2):833–841

Yoo S, Li X, Wu Y, Liu W, Wang X, Yi W (2014) Ammonia gas detection by tannic acid functionalized and reduced graphene oxide at room temperature. J Nanomater 2014

Zhang J, Hu JQ, Zhu FR, Gong H, O'shea SJ (2002) ITO thin films coated quartz crystal microbalance as gas sensor for NO detection. Sens Actuators B Chem 87(1):159–167

Zhang Z, Zou R, Song G, Yu L, Chen Z, Hu J (2011) Highly aligned SnO_2 nanorods on graphene sheets for gas sensors. J Mater Chem 21(43):17360–17365

Zhang J, Zhao C, Hu PA, Fu YQ, Wang Z, Cao W, Placido F (2013) A UV light enhanced TiO_2/graphene device for oxygen sensing at room temperature. RSC Adv 3(44):22185–22190

Zhang H, Feng J, Fei T, Liu S, Zhang T (2014a) SnO_2 nanoparticles-reduced graphene oxide nanocomposites for NO_2 sensing at low operating temperature. Sens Actuators B Chem 190:472–478

Zhang H, Wang L, Zhang T (2014b) Reduced graphite oxide/SnO_2/Au hybrid nanomaterials for NO_2 sensing performance at relatively low operating temperature. RSC Adv 4(101):57436–57441

Zhang D, Liu J, Chang H, Liu A, Xia B (2015a) Characterization of a hybrid composite of SnO_2 nanocrystal-decorated reduced graphene oxide for ppm-level ethanol gas sensing application. RSC Adv 5(24):18666–18672

Zhang D, Liu A, Chang H, Xia B (2015b) Room-temperature high-performance acetone gas sensor based on hydrothermal synthesized SnO_2-reduced graphene oxide hybrid composite. RSC Adv 5(4):3016–3022

Zhou L, Shen F, Tian X, Wang D, Zhang T, Chen W (2013) Stable Cu_2O nanocrystals grown on functionalized graphene sheets and room temperature H_2S gas sensing with ultrahigh sensitivity. Nanoscale 5(4):1564–1569

Zhou X, Wang X, Wang B, Chen Z, He C, Wu Y (2014a) Preparation, characterization and NH_3-sensing properties of reduced graphene oxide/copper phthalocyanine hybrid material. Sens Actuators B Chem 193:340–348

Zhou Y, Jiang Y, Xie G, Wu M, Tai H (2014b) Gas sensors for CO_2 detection based on RGO–PEI films at room temperature. Chin Sci Bull 59(17):1999–2005

Zhu L, Jia Y, Gai G, Ji X, Luo J, Yao Y (2014) Ambipolarity of large-area Pt-functionalized graphene observed in H_2 sensing. Sens Actuators B Chem 190:134–140

Zito CA, Perfecto TM, Volanti DP (2017) Impact of reduced graphene oxide on the ethanol sensing performance of hollow SnO_2 nanoparticles under humid atmosphere. Sens Actuators B Chem 244:466–474

Zou R, He G, Xu K, Liu Q, Zhang Z, Hu J (2013) ZnO nanorods on reduced graphene sheets with excellent field emission, gas sensor and photocatalytic properties. J Mater Chem A 1(29):8445–8452

Chapter 3
Corrosion Monitoring and Control in Aircraft: A Review

Rishi Kant, Pankaj Singh Chauhan, Geeta Bhatt and Shantanu Bhattacharya

Abstract There has been increasing trend for health monitoring of an aircraft to deal its life issues and to minimize the operational cost. The corrosion is one of the important factors which limits the life of an aircraft due to material loss and fatigue cracking. The development of corrosion monitoring systems and models can provide the solution for health monitoring and prediction for remaining life of an aircraft. This chapter provides a comprehensive review on different types of corrosion, corrosion monitoring, and corrosion control. The latest studies on aircraft corrosion have been discussed in chapter. The different techniques based on nondestructive testing methods such as ultrasonic, electromagnetic, radiographic, thermographic have been detailed out in the chapter. Further for corrosion control, different schemes of corrosion control and coating removal techniques have been discussed in detail.

Keywords Corrosion control · Aircraft · Health · Monitoring

3.1 Introduction

Corrosion of airframes is associated with degeneration mechanism which affects aircraft's structural integrity, reliability, and durability. Corrosion is a complex process that begins at microscopic length scale and its manifestation can be clearly visualized at macroscopic level is a major cause of material failure under severe

R. Kant (✉) · P. S. Chauhan · G. Bhatt · S. Bhattacharya
Microsystems Fabrication Laboratory, Department of Mechanical Engineering, Indian Institute of Technology Kanpur, Kanpur 208016, UP, India
e-mail: dsrishikant@gmail.com

P. S. Chauhan
e-mail: pankajs@iitk.ac.in

G. Bhatt
e-mail: geetabht@iitk.ac.in

S. Bhattacharya
e-mail: bhattacs@iitk.ac.in

S. Bhattacharya et al. (eds.), *Sensors for Automotive and Aerospace Applications*, Energy, Environment, and Sustainability,
https://doi.org/10.1007/978-981-13-3290-6_3

environmental conditions. Corrosion brings negative impacts such as aircraft unavailability, safety and financial cost for aircraft maintenance. The associated cost of corrosion is significant which make an impact on the economy of the nation. The United States spends almost $437 billion annually on maintenance and restoration of corroded structures (Winkleman et al. 2011). The trend of cost is rising as the aircrafts are generally exposed to extreme operating environment condition which negatively affects structural integrity and durability of aircrafts. Since, the damage caused by corrosion is cumulative in nature, if not detected at early stage, it may lead to fatigue failure in structure (Jaya et al. 2012) which results in the decrease in load carrying capacity of the aircraft. The aircrafts utilize high-strength aluminum alloys as structural materials which provide high strength to weight ratio, enhanced mechanical properties, improved corrosion resistance and formability. This chapter presents an overview of corrosion related to aerospace industry in introduction section and details out corrosion monitoring and control in subsequent section.

The alloying elements of aluminum significantly affect the corrosion properties which further may be responsible for initiation of various types of corrosion such as pitting, intergranular, galvanic, filiform, crevice, erosive corrosion, environmental assisted cracking, and fretting fatigue. Some of them have been detailed out in following manner.

3.1.1 *Pitting Corrosion*

Pitting corrosion is a localized corrosion of a metallic surface which remains restricted to small area or point that further grows to form cavities (Abood 2008). It frequently occurs in the structure due to corrosion attack. The entire surface of structure is affected by corroding medium which reacts with surface impurities and causes different isolated pits in the surface of structure. Pitting corrosion is responsible for localized stresses which stimulate crack formation in the structure. Pitting with the influence of chemicals can contribute in loss of ductility in the material at the pit surface. The material used in aircraft manufacturing such as high strength aluminum and steel are liable to influence by pitting corrosion under typical environmental condition, especially when protective coating gets broken. The common factors which provide start and the propagation to pitting corrosion are (Roberge 2008)

(i) Protective oxide film damage by localized chemical or mechanical factors.
(ii) Passive film breaks down due to acidity, high chloride concentrations and low dissolved oxygen.
(iii) The existence of non-uniformities (ex. nonmetallic inclusions) in the metallic structure.

Zaya (1984) proposed pitting theory and different stages of pit development which can be visualized in Fig. 3.1. It is divided into four stages. The unattacked surface covered by passive film is represented in stage 0. Stage 1 comprises of the breakage of thin passive film and subsequently the substrate dissolution starts. The

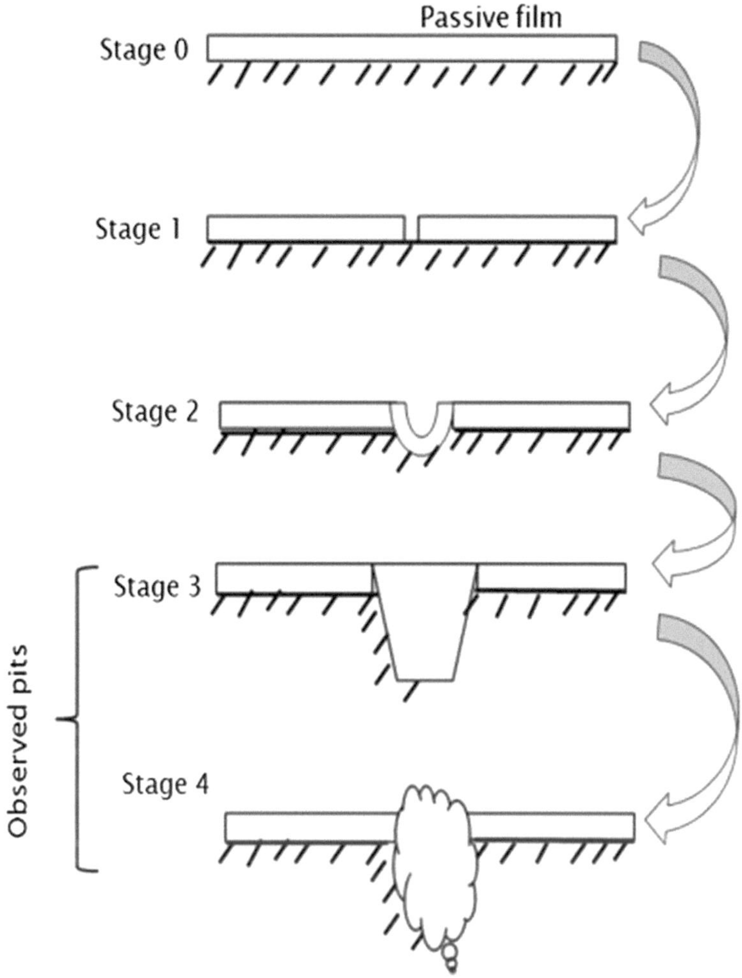

Fig. 3.1 Schematic diagram for pit development stages (Reprinted with permission from Bhandari et al. (2015). Copyright (2015) Elsevier.)

initiated pit starts to grow in stage 2, i.e., repassivation of layer cannot take place furthermore. The disintegration of the substrate starts to grow and larger pit (of 1–10 μm) can be observed under microscope. The shape of pit can be visualized in hemisphere or polyhedron form. Finally in stage 4 pits can be visualized with naked eyes. These pits may also have irregular shape if they are covered with corrosion products during dissolution of metal (Szklarska-Smialowska 1986; Zaya 1984). But several other researchers proposed that stage 0 does not exist, this means that after immersion of film-coated substrate directly starts at stage 1 (Thompson et al. 1978).

Steel and aluminum alloys are used for manufacturing of aircrafts and marine systems. These metals have tendency to form pit on their surface in marine environmental conditions (Roberge 2008). The alloys coated with passive layer get broken in such environmental conditions which initiate degeneration of underlying

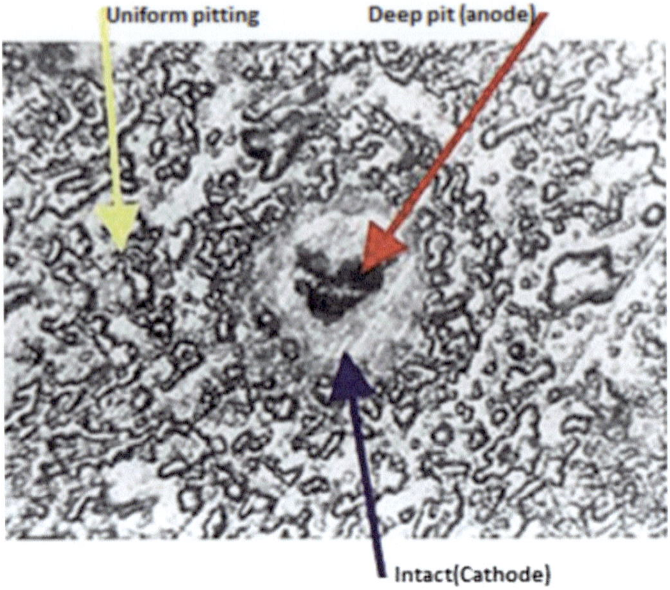

Fig. 3.2 Micrograph of a metal matrix with deep pit (Roberge 2008) (Reprinted with permission from Bhandari et al. (2015). Copyright (2015) Elsevier.)

material (Schiroky et al. 2013). For an instance, Fig. 3.2 is depicting details of deep pit corrosion on a metallic surface.

3.1.2 Intergranular Corrosion

Intergranular corrosion sometimes refers to intercrystalline corrosion or interdendritic corrosion which means between grains or crystals. It is form of corrosive attack which propagates along grain boundary (or interdendritic paths). Usually, it is identified by visualizing microstructure of material with the help of a microscope. The aircraft structures are more prone to this as they utilize high-strength aluminum alloys for manufacturing. Intergranular corrosion is a special type of exfoliation. By ASTM definition, exfoliation is defined as "corrosion that proceeds laterally from the sites of initiation along planes parallel to the surface, generally at grain boundaries, forming corrosion products that force metal away from the body of the material, giving rise to a layered appearance" (ASTM 1986).

The aluminum alloys of 7xxx series consist of zinc and copper elements which increases intergranular attack as zinc and copper produce galvanic couple among grain boundaries. This couple causes corrosion at grain boundaries which finally provides initiation point for further corrosion.

Usually, corrosion damage is identified by inspections which are complex, time-consuming and require disassembly of aircraft. But these inspections may include extra costs arising from incidental damage of structure and increased the possibility of corrosion due to inferior sealing (after repair). These cost associated with maintenance and repair of an aircraft can be reduced by employing structural health monitoring systems (SHMS). The structural health monitoring systems can detect broken coating, identify starting place of corrosion and characterize the environmental condition in specific area of an aircraft. The essential tools of SHMS are environment monitors, corrosion sensors in a network for providing precise and reliable information about corrosion, corrosion prediction models, and decision-making for maintenance.

3.2 Corrosion Monitoring

As we have discussed the corrosion may cause severe damage to the working, reliability, safety, and capital involved in aircraft. To avoid such a scenario it becomes extremely important to monitor, detect, and remove the possible causes of corrosion. This activity enables the preventive maintenance strategy to find out damages to the aircraft under scheduled inspections. Among various techniques of detection nondestructive techniques are widely used.

According to the American Society for Nondestructive testing (ASNT) the nondestructive testing (NDT) is defined as: "The determination of the physical condition of an object without affecting that object's ability to fulfill its intended function. Nondestructive testing techniques typically use a probing energy form to determine material properties or to indicate the presence of material discontinuities (surface, internal or concealed)". NDT techniques enable a continuous and safe monitoring of the aircraft. In health monitoring system of the aircraft, the NDT sensors are attached to the system or components to be monitored and give information about a particular physical or chemical property which is analyzed. The parameter's being monitored by NDT sensors are defined by the user according to the system properties such as material characteristics and working environment. The level of acceptability of a particular value is decided by the taking factor of safety into account. NDT techniques are classified according to their physical operating principle as follows.

3.2.1 Visual Testing (VT)

Visual testing is the most basic and easy technique where a trained inspector can detect the corrosion by using a magnifying glass or a flashlight. Hence, it is called the first line of defence against corrosion. This technique has limitation of accessibility to reach to enclosed part of the components. A device known as borescope is generally used in these circumstances. In borescope, a miniaturized camera is used which is attached to the optical cable and real-time images are viewed on screen by the inspector.

3.2.2 *Ultrasonic Testing (UT)*

The term ultrasonic is used for the frequencies above the human hearing capability. The normal values of ultrasonic frequencies used in ultrasonic techniques are 1–10 MHz. Ultrasonic waves can travel through the material easily irrespective of the type of material. The ultrasonic waves are reflected at the boundaries of the material, where the property of material such as density changes. Hence, at crack location the wave is reflected back and received at the source. The time taken for the travel of wave also gives the distance covered which provides the location of crack.

Figure 3.3 shows penetrant testing and ultrasonic testing on a specimen (length 13 mm and width 9 mm). The dye penetration can be clearly visualized at different location from point 1 to point 4 in Fig. 3.3a while in Fig. 3.3b a representation of cracks at different location can be visualized.

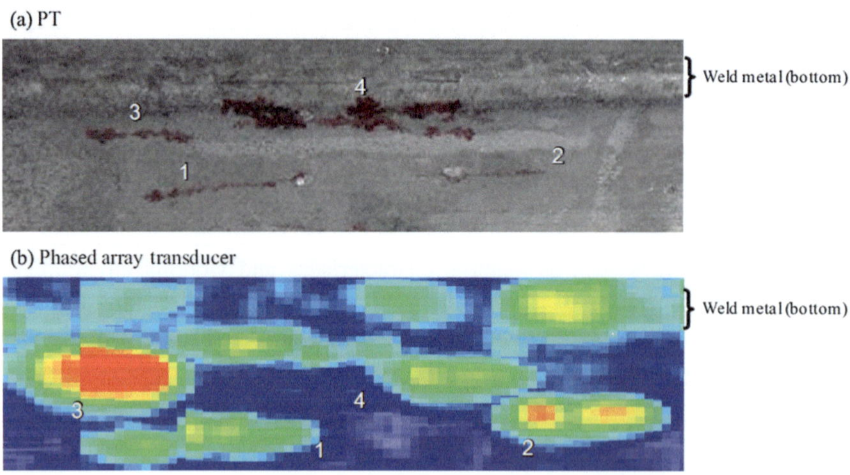

Fig. 3.3 Results of **a** the penetrant testing, **b** ultrasonic testing using the phased-array piezoelectric transducer (Reprinted with permission from Nakamura et al. 2016. Copyright (2016) Elsevier.)

3.2.3 *Thermographic Testing (TG)*

In this technique, thermal imaging infrared (IR) cameras are used to monitor the hot spots of component which is an indicator of fault. The energy is supplied to the test object and flow of heat through the material is monitored. And the change in heat flow at discontinuities present in the material is detected. In composite materials flash thermography technique is being successfully used to quantify defects. The following Fig. 3.4 represents the defects in this specimen. This technique can quantify defects (by calculating size of pixels) in quick time (30 s).

A comparison between two nondestructive testing (ultrasonic and thermographic testing) has been made in Table 3.1. The major difference between two techniques is the time taken to detect cracks. The ultrasonic testing method takes around 30 min while in thermographic method takes around 30 s to carry out detection process. Other differences have been details out in Table 3.1.

Since, these testing methods rely upon the geometry and sizes of testing specimen, so different method are used for distinct samples. Like, Ultrasonic testing is best suitable for small sample while thermographic technique is suited to all types of geometry. The optical method in comparison to ultrasonic testing has advantage as it can measure different geometrical discontinuities. But, thermographic method keeps restriction on thickness of specimen which is being tested for corrosion defects.

3.2.4 *Electromagnetic or Eddy Current Testing (ET)*

In this technique, a conducting coil carrying alternating current is placed close to the test object. In response the object generates eddy currents which are detected by

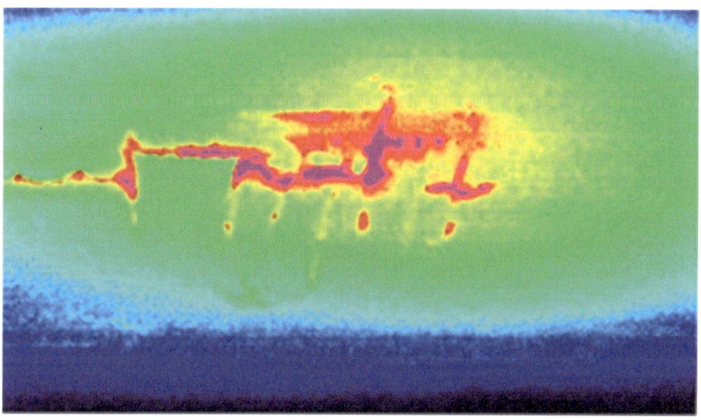

Fig. 3.4 InfraRed Thermography map of Specimen A (Reprinted with permission from Garnier et al. (2011). Copyright (2011) Elsevier.)

Table 3.1 Various characteristics of ultrasonic testing, thermographic testing (Reprinted with permission from Garnier et al. (2011). Copyright (2011) Elsevier.)

Testing method	Ultrasonic testing	Thermographic testing
Inspection type	Contact	Non-contact (optical)
Measurement	By mechanical vibration	By thermal radiation
Loading	Acoustic wave	Extended heating pulse, excitation of transient pulse
Output	Amplitude and time of flight of ultrasonic wave	Series of thermal images
Analysis	Quantitative and qualitative analysis by the ultrasonic amplitude	Quantitative and qualitative analysis by temperature distribution
Advantages	Depth of defect can be evaluated quickly	Quick evaluation of defect dimensions and control adapted of all the geometries types
Disadvantages	Gradual time of control and choice of different probe for different specimen	Defect depth not directly evaluated
Limitations	Depends upon size, shape and depth of discontinuity	Sample thickness is critical

a separate sensor. If a crack is present on or below the surface the induced eddy current change due to change in material's electromagnetic property. Thus the defect or cracks generated by corrosion can be detected by using this portable technique. Figure 3.5 shows the mechanism of magnetizing-based eddy current testing (MB-ECT).

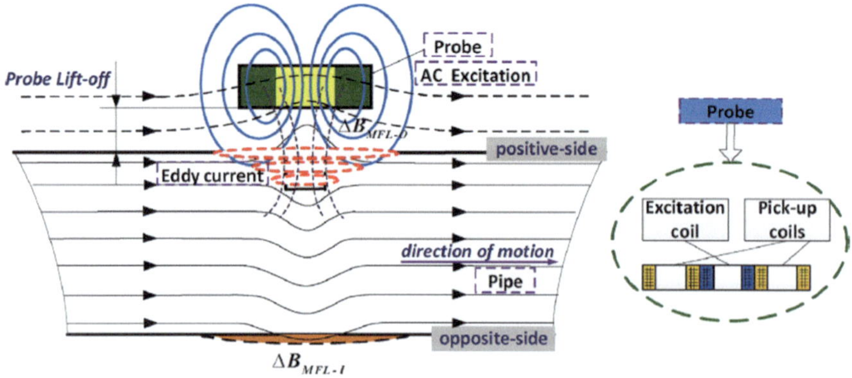

Fig. 3.5 The schematic diagram of magnetizing-based eddy current testing (MB-ECT) method and MB-ECT sensor (Reprinted with permission from Deng et al. (2018). Copyright (2018) Elsevier.)

3.2.5 Radiographic Testing (RG)

Radiographic technique is also a very common and old technique of surface deformity detection. It uses X-rays to generate the radiography images of the test object and the discontinuities are viewed. It is the very useful method used for casted, forged, and welded parts to identify the location of pores and cracks. The cracks generated by corrosion can be easily detected by this technique. This method has limitations in view of safety and generation of residual stresses due to the use of X-rays. Figure 3.6 shows the region of interest of a crack, root under cut and no disturbance in the signal with profile plots and cross-sections.

Above-mentioned testing methods are successfully employed in detection, monitoring, and control of the defects formed by corrosion and other conditions. In an aircraft, there are several small to large components which can be subjected to the different types of corrosion. Wings, engine components, fasteners, welded joints, surface coatings, etc., are such parts of an aircraft which are prone to the corrosion and needs to be monitored regularly.

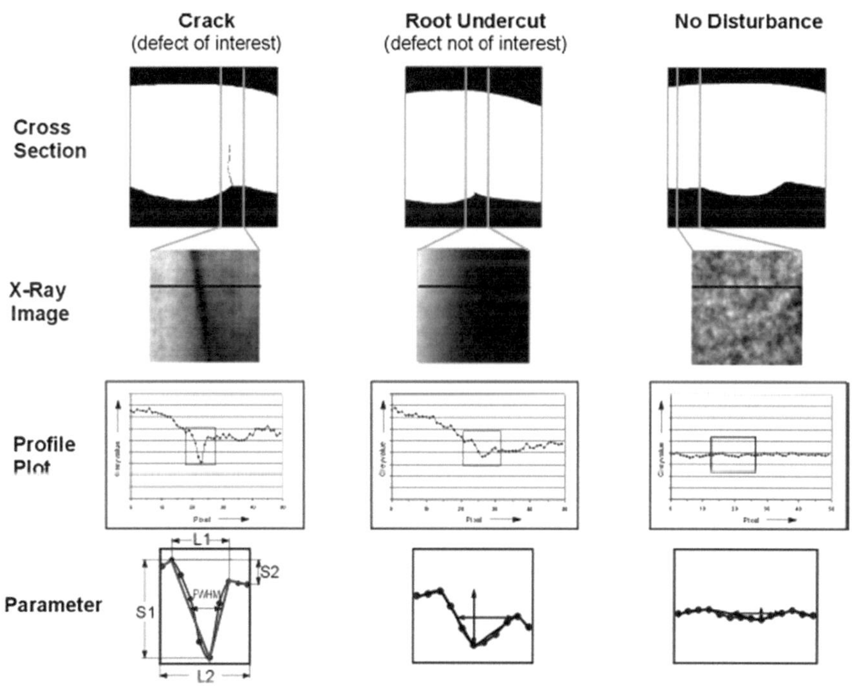

Fig. 3.6 Regions of interest of crack, of an undercut and of a region without disturbance with corresponding cross-sections and profile plots (Reprinted with permission from Perner et al. (2001). Copyright (2001) Elsevier.)

3.3 Corrosion Control

Due to the severe effects generated through corrosion, as discussed in earlier sections, corrosion control becomes a very important aspect to study. There are many issues related to corrosion control phenomenon and these are classified depending on the different aspects related to aerospace assembly and launching. The classifications are

(i) Different corrosion control schemes in aircrafts
(ii) Corrosion control in space launch vehicles
(iii) Techniques for coating removal.

3.3.1 *Different Corrosion Control Schemes in Aircrafts*

Aluminum alloy is the most used alloy for the aircraft industry due to the benefits it provides, like higher strength to weight ratio, as compared to other alloys. It contains many different metals like zinc, copper, magnesium, etc., to enhance the strength of the alloys but the inclusion of these metals also increases the susceptibility of the alloy towards corrosion. Additionally when there is any sort of relative movement between two parts, the friction and the corrosion tendency increase. Hence to take care of the issues generated through doping and relative motion, the need to restrain these effects increases. For the reduction of friction and corrosion in the aircraft raised through relative motion of parts, greases (lubricating material) are applied on the interconnecting surfaces. The greases contain lubricating base oil, a thickener and performance enhancing additives (Totten et al. 2003). This is the basic prevention method for reducing corrosion, while apart from greasing, the very important aspect to corrosion control is aircraft coating. Aircraft coatings consist of three layers, conversion coating (chemically grown layer to enhance bonding strength of coating), primer and top coating. Inhibitors like hexavalent chromium are also added to conversion coating/primer to provide effective coating, but inhibitors cost are high. Along with the coating, there are other different techniques which are used to protect the areas where coating is not useful to protect corrosion.

Corrosion prevention compounds are the materials which are used to inhibit new corrosion sites as well as to restrain the already grown corrosion sites. As an application of these compounds to aircraft can be done on site itself, these compounds are fetching great interest in themselves (Hinton et al. 1996). These compounds generally involve two main components, one is organic solvent (aliphatic hydrocarbon) and the other is film former (low-molecular-weight hydrocarbons). Organic solvent acts as a carrier and helps the compound to dispense properly by maintaining required viscosity depending on the application. Figure 3.7 shows the

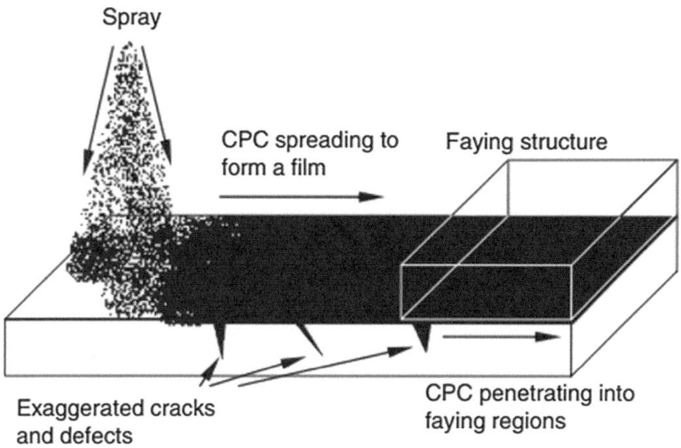

Fig. 3.7 Schematic illustrating the regions in which corrosion prevention compound provide protection (Reprinted with permission from Benavides (2009). Copyright (2009) Elsevier.)

schematic to illustrate the regions in which corrosion prevention compound can provide protection.

The other method of protection of aircraft surface from corrosion is to modify the surface. For surface modification, techniques like ion implantation and laser treatment are used.. Laser treatment has successfully shown increase in corrosion resistant (Anthony and Cline 1978) to steel surface and it is done through laser surface alloying or melting (followed by non-equilibrium cooling) (Watkins et al. 1997). Hence the different microstructure that gets generated through the melting/alloying and cooling process is observed to be corrosion resistant. Another important surface modification process is ion implantation. In this technique, a positively charged ions beam is impinged on the surface and after collision of ions with the atoms of the substrate, a uniformly doped layer is formulated. The doping of different materials like molybdenum (Al-Saffar et al. 1980) and other metals has shown improved corrosion resistant. The following Fig. 3.8 shows the schematic of ion implantation process.

3.3.2 Corrosion Control in Space Launch Vehicles

As the launch pad/launch vehicles are the inherent part of the aerospace system, the corrosion in launch pad structures, related hardware and facilities remains a challenge. Out of these, material selection, drainage/purging system and coating system for different parts of launch pad structure to control corrosion have been discussed.

The corrosion control of space shuttle orbiter is based on the vision of no failure with respect to structure for at least 10 years or 100 mission life (Benavides 2009). For this constraint, the material selection has always been the most important factor

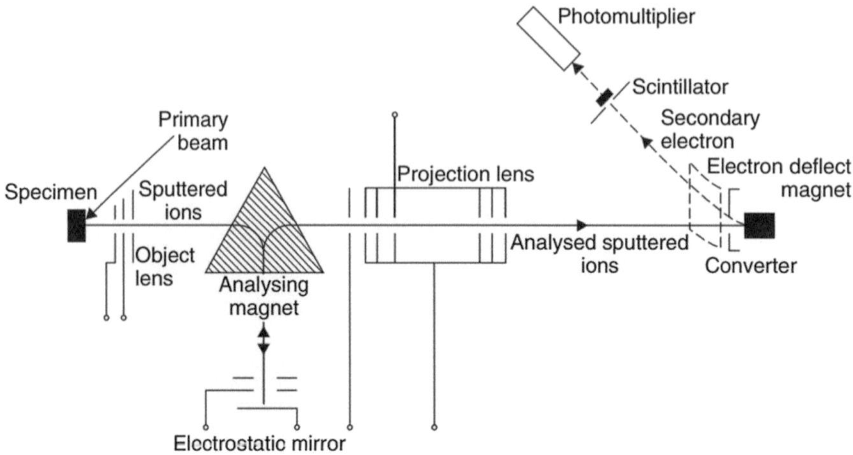

Fig. 3.8 Schematic of ion implantation process (Reprinted with permission from Benavides (2009). Copyright (2009) Elsevier.)

to consider. Hence to reduce the level of corrosion, the materials which are used for launch pad structure are 'A' rating materials (materials highly resistant to corrosion in normal environmental conditions) which are further required to meet MSFC-SPEC-250 (NASA 1977), class II requirement (NASA specification for defining protective finishes for space structures and related equipment). MSFC-SPEC-522 (NASA 1987) is the guideline which has to be followed by chosen metals and these are related to the rating of material depending on stress corrosion cracking. The other restrictions which are required to be followed are defined through Standard Design Manual. It prohibits using galvanically dissimilar metals in contact without suitably protecting them against corrosion. It recommends sealing faying surfaces of dissimilar metals against water inclusion or suggests using a separation layer of corrosion-inhibiting epoxy or room temperature vulcanized silicon rubber. It recommends installing fasteners wet with chromated primer epoxy. The primer epoxy (Super Koropon primer) has also been studied for its effect in aging process of coating (Lomness and Calle 2006) and it has proved efficient in reducing corrosion in this manner also.

The drainage system, which consists of many through holes and vacuum lines, is designed in a manner to not accumulate the water in open orbiter structure. Passive gravity as well as active vacuum line, where passive gravity does not work, is utilized to avoid water accumulation. Additionally the purge system, consisting of a series of onboard ducts to allow purging of un-pressurized compartments is implemented with dry nitrogen gas system to purge the interior spaces of the orbiter vehicles. The different zones of launch pad structure undergo different working environments with respect to direct/indirect rocket engines exhaust impingement, acid deposition, and temperature. NASA STD-5008 (NASA 2011) Qualified Product List provides standard for choosing corrosion protection coatings for launch structures. The coating of carbon steel launch structures is done with zinc primer in which zinc primers are coated with an inorganic topcoat above 30–34 m and for high heat area (at 30 m) an ablation topcoat is used. The wing covers are coated with epoxy polyurethane.

3.3.3 Techniques for Coating Removal

The major source of saving aerospace parts from corrosion is to coat. Aerospace coating is exposed to extreme working conditions from −40 to +40 °C (Benavides 2009). The coatings are also exposed to severe chemicals, sands and mechanical damages like maintenance through some tool/machine. As the coating undergoes various conditions which distort it, it would not be eligible for serving the purpose it is intended to. Typical life of coating is 5–6 years, while for that of the aircraft is more than 20 years. So refurbishing is always a required option to enhance the quality of coating and its respective efficiency. When the coating becomes cracked or stripped, the requirement of removing the original coating and applying new coating arise. However there are various factors which affect the removal of coating like hardness of coating (as it is difficult to remove harder coatings), limit of adhesion of coating (lower adhesion helps easy removal), thickness of coating (thick coatings are tougher to remove) and color of coating for some removal methods.

Scuff sanding, the manual sandpaper based abrasion method is the most common method to partially remove the coating. The other coating removal processes are based on different mechanisms like molecular dissociation, which uses chemical strippers of different pH, thermal methods, which uses robotic laser or heat lamps (CO_2) or impact methods. The impact method uses either high pressure water (up to 4000 bar) or particles projected towards surface at speed of the range 100–200 m/s (e.g., plastic media (dry stripping through different crushed plastic media) (USMilitary 1998), sodium bicarbonate (can be dry or wet mode), sponge media (dry stripping process using pliant matrix material) etc.). Depending upon the requirement, different method is selected.

3.4 Conclusion

Determination of location of corrosion in an aircraft structure is important to identify defects caused by corrosion or some other means. Corrosion defects associated with aircraft structure has been outlined in this chapter. The costs associated with corrosion, different types of corrosion, corrosion monitoring and corrosion control schemes have been discussed here in detailed manner. Different types of nondestructive techniques like visual, ultrasonic, thermographic, radiographic, and electromagnetic have been discussed in depth for corrosion monitoring in the structure. The comparisons among different nondestructive techniques have also been detailed here. The different schemes related to corrosion control have been explained in detail. In summary, this chapter provides latest review on corrosion attack types, corrosion monitoring, and control in the metallic structure of aircraft.

References

Abood TH (2008) The influence of various parameters on pitting corrosion of 316L and 202 stainless steel. Department of Chemical Engineering of the University of Technology, University of Technology

Al-Saffar AH, Ashworth V, Bairamov AKO et al (1980) The effect of molybdenum ion implantation on the general and pitting corrosion behaviour of pure aluminium and a high strength aluminium alloy. Corros Sci 20:127–144. https://doi.org/10.1016/0010-938X(80)90116-X

American Society for Testing and Materials (ASTM) (1986) Compilation of ASTM standard definitions, 6th edn. American Society for Testing and Materials, Philadelphia

Anthony TR, Cline HE (1978) Surface normalization of sensitized stainless steel by laser surface melting. J Appl Phys 49:1248–1255. https://doi.org/10.1063/1.325015

Benavides S (2009) Corrosion control in the aerospace industry. Elsevier

Bhandari J, Khan F, Abbassi R, Garaniya V, Ojeda R (2015) Modelling of pitting corrosion in marine and offshore steel structures—a technical review. J Loss Prev Process Ind 37:39–62

Deng Z, Kang Y, Zhang J, Song K (2018) Multi-source effect in magnetizing-based eddy current testing sensor for surface crack in ferromagnetic materials. Sens Actuators A Phys 271:24–36. https://doi.org/10.1016/j.sna.2018.01.009

Garnier C, Pastor ML, Eyma F, Lorrain B (2011) The detection of aeronautical defects in situ on composite structures using non destructive testing. Compos Struct 93:1328–1336. https://doi.org/10.1016/j.compstruct.2010.10.017

Hinton B, Trathen P, Haberecht P, Bushell P (1996) Prevention and control of corrosion on aircraft structure with corrosion prevention componds. In: 4th international aerospace corrosion and control symposium, Jakarta, Indonesia

Jaya A, Tiong UH, Clark G (2012) The interaction between corrosion management and structural integrity of aging aircraft. Fatigue Fract Eng Mater Struct 35(1):64–73

Lomness JK, Calle LM (2006) Comparison of the chromium distribution in new Super Koropon Primer to 30 year old Super Koropon using focused ion beam/scanning electron microscopy. KSC Corros Technol Lab 168–286

Nakamura N, Ashida K, Takishita T, Ogi H, Hirao M (2016) Inspection of stress corrosion cracking in welded stainless steel pipe using point-focusing electromagnetic-acoustic transducer. NDT E Int 83:88–93. https://doi.org/10.1016/j.ndteint.2016.06.005

NASA (1977) MSFC-SPEC-250A (1977), general specifications for protective finishes for space vehicle structures and associated flight equipment

NASA (1987) MSFC-SPEC-522 (1987), design criteria for controlling stress corrosion cracking

NASA (2011) NASA-STD-5008, protective coatings of carbon steel, stainless steel, and aluminum on launch structures, facilities, and ground support equipment

Perner P, Zscherpel U, Jacobsen C (2001) A comparison between neural networks and decision trees based on data from industrial radiographic testing. Pattern Recognit Lett 22:47–54. https://doi.org/10.1016/s0167-8655(00)00098-2

Roberge PR (2008) Corrosion engineering: principles and practice. McGraw-Hill, New York

Schiroky G, Dam A, Okeremi A, Speed C (2013) Pitting and crevice corrosion of offshore stainless steel tubing

Szklarska-Smialowska Z (1986) Pitting corrosion of metals. National Association of Corrosion Engineers

Thompson G, Furneaux R, Wood G, Richardson J, Goode J (1978) Nucleation and growth of porous anodic films on aluminium

Totten GE, Westbrook SR, Shah RJ (2003) Fuels and lubricants handbook: technology, properties, performance, and testing, Volume 1_Chapter9: Additives and additives chemistry. In: Fuels and lubricants handbook: technology, properties, performance, and testing, vol 1, p chap 5

USMilitary (1998) MIL-P-85891A, plastic media for removal of organic coatings

Watkins K, McMahon M, Steen W (1997) Microstructure and corrosion properties of laser surface processed aluminium alloys: a review. Mater Sci Eng A 231:55–61. https://doi.org/10.1016/S0921-5093(97)00034-8

Winkleman A, Svedberg EB, Schafrik RE, Duquette DJ (2011) Preventing corrosion from wearing our future away. Adv Mater Process 169:26–31

Zaya PGR (1984) Evaluation of theories for the initial stages of pitting corrosion. McMaster University, McMaster

Chapter 4
Energy Harvesting Techniques for Powering Wireless Sensor Networks in Aircraft Applications: A Review

Poonam Sundriyal and Shantanu Bhattacharya

Abstract In the recent years, wireless sensors (WS) have utilized for various structural applications such as structural health monitoring and flight tests. The conventional battery system to power these sensors is inefficient due to several limitations including heavyweight, large size, less power output, short cycle life, and requirement of their replacements. Therefore energy harvesting systems are gaining much research attention to build the self-powered WS. The energy harvesting unit can directly extract the energy from the local environment such as pressure, vibration, wind, thermal gradients and the solar source. The reported energy harvesting methods are piezoelectric, triboelectric, thermoelectric, electromagnetic, and magnetostrictive. This review focuses on the different energy harvesting techniques and challenges of their integration with the WS for the aircraft applications.

Keywords Energy harvesting · Wireless sensors · Aircraft structures
Structural health monitoring

4.1 Introduction

An aircraft structure consists of a large number of real-time sensors for the proper operation of the different systems. These systems include aircraft engine control system, flight control system, structure health monitoring system, and the cabin

P. Sundriyal (✉) · S. Bhattacharya (✉)
Department of Mechanical Engineering, Indian Institute of Technology Kanpur, Kanpur 208016, UP, India
e-mail: poonams@iitk.ac.in

S. Bhattacharya
e-mail: bhattacs@iitk.ac.in

P. Sundriyal · S. Bhattacharya
Microsystems Fabrication Laboratory, Indian Institute of Technology Kanpur, Kanpur 208016, UP, India

S. Bhattacharya et al. (eds.), *Sensors for Automotive and Aerospace Applications*, Energy, Environment, and Sustainability,
https://doi.org/10.1007/978-981-13-3290-6_4

environment control system. The conventional systems, which rely on the wired connections, are absolute due to the complex circuitry, heavyweight, and short life (Hespanha et al. 2007; Dilhac and Bafleur 2014). Also, the harsh environmental conditions lead to the degradation and failure of the wired connections. According to a report, the U. S. Navy lost six aircraft over a ten year period and cancelled 1000 missions each year due to the wiring problems (Collins 2006). Therefore, the wireless sensor network (WSN) has attracted the considerable attention to improving the aircraft system efficiency by reducing the system weight, improving the fuel efficiency, and increasing the system flexibility. Also, the use of WSN significantly reduces the overall cost of the system.

Aircraft maintenance is one of the major challenges faced by the aircraft industries. Aircraft tend to develop cracks due to substantial changes in loading conditions, humidity, temperature, speed, and pressure. One of the essential tasks is the prediction, monitoring, and prevention of the structural damages. The structural health monitoring (SHM) is used to predict the damage and failures in the structures (Zhao et al. 2007; Ciang et al. 2008). Integration of the WSN in the SHM system is recently adopted by the various aircraft manufacturing industries (Wang and Yuan 2007). However, powering the WSN is the main problem for their efficient function. A WSN consists of the WS and an independent energy source. In most of the systems, batteries are used to power the WSN. However, the batteries possess the inherent problems such as short lifetime, limited power density, leakage problem, and massive weight. These issues necessitate the replacement of hundreds of the discharged batteries which increases the maintenance cost and the environmental pollution. Notably, the batteries fail at the high altitudes due to the temperature rise (>60 °C) in some areas which may cause safety problems like fire, thermal runaway, and the structure failures (Vanhecke et al. 2015). Therefore, the batteries should be eliminated from the aircraft structures, and these should be replaced by the self-powered energy harvesting devices (Mitchell 2007). Although, the stand-alone use of the energy harvesting component is not sufficient and integration of an energy storage component is essential to store the excess harvested energy and to maintain the energy supply during less power generation conditions (Dilhac and Bafleur 2014).

Supercapacitors are the efficient energy storage devices that are widely adopted in the various energy storage applications due to their long life, high energy and power density, environment friendliness, and high stability (Gogotsi 2014; Sundriyal and Bhattacharya 2017a, b). The recent progress in the area of flexible and microscale supercapacitors has further favored its integration with the other electronic devices such as energy harvesting devices (nano-generators), (Zi et al. 2015; Zhong et al. 2017; Pu et al. 2018) sensors (Sun et al. 2018; Thostenson et al. 2018; Yun et al. 2018), and other electronic devices (Ferdous et al. 2016). Till date, the reported energy harvesting techniques are piezoelectric (Ma et al. 2018), triboelectric, thermoelectric (Samson et al. 2010), electromagnetic (Costanzo et al. 2014), and magnetostrictive (Jafari et al. 2017). With the increased focus on the flexibility for the advanced electronic devices, the integrated system should be flexible enough which can be easily mounted on the complexly shaped surface of the aircraft (Cheng et al. 2015; Liu et al. 2015; Sundriyal and Bhattacharya 2018). Also, with the rapid growth of the electronics waste and the raised environmental pollution, more focus should be devoted for the green electronics based devices for the sustainable future (Irimia-Vladu 2014; Patel et al. 2017; Sundriyal and Bhattacharya 2017a, b; Xu et al. 2017). This review demonstrates state of the art in the area of integrated energy harvesting devices and the supercapacitors for application in WSN. It also focuses on the various strategies used in the fabrication of the integrated energy storage-energy harvesting devices and the associated challenges for their utilization in the WSN for the aircraft applications.

4.2 Energy Sources in the Aircrafts

The aircraft consists of the various energy sources which can be utilized by the energy harvesting techniques. These energy sources include vibrations, temperature differences, pressure differences, and strains. The details of the energy sources are as follows.

4.2.1 Vibrations

Both internal and external mechanical vibrations occur in an aircraft structure. Figure 4.1 shows the distribution of the vibration levels in aircraft structures. The major internal source of the vibrations is the propulsion system. The generated vibrations vary in range and depend on the associated vibration source and the location of the affected area (Le et al. 2015). One of the previous studies reports the rotor generated vibration on the flight officer seat shows an acceleration of 0.8 m^2/s at a frequency of 73.6 Hz (Smith 2006). The components near the engine and gearbox experiences more vibration due to the multicomponent frequencies. For example, a gearbox system shows acceleration of 3.5 g at 80 Hz (Dickerson 2011). Turbulence due to the air is the main source of external vibrations, and the vibration frequencies depend on the altitude of the aircraft (Le et al. 2015).

4.2.2 Thermal Gradients

Another important source of energy in the aircraft is the thermal gradients which are always present in the aircraft. Figure 4.2 displays the thermal gradients exist in the different parts of the airplanes and helicopters. Literature shows that the military aircraft have a thermal gradient of $\sim$120 °C in the wing which is around 10–20 °C for the normal airlines (Tendeland and Schlaff 1948; Kordes and Reed 1962). The other components also show the temperature differences such as the rotor blade and

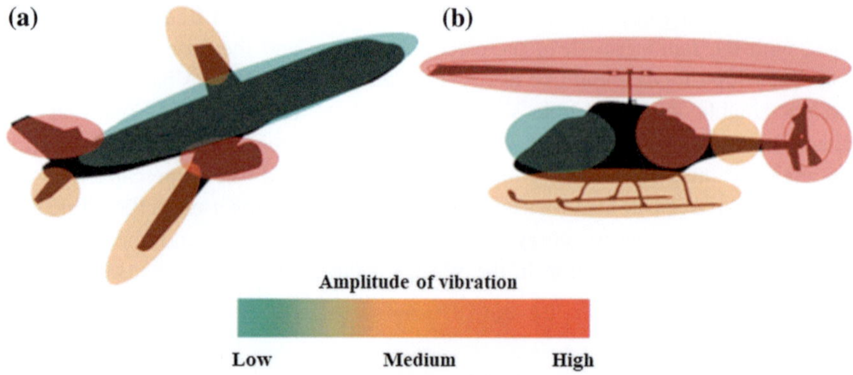

Fig. 4.1 Distribution of the vibration levels at different areas of the: **a** airplane and **b** helicopter structures. Reproduced from Le et al. (2015) with permission from the Elsevier

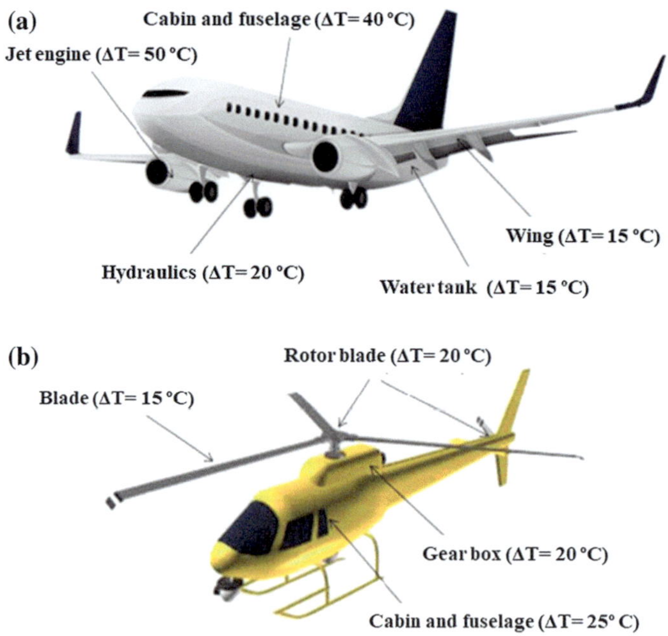

Fig. 4.2 Distribution of the thermal gradients at different areas of the: **a** airplane and **b** helicopter structures. Reproduced from Le et al. (2015) with permission from the Elsevier

Fig. 4.3 The energy powering components of a solar-based aircraft. Reproduced from Gao et al. (2015) with permission from the Elsevier

gearbox have the average thermal gradient of 20 °C (Ducharme 2006), the hydraulic system situated in the fuselage shows a thermal gradient of 40 °C (Pearson et al. 2012), and some thermal gradient also occurs in the cabin (Ducharme 2006).

4.2.3 Solar Energy

Solar energy is a potential source of energy which provides excellent power output (Raghunathan et al. 2005; Alippi and Galperti 2008). Figure 4.3 displays an aircraft with the solar panels. However, some limitations such as changing weather conditions (absence of sunlight during the night and cloudy weather), low efficiency, variation in the incident angles throughout the day, and the complex structure hinder its full use for the energy harvesting techniques (Alippi and Galperti 2008; Gao et al. 2015). Therefore, the solar-powered systems must be designed properly to get the required outcomes (Klöckner et al. 2012).

All the mentioned energy sources namely vibration, thermal gradients, and the solar are the promising sources for powering a WS for different applications of the aircraft structures.

4.3 Energy Harvesting Techniques

The energy can be harvested from different environmental sources based on various techniques. The potential energy harvesting techniques for an aircraft are as follows.

4.3.1 Vibration Energy Harvesting Techniques

Vibration energy harvesting techniques have gained great attention to power the small electronic devices. It converts the mechanical vibrations to the electrical energy based on different energy conversion methods (Wei and Jing 2017). These methods include piezoelectric, electromagnetic, and electrostatic conversion methods (Erturk 2009). The brief introduction of these methods and state of the art for their application in aircraft structure are as follows.

4.3.1.1 Piezoelectric Technique

The piezoelectric generators are used to convert the vibration induced mechanical strain into the electrical charge. The piezoelectric energy harvesters have greatly explored for integrating into the different microelectronic devices, and it is one of the widely used energy harvesting methods for the aircraft applications. The schematic diagram (as shown in Fig. 4.4) shows the mechanism of a piezoelectric converter. In the no stress state, the generator shows the neutral state with no flow of the electrons. After the applied compression stress, the generated electrons will flow from the top toward the bottom electrode through an external circuit. The collected electrons on the bottom electrode will balance the potential which generates the positive voltage and the current signals. The generated potential and current signals instantly vanish with the removal of the compression stress.

For the piezoelectric based harvesting, two configurations have been used. In the first configuration, the piezoelectric component is directly mounted on the structure. While on the other one, an intermediate mechanical component is used which provides easier operation. However, matching the resonance frequency modes of the intermediate component and the structure is a challenge (Pearson et al. 2012). In the case of direct mounting, the host structure transforms the input energy into the mechanical energy, which is further converted to the electrostatic energy using the piezoelectric component. It provides a high energy harvesting efficiency over a wide range of frequencies (Lallart 2008). On the other hand, the indirect mounting configurations have the lower energy harvesting efficiency, due to the transfer of some part of the host structure energy to the intermediate mechanical component. However, the benefit of the indirect mounting is the flexible and independent design of the host structure which leads to the easy maintenance of the system. The following points should be considered while designing a piezoelectric generator (Le et al. 2015):

- The piezoelectric element should be properly positioned near the maximum stress/strain locations (for direct mounting) or the maximum acceleration locations (for indirect mounting).
- Selection of a proper piezoelectric component which shows high coupling coefficients.

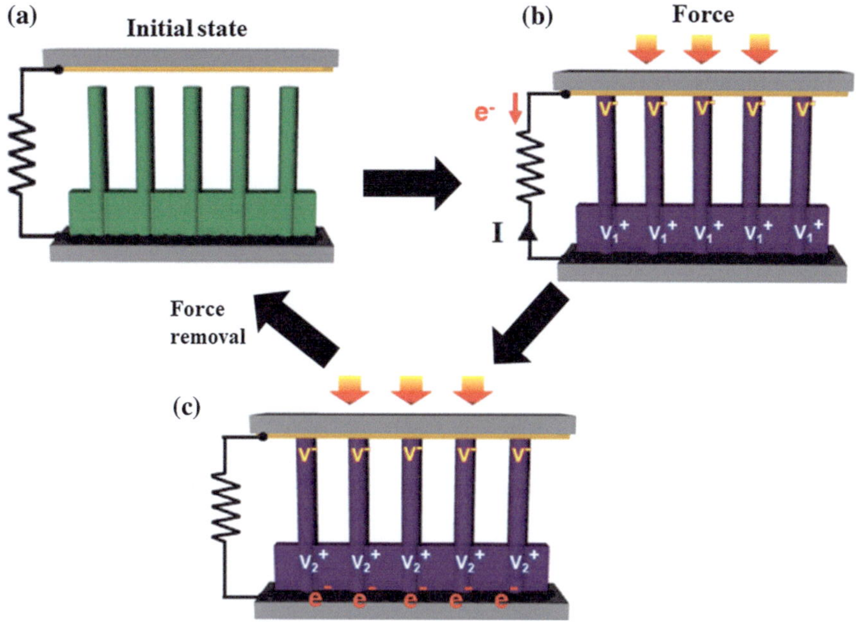

Fig. 4.4 Schematic of a piezoelectric generator in different states: **a** initial state; **b** under externally applied force; **c** the generated positive potential due to the presence of some electrons near the bottom electrode. Reproduced from Kumar et al. (2011) with permission from the American Chemical Society

A huge amount of energy can be obtained at the wide frequency range by using the structure optimization. The reported methods to achieve the optimum performance are the selection of a proper geometry (Zhu et al. 2009), nonlinear structures (Blystad et al. 2010), using several cantilevers with different lengths (Shahruz 2006), and using variable resonance frequencies (Lallart et al. 2010). The piezoelectric methods are widely adopted for powering the sensors in the aircraft structures. Arm et al. (2010) coupled the piezoelectric component to the pitch link (a rod to control the rotor's angle of attack and bears huge dynamic loads) of a helicopter. The harvested strain energy was sufficient to power a WS node. The integrated energy harvester—WS tracked the load records of the helicopter components to decide the conditional maintenance, to reduce the testing costs, and to improve the safety of the aircraft.

Dumas et al. also used a piezoelectric component with the wireless network system for damage detection in the aircraft (Dumas et al. 2011). The degree of damage was checked by the comparative analysis of the generated ultrasonic waves and the stored reference signals. They also validated their experimental data with the finite elemental results and provided a reference for the design of a structure monitoring system.

Montheard et al. have presented an aeroacoustic energy harvesting system to power a WS (Monthéard et al. 2014). The energy harvesting component was installed on the outer side of the aircraft to capture the air turbulence energy. The device successfully generated 2 mW power from 0.5 Mach air flow speed. Jong et al. have demonstrated a piezoelectric energy harvesting device mounted on the damper rod to achieve a high energy output of 7.5 W per damper (de Jong et al. 2011). This energy is sufficient to power the several sensing nodes. Despite the several advantages of the piezoelectric energy harvesters, its limitations are charge leakage, aging, depolarization, and brittleness.

4.3.1.2 Electromagnetic Technique

The electromagnetic generator converts the motion of a permanent magnet to produce the voltage across the coil terminals. The generated voltage is further used to power an electric circuit (Lallart 2008). A lot of electromagnetic energy harvesters have reported for several applications. Beeby et al. have demonstrated a miniaturized (0.15 cm^3 volume) electromagnetic generator using four magnets mounted on a cantilever and a wound coil. The optimized device produced a power output of 46 μW at a load of 4 kΩ. The device also produced a voltage of 428 mV with 2300 turns of the coil (Beeby et al. 2007). Jones et al. have designed a micro-generator with two magnets mounted to a coil and the cantilever. This device produced a peak power of 3.9 mW with an average power of 157 μW (Glynne-Jones et al. 2004).

Some literature also reports the electromagnetic energy harvesting from the low-frequency vibrations of the helicopters during their flight operations (Hadas et al. 2012a, b). The output power of the reported generators was sufficient to power the WSN. Hadas et al. have developed a mechatronic device based on the combination of the electromagnetic harvester and a resonance mechanism to capture the low vibrations of the helicopters (Hadas et al. 2012a, b). The resonance mechanism was properly designed to tune the frequency and the excited vibration levels. The generator is used up to the 17 Hz frequency levels, and nonlinear effect of the level of vibration on the output power was reported in the study. The generator displayed the output power as 2–25 mW with the vibration range of 0.1–1 G. Therefore, the different power output can be obtained according to the actual requirement of the various sensors.

These micro-generators are also explored for the practical applications and health monitoring of the aircraft structures. Previously, the MicroStrain has checked the feasibility of these generators by monitoring the pitch link stresses to find the components life of the Apache Helicopter. The pitch link strain measurement is a complicated process with the conventional wired sensors, and therefore WS are more meaningful for such applications (Leamon et al.; Le et al. 2015).

The MicroStrain also developed a specific design of the electromagnetic energy harvester for regulating the UH-60 rotor head. It is one of the major parts of an aircraft which contains many critical components and mounting a high-performance

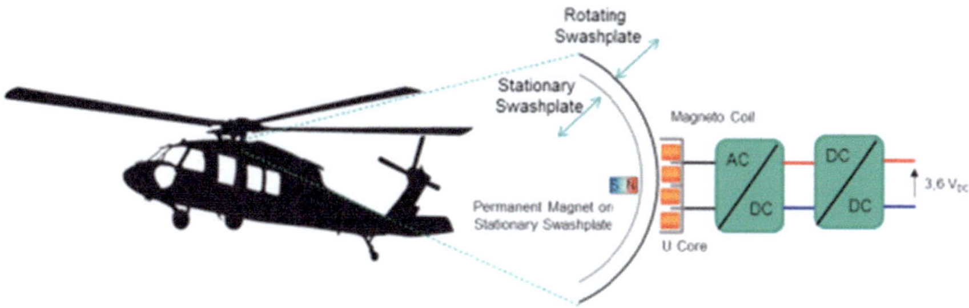

Fig. 4.5 Configuration of the HV-60 spin harvester. Reproduced from Le et al. (2015) with permission from the Elsevier

energy harvester and WS on it will be useful for its proper maintenance. The reported generator was benefited from the relative motion between the rotating and stationary sides of the swashplate. The permanent magnets were fitted to the stationary side while the power electronics and the power-generating coils were mounted on the moving side of the swashplate (as shown in Fig. 4.5). The energy was produced due to the relative motion between the permanent magnets and the coils. For the one magnet and four coils combination, and at 3 m/s relative velocity, the power output was produced in a range of 1–60 mW. This power output was further increased from 9.2 to 500 mW as the relative velocity reached at 9 m/s (Leamon et al.; Le et al. 2015).

Therefore, a lot of electromagnetic generators are reported for powering the sensor networks. However, their output performance is limited by the several challenges, which are as follows:

- The complexity of the coil fabrication to achieve high performance.
- Less distance between the magnets and the coil.
- Fabricating the small-sized magnets.

4.3.1.3 Electrostatic Technique

In electrostatic conversion technique, the vibration causes the change in the capacitance which results in a voltage increase in a constant current system or a current increase in a constant voltage system (Wei and Jing 2017). These generators need an extra charging unit for the initial operation. The satisfactory outcome of these devices for the practical applications requires further improvements to overcome several drawbacks such as thermal stability, mechanical efficiency, and structure design.

Table 4.1 displays the comparison of the three different conversion methods (Le et al. 2015) and Fig. 4.6 exhibits the data plot of the power density as a function of the applied frequency (Le et al. 2015). The frequency range for each technique is also added to the plot.

Table 4.1 Comparison of the vibration-based energy harvesting techniques

	Piezoelectric	Electromagnetic	Electrostatic
Advantages	• High output voltage • Large temperature range • Easy to use • Robust	• High output current • Long life • Robust	• High output voltage • Simple integration • Capturing low frequencies
Disadvantages	• The conversion properties depend on the piezoelectric element	• Low output voltage • Heaviness • Electromagnetic compatibility problems • Low efficiency	• The requirement of a polarization source • Poor mechanical guiding • Complicated power circuit management

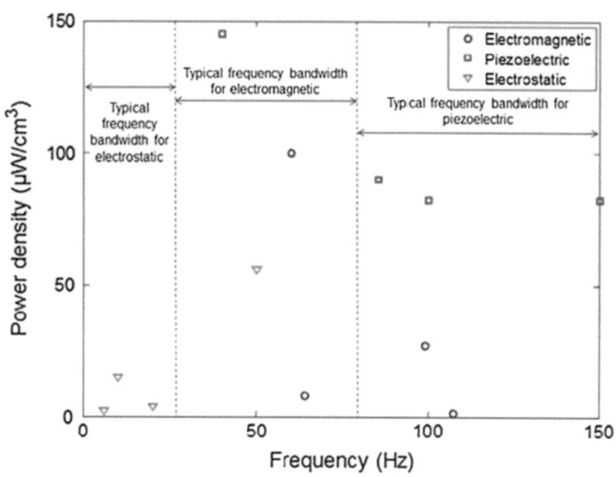

Fig. 4.6 Power density versus frequency for different vibration-based energy harvesters. Reproduced from Le et al. (2015) with permission from the Elsevier

4.3.2 *Thermal Energy Harvesting Techniques*

Another potential technique, for the self-powering sensors in the aircraft application, is the conversion of thermal gradients into the electrical energy. The most common thermal energy harvesters are the thermoelectric generators (TEG), which are based on the Seebeck effect. The energy harvesting using TEG is a mature technology which is widely adopted for the aircraft applications as aircraft have several thermal gradients during the flight conditions (as already explained earlier in Sect. 2.2). A good TEG should have a low thermal conductivity to minimize the thermal losses, the high Seebeck coefficient to generate the high voltage output, and high electrical conductivity to minimize the thermal noise (Carlson et al. 2010). The mechanism of the TEG is based on the Seebeck effect that converts the thermal gradients into the electrical energy. Heat flow arises due to the temperature

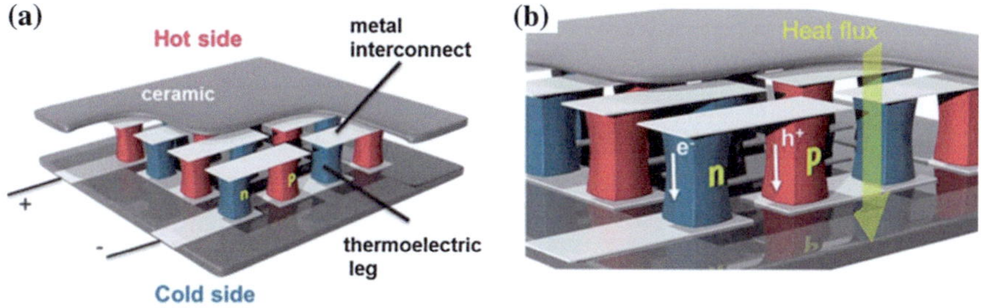

Fig. 4.7 **a** and **b** Schematic of the thermoelectric components with the n-type and p-type layers. Reproduced from Bubnova et al. (2012) with permission from the Royal Society of Chemistry

difference between the opposite parts of a conducting material and simultaneously charge flows due to the diffusion of energy carriers from high to low concentration regions. TEG is constructed with the n-type and p-type junctions with various series and parallel connections (as shown in Fig. 4.7). The n-type and p-type materials are electrically connected at the high-temperature junction allows heal flow to carry the high charge carriers of each material to the low-temperature side, which produces a voltage difference across the base electrodes (Carlson et al. 2010; Crispin 2012). The voltage and power generation is proportional to the thermal gradient.

Pearson et al. have reported a combined TEG generator and piezoelectric material mounted on an aircraft to harvest the energy from the thermal gradients (up to 50 °C) and the vibrations (20–400 Hz), respectively. Temperature data were obtained from the thermocouples located at different positions of the aircraft as shown in Fig. 4.8. The simulation shows that the power was dependent on the position and orientation of the energy harvesters (as detailed in Table 4.2). The generated output power range for the TEG was in the range of 5–30 mW. The combined energy harvesters show a power output range of 1–180 mW. The generated power was sufficient to power the WS (Pearson et al. 2012).

Samson et al. have tested a self-powered WS system containing an energy management unit, a TEG generator, and a WS for aircraft applications. They reported the improvement in the energy output by 14% as compared to other TEGs and use capacitors for energy storage. The overall efficiency of the system was improved by 50% and the total power output produced was ~3.2 V (Samson et al. 2011). In another study, the performance of the TEG was improved by using phase change materials. Phase change material was used to increase the thermal gradient. The TEG harvester containing two cavities filled with the phase change materials was attached to the fuselage. The temperature difference was studied during take-off and landing (Elefsiniotis et al. 2014). Therefore, they proposed a method to increase the temperature difference by applying phase change materials in the TEG systems to improve their output power. The TEGs are the promising devices to power the WS in the aircraft however their low efficiency is the major problem for their full realization. Table 4.3 shows the advantage and limitation of the TEG based energy harvester systems.

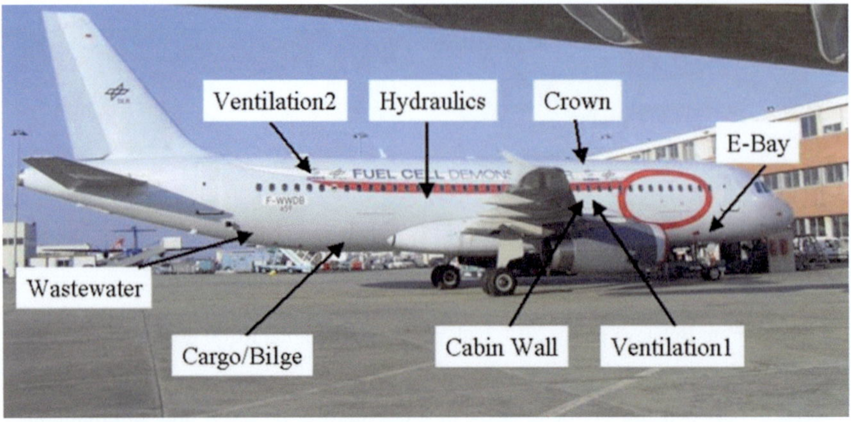

Fig. 4.8 Thermocouple locations for the thermoelectric generator output power simulation. Reproduced from Pearson et al. (2012) with permission from the IOP Science

Table 4.2 Simulation results for the generated power output for temperature gradients between different parts of the aircraft (Pearson et al. 2012)

S. no.	Aircraft location	Peak temperature gradient (°C)	Peak power (mW)	Average power (mW)
1	E-bay fuselage skin E-bay primary insulation	35	18.72	6.42
2	Cabin wall fuselage skin Cabin wall primary insulation	30	13.36	3.97
3	Cabin wall fuselage skin Cabin wall secondary insulation	40	30.06	11.70
4	Hydraulic pipeline 1 Hydraulic pipeline 2	20	7.97	3.07
5	Waste water tank Waste water ambient	15	5.46	2.99
6	Cargo skin Cargo primary insulation	40	34.15	22.58

4.3.3 Solar Energy Harvesting Techniques

Solar energy harvester uses photovoltaic (PV) cells, which converts the sun's energy into the electricity. When the sun rays hit the PV cell, photons are absorbed by the semiconducting material layer of the PV cell (Gao et al. 2015). The photon

Table 4.3 Advantage and challenges for the existing TEGs

	Thermoelectric energy generators
Advantages	• Availability and a wide range of the materials • Mature technology • Robust • Commercially available circuits
Disadvantages	• Very low efficiency ($\sim$10%) • Heavy • Require constant thermal gradient. The temporary fluctuations in the temperature results difficulty in the energy harvesting

Table 4.4 Comparison of the power densities of the different energy harvesting methods (Roundy 2003)

S. no.	Technique	Power density ($\mu W/cm^3$) 1-year time	Power density ($\mu W/cm^3$) 10-year time
1	Vibrations	200	200
2	Thermal gradient	15 at 10 °C temperature difference	15 at 10 °C temperature difference
3	Solar	15000 (direct sun) 150 (cloudy day)	15000 (direct sun) 150 (cloudy day)

energy produces free electrons which flow as current in the presence of an electric field. This method provides the maximum relative power density as compared to the other methods (as displayed in Table 4.4). However, the solar-based systems are limited due to the unavailability of the sunlight during the night time and the cloudy weather and the low efficiency of these systems.

4.4 Recent Progress in the Development of the Integrated Energy Harvesting, Energy Storage, and Sensor Hybrid Devices

The previous research focus was on the development and the improvement of the individual components such as energy harvesting devices, energy storage devices, and the WS. However, their integration at a limited surface area causes problem both regarding the physical mounting and the output performance. The post-coupling of the different devices in a system also increases the weight of the system and creates proper assembly problems. Therefore, the recent research is focused on the development of the integrated hybrid systems consisting energy harvesting, energy storage, and sensing in a single device (Hu et al. 2011; Fu et al.

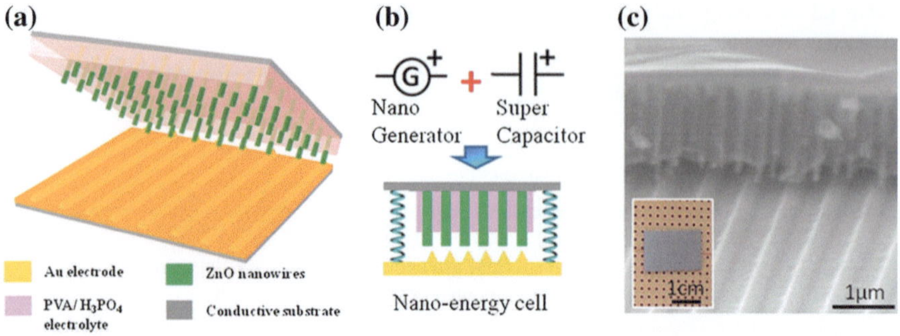

Fig. 4.9 **a** Schematic of the integrated piezoelectric nanogenerator-supercapacitor hybrid device; **b** working mechanism and cross section of the device; and **c** the cross-sectional SEM image of the device (inset shows a photograph of the hybrid device). Reproduced from Wang et al. (2016) with permission from the Elsevier

2013; Xu et al. 2015; Hwang et al. 2016; Wang et al. 2016; Iezzi et al. 2017; Park et al. 2018). The various harvester systems such as the piezoelectric generator, thermoelectric generator, photoelectric generator, and others are recently reported with the integrated supercapacitor and the sensing units.

4.4.1 Hybrid Nanogenerator and Supercapacitor Device

Wang et al. have presented a hybrid piezoelectric nanogenerator and supercapacitor device to power miniaturized electronic devices (Wang et al. 2016). The ZnO nanowires were used as an energy harvesting component and ZnO nanowires with PVA–H_3PO_4 solid electrolyte served as a supercapacitor component in a single device. The mechanical to electrical energy conversion efficiency of the hybrid device was ten times higher than the energy harvesting device alone. The design and working mechanism of the hybrid device are shown in Fig. 4.9a, b. Figure 4.9c shows the cross-sectional SEM images of the SEM images of the device and Fig. 4.10 shows the process of device functionality and performane.

4.4.2 Hybrid Nanogenerator and Sensor Device

Hwang et al. have developed an integrated piezoelectric energy harvester and a self-powered WS system in a single device (Hwang et al. 2016). The developed flexible device shows the high performance and the excellent flexibility of the device enables it to be mounted on any surface. The piezoelectric component generated an open circuit voltage of 200 V and a short circuit current of 35 μA. It was able to charge a supercapacitor which powered the integrated temperature

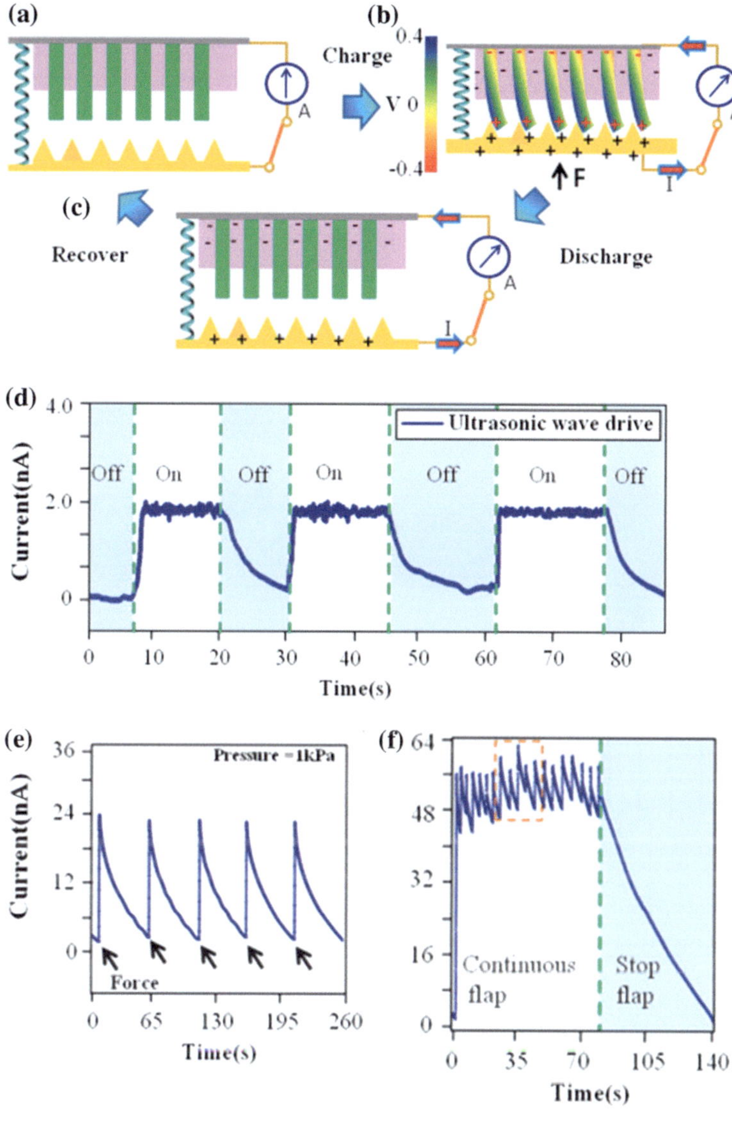

Fig. 4.10 The process of the device functionality and its performance. **a** Device in a neutral state; **b** device under external applied load on the bottom electrode; **c** device recovery with removal of the applied load; **d** the closed circuit current output of the device as it gets mechanical energy input from an ultrasonicator at 42 kHz frequency; **e** the closed circuit current output of the device at 1 kPa pressure; and **f** the closed circuit current output of the device under 15 kPa pressure, showing device charging with the help of harvesting component. Reproduced from Wang et al. (2016) with permission from the Elsevier

sensor node. Hu et al. also demonstrated an integrated device with a piezoelectric nano-generator, a supercapacitor, and WS components (Hu et al. 2011). They also checked its feasibility for long-range data transmission. The piezoelectric nano-generator was a cantilever structure with a flexible substrate, electrodes, and the ZnO nanowires as the piezoelectric material. The achieved output voltage was 10 V with an output current of 0.6 μA at 0.12% straining. The wireless signals were detected at 5–10 m distance using a radio. Figure 4.11 shows different components and the prototype of the of a self-powered sensor device.

Park et al. have studied a hybrid pressure/temperature/strain sensor and solid-state supercapacitor using polydimethylsiloxane coated polypyrrole-graphene foam

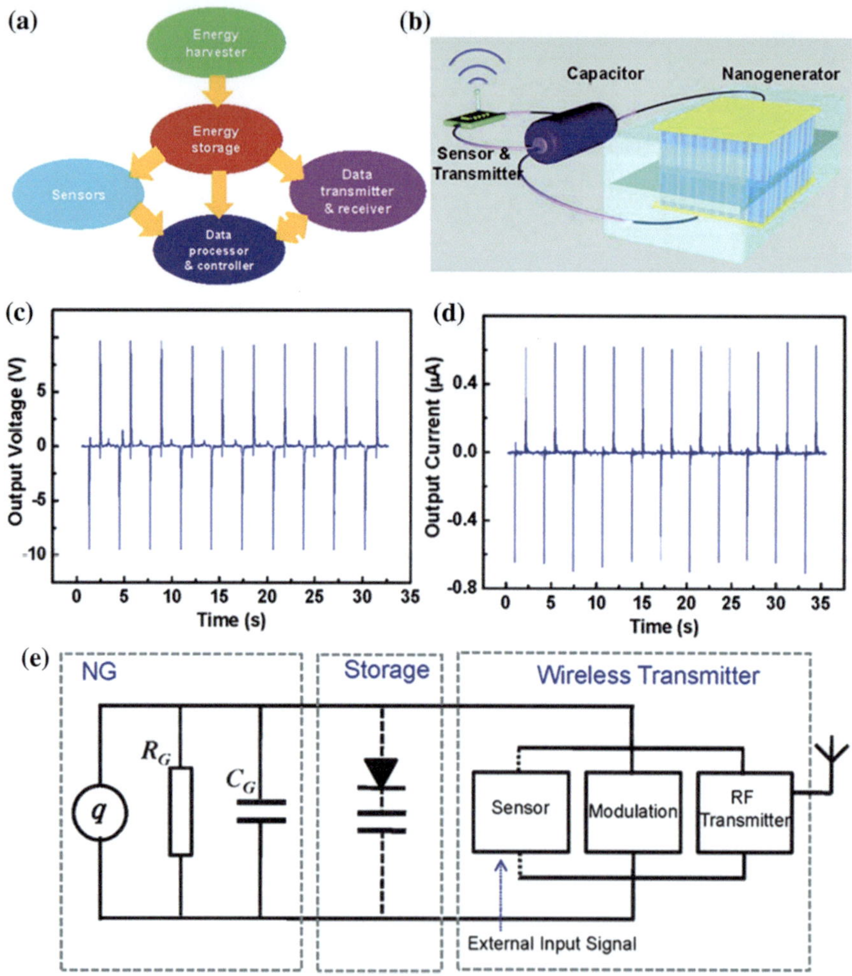

Fig. 4.11 **a** Schematic of an integrated self-powered device with its different modules; **b** prototype of the integrated system; **c** output voltage performance of the nanogenerator; **d** output current performance of the nanogenerator; and **e** schematic of the integrated device. Reproduced from Hu et al. (2011) with permission from the American Chemical Society

nanocomposite (Park et al. 2018). The developed sensor displays fast response, high sensitivity, and long life up to 10,000 pressure loading cycles. An integrated device was constructed by using the multifunctional sensors and the supercapacitor on a flexible substrate. The sensors were operated by the supercapacitor, which shows successful integration of sensor unit with the energy storage component. Iezzi et al. fabricated a screen printed, flexible, and inexpensive thermoelectric generator and a sensor unit based on the silver and nickel inks (Iezzi et al. 2017). The generated power output of the thermoelectric generator was 308 μW. It was sufficient to power a temperature sensor unit. The shape and size versatile screen printed thermoelectric generator and sensor unit was integrated into steam pipe insulation. Figure 4.12 shows images of the printed thermoelectric generator elements and their mounting on a heat pipe.

4.4.3 Hybrid Solar Cell and Supercapacitor Device

Xu et al. have presented an integrated solar cell and a supercapacitor to examine its combined performance (Xu et al. 2015). The solar cell was based on the $CH_3NH_3PbI_3$ while the supercapacitor was based on the polypyrrole material. The developed device shows 10% energy storage efficiency and an output voltage of 1.45 V under AM 1.5 G illumination. In another literature, Scalia et al. proposed an integrated flexible device with TiO_2 nanotube-based solar cell and a graphene-based supercapacitor. Electrochemical analyses were performed to check the performance of the device. They obtained a maximum energy conversion and storage efficiency

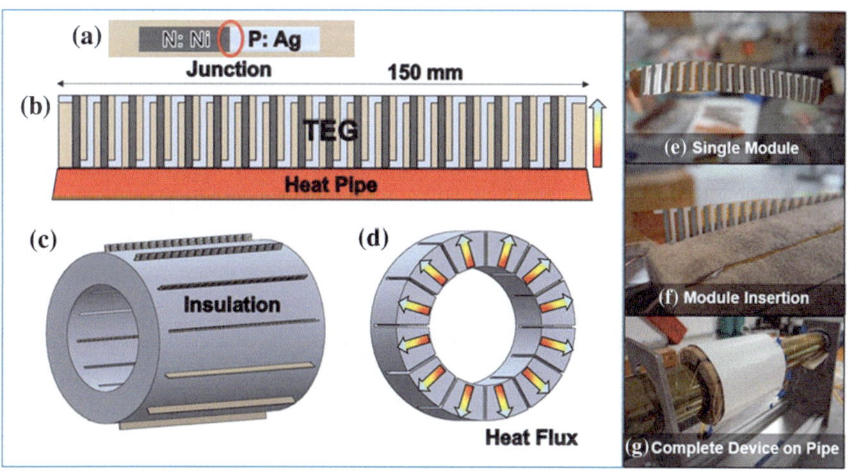

Fig. 4.12 **a** Screen printed p and n-type junctions; **b** printed thermoelectric module; **c** and **d** inserted thermoelectric generator with the heat pipe; **e**–**g** photographs of the device. Reproduced from Iezzi et al. (2017) with permission from the Elsevier

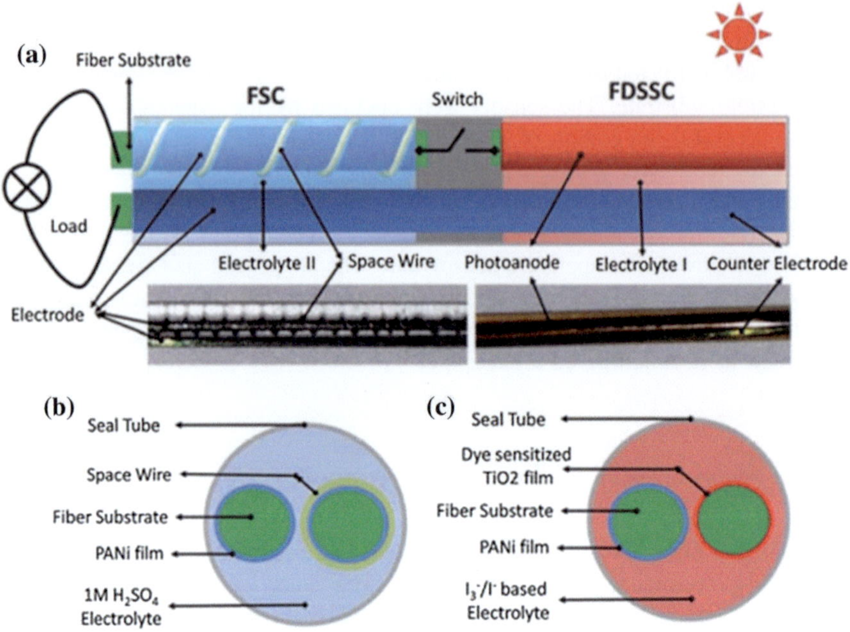

Fig. 4.13 **a** Schematic and photographs of the fiber-based integrated FDSSC and FSC; **b** cross-sectional image of the FSC; and **c** cross-sectional image of the FDSSC. Reproduced from Fu et al. (2013) with permission from the Royal Society of Chemistry

of 1.46%. Fu et al. have reported an innovative integrated dye-sensitized solar cell and supercapacitor hybrid device using the fiber type configuration (Fu et al. 2013). The polyaniline coated stainless steel wire was used as the electrode for both dye-sensitized solar cell (FDSSC) and supercapacitor (FSC) while 1 M H_2SO_4 was chosen as an electrolyte (as shown in Fig. 4.13) The output voltage and current can be easily altered for different power needs by constructing the different series or parallel connections. Figure 4.14 a shows the J-V curves of two dye-sensitized solar cells and also their series and parallel connection results. The maximum short circuit current was 2.91 mA with the parallel connection, and the highest short circuit voltage was 1.26 V with the series connection. The galvanostatic charge–discharge curves display a maximum voltage of 1.2 V for the series connected fibers (Fig. 4.14b). Figure 4.14c shows the working principle diagram of the integrated solar cell and a supercapacitor and Fig. 4.14d displays the solar charge and supercapacitor discharge process of the hybrid device. This study reports an energy conversion efficiency of 2.1%. The attractive feature of the fiber-based components is that they can be used to form the different size and shape of the large area components by using weaving methods. Therefore, the integrated devices on the fiber are potential candidates for the next generation flexible, miniaturized, and lightweight electronic systems.

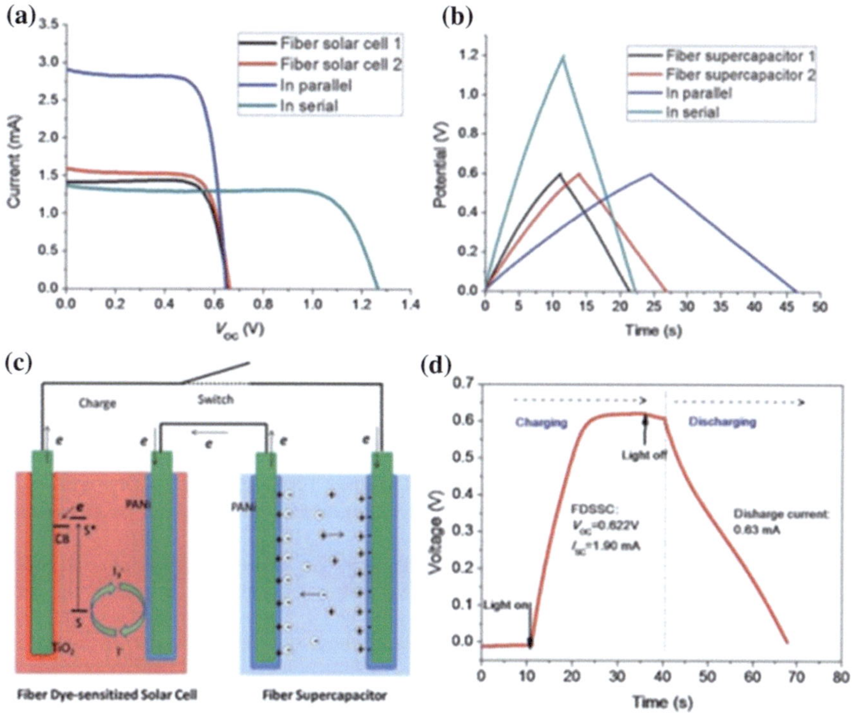

Fig. 4.14 **a** The current–voltage curves of the two FDSSCs connected in series and parallel configuration; **b** the galvanostatic charge–discharge curves of the two FDSSCs; **c** schematic of the working mechanism of the hybrid FDSSC-FSC device; and **d** charging profile of the FDSSC-FSC and discharging profile. Reproduced from Fu et al. (2013) with permission from the Royal Society of Chemistry

4.5 Conclusion

In summary, the energy harvesting techniques are effective methods for operating the WS network in an aircraft system. The energy requirement for various sensors used in the aircraft applications can be fulfilled by the own environmental conditions of the aircraft, which can effectively replace the wired and battery systems. The major energy sources in the aircraft are vibration, thermal gradient, and the solar energy. This paper reviews the currently used energy harvesting methods in the aircraft and their potential for the self-powered WS. Recently, some integrated energy harvester, energy storage, and WS have been reported for the efficient hybrid devices. However, their realization is limited to the research stage and their utilization for the real-world problem is a big challenge.

References

Alippi C, Galperti C (2008) An adaptive system for optimal solar energy harvesting in wireless sensor network nodes. IEEE Trans Circuits Syst I Regul Pap 55(6):1742–1750

Arms SW, Townsend CP, Churchill DL, Hamel MJ (2010) Energy harvesting, wireless structural health monitoring system, Google Patents

Beeby SP, Torah R, Tudor M, Glynne-Jones P, O'donnell T, Saha C, Roy S (2007) A micro electromagnetic generator for vibration energy harvesting. J Micromech Microeng 17(7):1257

Blystad L-CJ, Halvorsen E, Husa S (2010) Piezoelectric MEMS energy harvesting systems driven by harmonic and random vibrations. IEEE Trans Ultrasonics Ferroelectr Freq Control 57(4)

Bubnova O, Chrispin X (2012) Towards polymer-based organic thermoelectric generators. Energy Environ Sci 5:9345–9362

Carlson EJ, Strunz K, Otis BP (2010) A 20 mV input boost converter with efficient digital control for thermoelectric energy harvesting. IEEE J Solid-State Circuits 45(4):741–750

Cheng T, Zhang Y, Lai WY, Huang W (2015) Stretchable thin-film electrodes for flexible electronics with high deformability and stretchability. Adv Mater 27(22):3349–3376

Ciang CC, Lee J-R, Bang H-J (2008) Structural health monitoring for a wind turbine system: a review of damage detection methods. Meas Sci Technol 19(12):122001

Collins J (2006) The challenges facing US navy aircraft electrical wiring systems. In: Proceedings of the 9th annual aging aircraft conference

Costanzo A, Dionigi M, Masotti D, Mongiardo M, Monti G, Tarricone L, Sorrentino R (2014) Electromagnetic energy harvesting and wireless power transmission: a unified approach. Proc IEEE 102(11):1692–1711

Crispin DX (2012) Retracted article: towards polymer-based organic thermoelectric generators. Energy Environ Sci

de Jong P, Loendersloot R, de Boer A, van der Hoogt P (2011) Power harvesting in a helicopter lag damper. In: Proceedings of SMART 2011, University of Saarland

Dickerson SM (2011) CH-47D rotating system fault sensing for condition based maintenance, Air Force Inst of Tech Wright-Patters on AFB OH School of Engineering and Management

Dilhac J-M, Bafleur M (2014) Energy harvesting in aeronautics for battery-free wireless sensor networks. IEEE Aerosp Electron Syst Mag 29(8):18–22

Ducharme MB (2006) Heat stress of helicopter aircrew wearing immersion suit. Ind Health 44 (3):433–440

Dumas D, Lani F, Monnier T, Smaili R, Loyer J (2011) Damage detection in composite structures using autonomous wireless systems: simulation and validation. J Phys Conf Series, IOP Publishing

Elefsiniotis A, Kokorakis N, Becker T, Schmid U (2014) A thermoelectric-based energy harvesting module with extended operational temperature range for powering autonomous wireless sensor nodes in aircraft. Sens Actuators A 206:159–164

Erturk A (2009) Electromechanical modeling of piezoelectric energy harvesters, Virginia Tech

Ferdous RM, Reza AW, Siddiqui MF (2016) Renewable energy harvesting for wireless sensors using passive RFID tag technology: a review. Renew Sustain Energy Rev 58:1114–1128

Fu Y, Wu H, Ye S, Cai X, Yu X, Hou S, Kafafy H, Zou D (2013) Integrated power fiber for energy conversion and storage. Energy Environ Sci 6(3):805–812

Gao X-Z, Hou Z-X, Guo Z, Chen X-Q (2015) Reviews of methods to extract and store energy for solar-powered aircraft. Renew Sustain Energy Rev 44:96–108

Glynne-Jones P, Tudor MJ, Beeby SP, White NM (2004) An electromagnetic, vibration-powered generator for intelligent sensor systems. Sens Actuators A 110(1–3):344–349

Gogotsi Y (2014) Materials science: energy storage wrapped up. Nature 509(7502):568

Hadas Z, Kurfurst J, Ondrusek C, Singule V (2012a) Artificial intelligence based optimization for vibration energy harvesting applications. Microsyst Technol 18(7–8):1003–1014

Hadas Z, Vetiska V, Singule V, Andrs O, Kovar J, Vetiska J (2012b) Energy harvesting from mechanical shocks using a sensitive vibration energy harvester. Int J Adv Rob Syst 9(5):225

Hespanha JP, Naghshtabrizi P, Xu Y (2007) A survey of recent results in networked control systems. Proc IEEE 95(1):138–162
Hu Y, Zhang Y, Xu C, Lin L, Snyder RL, Wang ZL (2011) Self-powered system with wireless data transmission. Nano Lett 11(6):2572–2577
Hwang GT, Annapureddy V, Han JH, Joe DJ, Baek C, Park DY, Kim DH, Park JH, Jeong CK, Park KI (2016) Self-powered wireless sensor node enabled by an aerosol-deposited PZT flexible energy harvester. Adv Energy Mat 6(13):1600237
Iezzi B, Ankireddy K, Twiddy J, Losego MD, Jur JS (2017) Printed, metallic thermoelectric generators integrated with pipe insulation for powering wireless sensors. Appl Energy 208:758–765
Irimia-Vladu M (2014) "Green" electronics: biodegradable and biocompatible materials and devices for sustainable future. Chem Soc Rev 43(2):588–610
Jafari H, Ghodsi A, Azizi S, Ghazavi MR (2017) Energy harvesting based on magnetostriction, for low frequency excitations. Energy 124:1–8
Klöckner A, Schlabe D, Looye G (2012) Integrated simulation models for high-altitude solar-powered aircraft. In: AIAA modeling and simulation technologies conference
Kordes EE, Reed RD (1962) Structural heating experiences on the X-15 airplane, National Aeronautics and Space Administration Washington DC
Kumar B et al (2011) Controlled growth of semiconducting nanowire, nanowall, and hybrid nanostructures on graphene for piezoelectric nanogenerators. Acs Nano 5.5: 4197–4204
Lallart M (2008) Amélioration de la conversion électroactive de matériaux piézoélectriques et pyroélectriques pour le contrôle vibratoire et la récupération d'énergie-Application au contrôle de santé structurale auto-alimenté, INSA de Lyon
Lallart M, Anton SR, Inman DJ (2010) Frequency self-tuning scheme for broadband vibration energy harvesting. J Intell Mater Syst Struct 21(9):897–906
Le MQ, Capsal J-F, Lallart M, Hebrard Y, Van Der Ham A, Reffe N, Geynet L, Cottinet P-J (2015) Review on energy harvesting for structural health monitoring in aeronautical applications. Prog Aerosp Sci 79:147–157
Leamon TM, Townsend CP, Churchill DL, Hamel MJ Energy harvesters for rotorcraft wireless sensor networks
Liu Z, Xu J, Chen D, Shen G (2015) Flexible electronics based on inorganic nanowires. Chem Soc Rev 44(1):161–192
Ma D, Lan G, Xu, M. Hassan and W. Hu (2018). SEHS: simultaneous energy harvesting and sensing using piezoelectric energy harvester. internet-of-things design and implementation (IoTDI). In: 2018 IEEE/ACM third international conference on, IEEE
Mitchell B (2007) Energy harvesting applications and architectures at Boeing commercial airplanes. NanoPower Forum, San Jose
Monthéard R, Airiau C, Bafleur M, Boitier V, Dilhac J-M, Dollat X, Nolhier N, Piot E (2014) Powering a commercial datalogger by energy harvesting from generated aeroacoustic noise. J Phys Conf Series, IOP Publishing
Park H, Kim JW, Hong SY, Lee G, Kim DS, Oh JH, Jin SW, Jeong YR, Oh SY, Yun JY (2018) Microporous polypyrrole-coated graphene foam for high-performance multifunctional sensors and flexible supercapacitors. Adv Func Mat 1707013
Patel V, Sundriyal P, Bhattacharya S (2017) Aloe vera versus poly (ethylene) glycol-based synthesis and relative catalytic activity investigations of ZnO nanorods in thermal decomposition of potassium perchlorate. Part Sci Technol 35(3):361–368
Pearson MR, Eaton MJ, Pullin R, Featherston CA, Holford KM (2012). Energy harvesting for aerospace structural health monitoring systems. J Phys Conf Series, IOP Publishing
Pu X, Hu W, Wang ZL (2018) Toward wearable self-charging power systems: the integration of energy-harvesting and storage devices. Small 14(1):1702817
Raghunathan V, Kansal A, Hsu J, Friedman J, Srivastava M (2005) Design considerations for solar energy harvesting wireless embedded systems. In: Proceedings of the 4th international symposium on Information processing in sensor networks, IEEE Press
Roundy SJ (2003) Energy scavenging for wireless sensor nodes with a focus on vibration to electricity conversion. University of California, Berkeley, CA

Samson D, Kluge M, Becker T, Schmid U (2010) Energy harvesting for autonomous wireless sensor nodes in aircraft. Proc Eng 5:1160–1163

Samson D, Kluge M, Becker T, Schmid U (2011) Wireless sensor node powered by aircraft specific thermoelectric energy harvesting. Sens Actuators A 172(1):240–244

Shahruz S (2006) Design of mechanical band-pass filters for energy scavenging. J Sound Vib 292 (3–5):987–998

Smith SD (2006) Seat vibration in military propeller aircraft: characterization, exposure assessment, and mitigation. Aviat Space Environ Med 77(1):32–40

Sun F, Chaudhry ZA, Sarlashkar A (2018) Self-powered multi-functional structural health monitoring sensor, Google Patents

Sundriyal P, Bhattacharya S (2017a) Inkjet-printed electrodes on A4 paper substrates for low-cost, disposable, and flexible asymmetric supercapacitors. ACS Appl Mater Interf 9(44):38507–38521

Sundriyal P, Bhattacharya S (2017b) Polyaniline silver nanoparticle coffee waste extracted porous graphene oxide nanocomposite structures as novel electrode material for rechargeable batteries. Mat Res Exp 4(3):035501

Sundriyal P, Bhattacharya S (2018) Inkjet-printed sensors on flexible substrates. In: Environmental, chemical and medical sensors, pp 89–113. Springer

Tendeland T, Schlaff BA (1948) Temperature gradients in the wing of a high-speed airplane during dives from high altitudes

Thostenson J, Li Z, Kim C, Ajnsztajn A, Parker C, Liu J, Peterchev A, Glass J, Goetz S (2018) Integrated flexible conversion circuit between a flexible photovoltaic and supercapacitors for powering wearable sensors. J Electrochem Soc 165(8):B3122–B3129

Vanhecke C, Assouère L, Wang A, Durand-Estèbe P, Caignet F, Dilhac J-M, Bafleur M (2015) Multisource and battery-free energy harvesting architecture for aeronautics applications. Power Electron IEEE Trans 30(6):3215–3227

Wang F, Jiang C, Tang C, Bi S, Wang Q, Du D, Song J (2016) High output nano-energy cell with piezoelectric nanogenerator and porous supercapacitor dual functions—a technique to provide sustaining power by harvesting intermittent mechanical energy from surroundings. Nano Energy 21:209–216

Wang L, Yuan FG (2007) Energy harvesting by magnetostrictive material (MsM) for powering wireless sensors in SHM. In: Sensors and smart structures technologies for civil, mechanical, and aerospace systems 2007, International Society for Optics and Photonics

Wei C, Jing X (2017) A comprehensive review on vibration energy harvesting: modelling and realization. Renew Sustain Energy Rev 74:1–18

Xu W, Huang LB, Wong MC, Chen L, Bai G, Hao J (2017) Environmentally friendly hydrogel-based triboelectric nanogenerators for versatile energy harvesting and self-powered sensors. Adv Energy Mat 7(1):1601529

Xu X, Li S, Zhang H, Shen Y, Zakeeruddin SM, Graetzel M, Cheng Y-B, Wang M (2015) A power pack based on organometallic perovskite solar cell and supercapacitor. ACS Nano 9 (2):1782–1787

Yun J, Song C, Lee H, Park H, Jeong YR, Kim JW, Jin SW, Oh SY, Sun L, Zi G (2018) Stretchable array of high-performance micro-supercapacitors charged with solar cells for wireless powering of an integrated strain sensor. Nano Energy 49:644–654

Zhao X, Qian T, Mei G, Kwan C, Zane R, Walsh C, Paing T, Popovic Z (2007) Active health monitoring of an aircraft wing with an embedded piezoelectric sensor/actuator network: II. Wireless approaches. Smart Mater Struct 16(4):1218

Zhong Y, Xia X, Mai W, Tu J, Fan HJ (2017) Integration of energy harvesting and electrochemical storage devices. Adv Mat Technol 2(12):1700182

Zhu D, Tudor MJ, Beeby SP (2009) Strategies for increasing the operating frequency range of vibration energy harvesters: a review. Meas Sci Technol 21(2):022001

Zi Y, Lin L, Wang J, Wang S, Chen J, Fan X, Yang PK, Yi F, Wang ZL (2015) Triboelectric–pyroelectric–piezoelectric hybrid cell for high-efficiency energy-harvesting and self-powered sensing. Adv Mater 27(14):2340–2347

Chapter 5
Embedded Sensors for Health Monitoring of an Aircraft

Sudarsana Jena and Ankur Gupta

Abstract Modern aircraft systems are system of systems involving multidisciplinary engineering viz., aeronautical/aerospace, RF, computer science, electrical, electronics, mechanical and electromechanical, etc. The maintenance of such a system becomes complex when a system is to be operational on the 24 × 7 basis. Commercial aircraft travel thousands of miles everyday and restless schedule of takeoff, landing, and turn around to do it again. Similarly, military aircraft undergo rigorous and continuous operations during war and trial exercises. In both the cases, human lives are directly at risk if the aircraft maintenance is not proper. In addressing the aircraft system requirements, the normal tendency is to deal primarily with those elements of the aircraft system that relates directly to the aircraft performance for flying. At the same time, very little attention is given to periodic maintenance and aircraft health monitoring system until the system fails and needs breakdown maintenance. This breakdown maintenance consumes more resources in terms of time, cost, and manpower. With the advent of different embedded smart sensors, online health monitoring of such complex aircraft systems can be possible. The health of the different components of the aircraft systems can be monitored continuously and necessary preventive action can be taken immediately before it fails. This paper introduces the use of various embedded smart sensors to detect and monitor the health issues or failure of different components of the aircraft, ultimately enabling proactive maintenance to prevent aircraft subsystems/components from breakdowns.

Keywords Embedded sensors · Health monitoring · Intelligent maintenance

---

S. Jena · A. Gupta (✉)
School of Mechanical Sciences, Indian Institute of Technology Bhubaneswar, Bhubaneswar 752050, Odisha, India
e-mail: ankurgupta@iitbbs.ac.in; ankurgupta.k10@gmail.com

S. Bhattacharya et al. (eds.), *Sensors for Automotive and Aerospace Applications*, Energy, Environment, and Sustainability,
https://doi.org/10.1007/978-981-13-3290-6_5

5.1 Introduction

In present scenario, importance of aircraft availability on 24 × 7 basis cannot be overemphasized, since it plays a vital role for transportations media and military operation. Maintenance of such a complex aircraft system is becoming a challenge when the system is to be operational on the 24 × 7 basis. Flight working in a very harsh environment which affects its health due to large variations in pressure, temperature, humidity, speed, and with different load conditions. As a result, it will undergo severe stress and failure like cracks, corrosion, creep, delamination, etc. will occur in aircraft which may severe if it is not monitored and corrected in time.

Aircraft system consists of complex subsystems like engines to provide thrust, fuselage to provides volume, wings to generate lift, vertical tail to provide directional stability and control, horizontal tail to provide longitudinal stability and control and control systems to provide flight monitoring and control, etc. as shown in Fig. 5.1.

The general practice of maintenance is either reactive, i.e., fixing or replacing equipment after it gets down in terms of efficiency or it is down to business, i.e., it is facing degradation against its schedule work, with no input from the subsystem itself, and servicing component on a routine schedule whether service is actually needed or not. Both scenarios are extremely wasteful. To an operator, it frequently appears that system fall short unexpectedly but, in fact, the system typically faces a

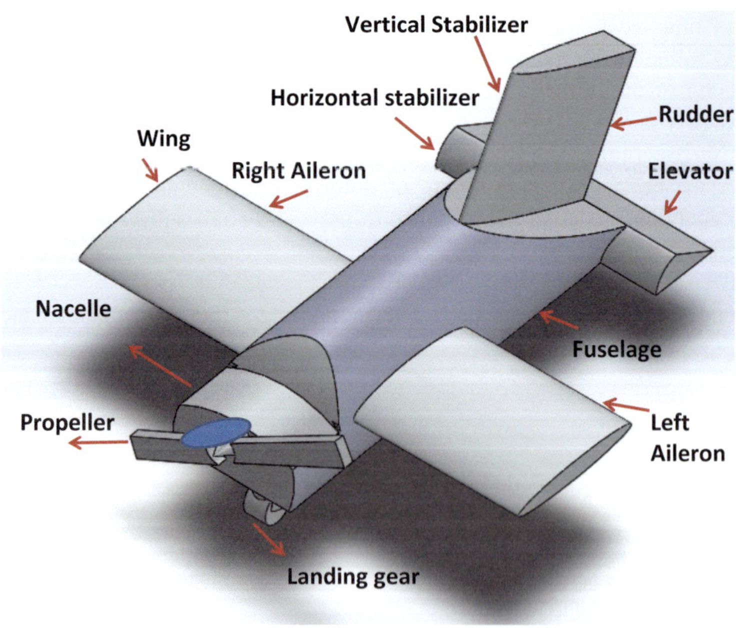

Fig. 5.1 Overview of various components in aircraft systems (not to scale)

quantifiable process of dreadful conditions before they fall short. Today, those dreadful conditions are mostly imperceptible to an operator, even though an immense agreement of technology has been developed that could formulate such information noticeably.

Many sophisticated sensors and computerized equipment are capable of delivering data about the component's status and performance. A commercial or a military aircraft consists of many critical control systems, viz., wings control system, vertical and horizontal tail control system, engine control system and diverse safety monitoring systems such as engine and structural health monitoring, humidity monitoring, cabin temperature, pressure, and in-flight entertainment system, etc. Different interconnectors are used to establish the communications between components and control unit. When such communication interconnects are come across with the power source or any high signal source, then interference occurs which creates a buzz and interrupted the communication. These systems require a huge figure of real-time-based sensors for their optimal operation. Existing systems, which are based on wired connections are multifaceted, complicated to route, heavy, and prone to harm and dreadful conditions due to wear.

More preferably sensors are made of embedded systems in order to achieve many advantages such as performance is good as it is dedicated to one function, less weight and size, use fewer recourses for processing, faster to load, etc. Embedded systems often use a real-time computing system. It consists of two parts; one is called firmware and other is hardware. Firmware is specific software stored in the flash ROM of a hardware device, while ROM is "read-only memory," and flash ROM can be erased and rewritten because it is actually a type of flash memory. Nowadays, embedded systems are very commonly used to control many devices (Massa 2006). More than 98% of microprocessors are designed and manufactured based on the embedded system requirements (Barr 2009).

5.2 Research on Embedded Sensors

Embedded sensors are mainly used in composite materials as these materials are manufactured by layered fabrication methods. MEMS-based and fiber optic sensors are more popular day by day. On recent development, wireless sensor network (WSN) embedding and fiber bragg grating (FBG) sensor embedding have become more popular due to their many advantages. Various sensors used for embedding are discussed in the following paragraphs.

5.2.1 MEMS-Based or Thin-Film Sensor Embedding

Small-size piezoelectric sensing elements are embedded below the surface of interest in order to check the interface loads of tooling workpiece. The disturbance of the surface and process should be minimal (Du and Klamecki 1993, 1999) as stated by Du et al. The sensor output is characterized based on its sensor depth from the surface level, the distance between load location to sensor location, and the spacing between sensor to sensor. A suitable mechanical model was developed to find out the sensor output.

As per Friswell and Inman (2000), structures having more number of embedded sensors are mostly used to identify damage location and model validation. Such arrangements are more useful, as it provides seamless information about the structure. Based on this concepts and design, many sensors have been developed such as accelerometers, PVDF film sensors, strain gauges, piezoceramic patches, and optical fiber sensors. A structure manned with such sensors are called smart structure whose health can be continuously monitored. Such a system is validated based on hypothesis testing model by Friswell et al.

The present research is focused on applications of shape memory alloy (SMA) to produce smart sensors for health monitoring and control of crack growth. A test was conducted on embedded SMA foil sensors and actuators by Ogisu et al. in order to study the crack growth on CFRP laminates (Ogisu et al. 2000). The aim of this study is to control the damage growth in CFRP laminates. Similarly, an embedded MEMS-based strain measurement sensor was fabricated in fiber-reinforced laminated composite plates by Hautamaki et al. (2000). He evaluated the response of three different designs of embedded strain measurement sensors fitted in a composite plate by applying a load in uniaxial tension and bending. These three designs of strain measurement sensors were (i) a piezoresistive filament fabricated directly on the wafer, (ii) a rectangular cantilever beam, and (iii) a curved cantilever beam. The test results of these three designs were compared for their repeatability, sensitivity, and reliability study of respective design. The test was conducted on all devices to a uniaxial tension load up to 10,000 cycles and it was found that all devices have responded well at this load condition.

5.2.2 Wireless Sensor Embedding

Wireless sensor embedding is a current possibility to overcome the difficulty of using traditional wired network sensors in composite structures (Krantz and Belk 1997; Krantz et al. 1999; Hautamaki et al. 2000; Pereira et al. 2000). Wireless embedded microsensors are first used in 1997 by Krantz and Belk (1997), Krantz et al. (1999). Physical connections are not required in this types of sensors. They used micromachined sensors and its associated remote query device that established the communication without physical wire connections. The main part of this research is to develop an embeddable antenna which is capable to both receive and transmit the data.

Further, Hautamaki et al. used a wireless sensing network (Hautamaki et al. 2000) in a MEMS-based embedded sensor for measuring strain in composites. A MEMS-based sensor was developed with a driving element, sensing element, and a telemetry circuitry which were fabricated on a single silicon wafer. He used finite element methods (FEM) to analyze different designs before proceeding to manufacture.

Pereira et al. looked into the method of sensing in an intelligent way and established the wireless communication for intelligent sensing and wireless communications in ruthless environments (Pereira et al. 2000). This paper explained the current efforts to characterize and performance test of MEMS-based inertial sensors. It also introduced the use of embedded sensors without a battery in weapons. It discussed the requirements of the internal wireless communication sensors integrated into smart materials/electronics with a built-in capability which can communicate between telemetry and sensors. Embedded wireless telemetry enables wireless and stresses fewer communications between a MEMS-based sensor and processing unit. Research is further extended on the integration of smart sensors, actuator, and its control with the extraordinary capability to establish an embedded telemetry system.

5.2.3 Fiber Optic Sensor Embedding

Fiber optic sensor embedding is one of the dominant technologies (Udd 1995) in which research is going on last two decades to implement in structural composites. This is named as "fiber optic smart structures." Vital parameters of different structures are sensed by such sensors in different conditions such as lightweight, electromagnetic interference, resistance to ruthless environments, nonobtrusive embeddability and very high bandwidth capability. Health monitoring of a structure or a system can easily be established by means of a network-based embedded fiber optic sensors during its manufacturing and service as well. Use of this sensors enable the functions same as a human being can sense by his sensory systems like visual inspection, hearing, smell, taste, and touch and provides real-time data about the system health status. Apart from these advantages, fiber optic sensors have some well-known disadvantages such as high fabrication costs, high fragility, and are required to provide ingress and egress from the structure. The embedded fiber optic sensors used for detecting the damage developed in a composite structure are well known as damage sensors (Chang and James 1918; Sirkis et al. 1994). These sensors are used in aircraft to investigate the low-velocity impact on composite structure layer.

Presently, fiber bragg grating sensors are becoming more and more popular for monitoring the process, mainly it is used in strain and temperature measurements device (Jin et al. 1998; Murukeshan et al. 2000, 2001).

5.3 Sensors Used in Aircraft

There are many sensors used in aircraft to monitor the control of different components such as nose wheel steering, leading edge flap, trailing edge flap, rubber pedal, pilot's stick, spoilers, ailerons, tail elevators, rudder, etc. Figure 5.2 shows different sensors used in aircraft at a different locations in order to smooth control of aircraft during take up, flying, and landing.

Apart from these sensors, there are other sensors that are used for continuous monitoring of cabin environment condition such as temperature, pressure, and humidity. Airflow control is also monitored and regulated as required.

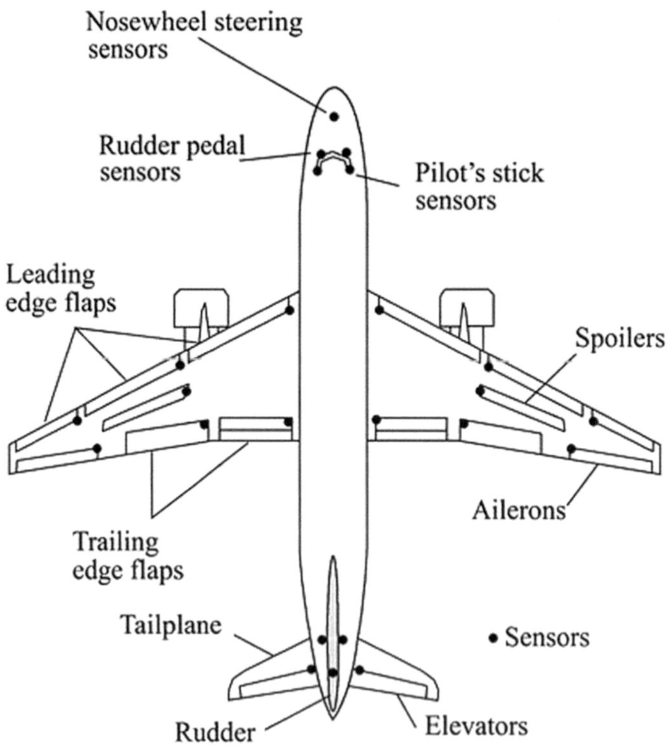

Fig. 5.2 Typical sensor locations of a commercial aircraft (reprinted with kind permission from (Collinson 2002))

A list of sensors used in aircraft are mentioned below:

(a) Airspeed (electronic/aneroid) sensor,
(b) Magnetic Compass,
(c) GPS (satellite/terrestrial),
(d) Radio Compass (NDB),
(e) Altitude (electronic/aneroid) sensor,
(f) Barometric Pressure (electronic/aneroid) sensor,
(g) Outside Air Temperature (C/F) sensor,
(h) Cabin air pressure (psi/hg) sensor,
(i) Cargo air pressure; all doors, bulkhead(s) sensor,
(j) Cabin temperature; bulkhead(s), all doors sensor,
(k) Cargo temperature sensor,
(l) Fuel temperature; fuel tanks, fuel pump sensor,
(m) Hydraulic Pressure (if equipped); brakes, flaps, spoilers, rudder, aileron, landing gear pump sensor,
(n) Weight sensors (if equipped)—landing gear,
(o) Fuel pressure (x number of engines) sensor,
(p) Fuel flow (x number of engines) sensor,
(q) Turbines; RPM (N1/N2), Inlet-turbine pressure, Temperature, fuel burn sensor,
(r) Voltmeter; cockpit, main bus, cabin, auxiliary power, cargo, engines, APU,
(s) Generator meters (engines, APU),
(t) Electricity Load (amp/h) sensors; flight deck, cabin, cargo,
(u) Fire sensors; cabin, cargo, engines, fuel, brakes, electronics bay,
(v) Carbon Dioxide sensor; cabin, cargo,
(w) Radar air traffic—TCAS, and
(x) Doppler radar; weather, lightning, downdraft (microburst).

5.4 Embedded Sensors Used for Health Monitoring Purpose in Aircraft

Aircraft structures and its subsystems necessitate sporadic and planned check and maintenance operations due to their unique operating conditions. Therefore, structural health monitoring is essential and it has an immense potential to diminish the costs associated with these operations.

Research has been made on all aspects of optimum requirements for monitoring the health of the aircraft. Promising technologies have explored the use of embedded smart sensors to predict and detect the issues on various critical components of the aircraft.

Following paragraphs present the use of embedded sensors for a specific purpose in aircraft.

5.4.1 Embedded Sensor/Actuator System for Aircraft Active Flow Separation Control

An airplane flies at constant altitude because of the balancing force called the force of lift, which is produced due to the pressure difference between upper and lower surfaces of the aircraft wings against the gravity force. During this process air sticks over the surface of the wing due to air viscosity which may cause the loss of lift. The force on the wing can be controlled (i.e., decrease and increase) by varying the angle of attack. The angle between the plane of the wing (i.e., airfoil chord) and the direction of airflow is called angle of attack. The large boundary layer division of airflow over the surface of the wings at low Mach numbers and high angles of attack is the main cause of loss of lift in aircraft. During the maneuver phase, the flow is unsteady which is responsible for the early flow of separation. This leads the aircraft to stall (loss of lift) at a lower angle of attack than the usual angle as per its design specifications. Flow separation plays an important role in aircraft operation. This has a big impact on the performance and safety of an airplane. So, this is important to predict and detect such a phenomenon before it really disturbs the wing operation. Network-based pressure sensors distributed over the wing profile can measure the fluctuations of pressure along the wing.

Pressure fluctuations can be measured by implementing a monitoring system in which fluctuations can be related by the root mean square (RMS) value of the signal received from transducer due to pressure variation. This signal will boost at the laminar separation bubble and reduces after attachment. In another way, it enhances as the flow accelerates and reduces at deceleration (Hummel et al. 2013). A reference aircraft ATR-72 was used as an experimental basis to understand the phenomena. ATR-72 was made by French-Italian aircraft manufacturer ATR. This aircraft has twin-engine turboprop short-haul regional airliner. The scenario of airflow over the wing surface at diverse flight conditions was simulated and analyzed using computational fluid dynamics (CFD). This analysis helped to obtain necessary information that says the exact parameters to be measured and optimized for sensor design and its location. These experimental results were validated on a conventional NACA0012 airfoil. In each configuration, characteristic of the secondary flow and the pressure fluctuation rate was estimated in both the cases of attached and separated flow. In this work, a capacitive-based pressure sensor was proposed to be the best-suited sensor to meet the estimated requirements of bandwidth at low-pressure fluctuations. In this experiment, the designer focused on the low budget and network-based sensors for large area application which are the key features of the design aspects. In proposed sensors, a deformable membrane and moving electrode (obtained from a circular copper-clad Kapton polyimide layer) coupled to a rigid copper-clad fiberglass (Fig. 5.3) were used as a fixed electrode. The deflection and stress distribution in a deferent shape of floating membranes were simulated by finite element method (FEM) simulations. The sensitivity of the pressure sensor over the applied pressure load was predicted using

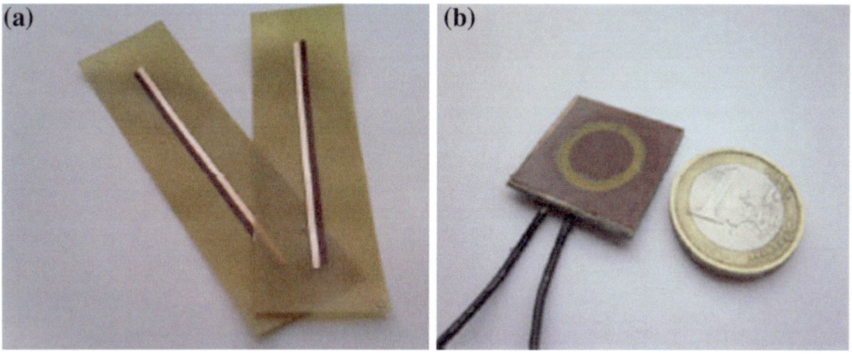

Fig. 5.3 Shows the DBD actuators (**a**) and **b** shows the pressure sensor with an exposed sensitive membrane (reprinted with kind permission from (Francioso et al. 2015))

this simulation. The same methodology was used to optimize the design of different capacitive-based pressure sensors and to calibrate it.

The experiment was conducted on the conventional NACA0012 airfoil. In this, the pressure variation was observed and recorded for the different positions of the sensors. All the pressure sensors were connected to a network and data was collected in a central computer system. Figure 5.4 (top) shows the absolute pressure variations by increasing the angle of attack from 4° to 18° and Fig. 5.4 (bottom) shows the absolute pressure variations with respect to a variation of sensor position varying from 0 to 0.14 m.

5.4.2 *Active Health Monitoring of an Aircraft Wing with Embedded Piezoelectric Sensor/Actuator Network*

Normally, ultrasound bulk wave and eddy current technology are very frequently used to identify the general health condition of the aircraft structure. Such a technique is very limited to point-by-point manner and it consumes a lot of time for checking the complete structure of aircraft. Sometimes, it is required to remove or disassemble some of the interior parts or components and also some inaccessible parts of the aircraft structure for regular visual inspection, which is very tedious and laborious job. In situ monitoring system has been developed which is relatively fast and capable to scan a large area at a time. It will give a quantitative and qualitative information about the aircraft health. It provides real-time information about defect types, location, severity, and possible solution for the issues. The health information is communicated to the operator by means of a wireless communication network. The ultrasonic guided waves are elastic by nature and it propagates in solids with boundaries (Lamb 1917; Viktorov 1970; Rose 1999). Such technique is used for nondestructive inspection (NDI) to detect different damages and its intensity of

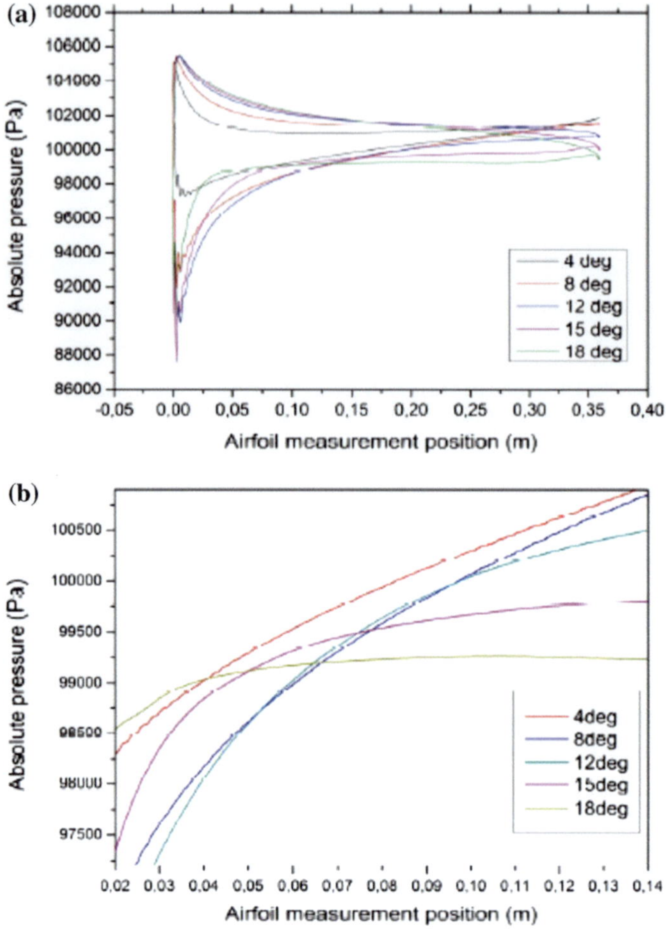

Fig. 5.4 Depicts the absolute pressure on a NACA0012, to increase the angles of attack (top), detail of 0–0.14 m position (bottom) (reprinted with kind permission from (Francioso et al. 2015))

various locations in aircraft structures (Mal and Chang 1998; Rose and Soley 2000; di Scalea et al. 2002). This technique has many advantages such as scanning capability is very fast, large area scanning, long-range inspection, low cost, and it can scan inaccessible location of the aircraft.

Presently, the technology brings a minute and conformal piezoelectric ceramics with wafer transducers which can be mounted on its surface or embedded leave in place on the aircraft structures. This uses guided waves for inspecting and continuous monitoring of the aircraft structure. A built-in piezoelectric transducer has been developed by Keilers and Chang (1995), Ihn and Chang (2004a, b) and Yang and Chang (2006a, b) which is used to detect the delamination layer of the composite plates, crack detection and its growth monitoring, stress in repair patches, reverted joints and bolted joint in aircraft. A product called SMART layer is now commercially available which is used to develop stemming effect from different sensors and actuators in a network-based system (Lin et al. 2001). A study was made by Giurgiutiu et al. (2003) on the effect of ultrasonic waves that propagate on a plate or a beam manned with an embedded piezoelectric wafer active sensor (PWAS). When the length of PWAS equals to the half of the wavelength of the Lamb wave mode, the maximum coupling is achieved while other modes are minimized. So, the change in the length of PWAS resulted in the tuning and setting of different wave modes (Giurgiutiu 2005). Matt et al. (2005) used guided wave in piezoelectric disks to inspect the composite wing to spar bonded joints. There are various applications of piezoelectric sensors in the practical field such as plates, sandwich beam structures, etc. as reported by Lin and Yuan (2001), Yang and Qiao (2005), and Annamdas and Soh (2006). But still, it is required to develop an effective algorithm to process the data acquired from sensors in order to minimize the bias or random errors and it provides an accurate detection of defects and its location with intensity.

The wings of the aircraft are a very complex structure in design. It consists of many things such as ribs, spars, reverted joints, skin panels, stiffeners, etc. Due to the ruthless working environment, the wings will be subjected to severe stress which may lead to a fatigue failure of the structure at rivet holes, spars, or stiffeners point. A detailed study was made by Xiao et al. (2005) on three different aircraft structures made of different materials such as CFRP, GLARE, and aluminum alloy. The study was focused on fatigue failure under different load conditions.

An experiment was conducted on an E-2 surveillance aircraft. A wing section of this aircraft was cut into different pieces for experimental purpose. Initially, a test was conducted with the angle beam Lamb wave transducer. A wave was propagated on the surface of the wing structure in order to evaluate the capabilities of wave propagation, attenuation, and scattering effect due to paint and rivet rows. It was observed that the use of paint has a negative effect on the return signal of the propagated wave. Then, a piezoelectric layer of 1-foot diameter was bonded to the inner part of the wing that resulted in a better generation and reception of the ultrasonic guided waves. In this experiment, the different defects were simulated in such way that signal effect due to loose rivets joint, crack in rivet holes, etc. were studied. An algorithm called RAPID was developed in order to understand and monitor the defect detection, its location and growth from the collected signals at the piezoelectric disk array (Gao et al. 2005).

Figure 5.5a shows an E-2 surveillance airplane wing section acquired from the US Navy for this study purpose. This section is made up of aluminum alloys and it is coated with paint for the avoidance of corrosion. Using plasma cutting tool, the upper skin of the wing structure was removed for easy mounting of sensors on this experiment.

Figure 5.5b represents a part of the removed skin piece of the wing section after cutting. Like this, many samples were produced for experimental purpose after removal of inner parts of the wing such as spars, stiffeners, ribs, etc.

A sample signal of magnitude 1.8 MHz S_0 Lamb wave signals was collected by the receiver transducer. This receiver transducer was located 20–200 mm distance

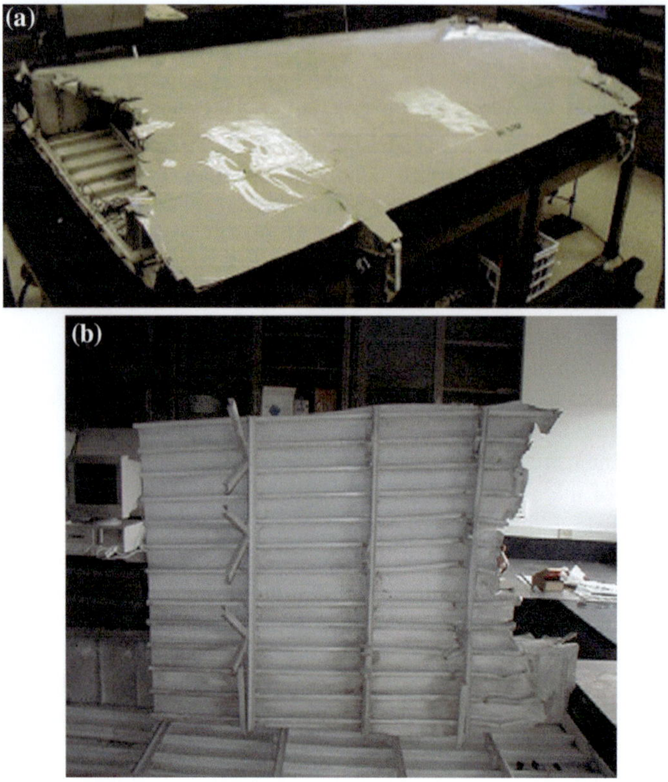

Fig. 5.5 **a** E-2 surveillance aircraft wing section; **b** the disassembled top panel of the wing (reprinted with kind permission from (Zhao et al. 2007))

Fig. 5.6 Represents the guided wave signals between rivet rows which show the wave attenuation with propagation distance (reprinted with kind permission from (Zhao et al. 2007))

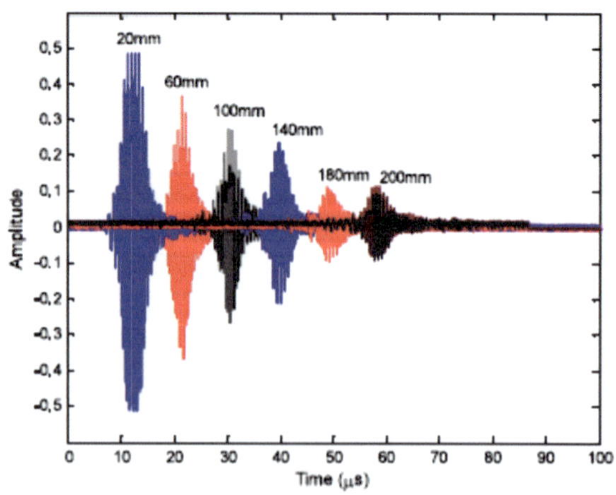

from the transmitter. It is quite observable that the amplitude of the wave signal kept on reducing with distance that varies from the transmitter as shown in Fig. 5.6 due to different effects such as beam broadening, multipath, and signal attenuation.

5.4.3 Optical Fiber Sensors for Aircraft Structural Health Monitoring

Optical fiber sensors have a great popularity in use for structure health monitoring system (SHM) in aircraft. SHM technology using optical sensors has a potential use in aerospace industries due to its inherent capabilities such as immune to electromagnetic radiation, less weight which is a very important factor in aircraft design, high sensitivity, accuracy, and resolution and moreover it is very suitable to be embedded into the composite structure of aircraft (Zhou and Sim 2002). Therefore, the use of optical fiber sensors (OFSs) for real-time health monitoring system (Güemes 2013) is very beneficial.

Mainly there are three approaches that are normally used for deployment of the optical sensor in the SHM system (Grattan and Sun 2000). These three approaches are given below:

(a) Single-point sensors (Claus et al. 1992): Used for detecting the pressure and temperature of the system or environments.
(b) Distributed sensing (Güemes et al. 2014): Measured data can be trapped at any point of an optical fiber.
(c) Quasi-distributed systems (Mrad and Xiao 2005): Used in a large structure with a number of single-point sensors.

5.5 Conclusion

This article describes the use of various embedded sensors to detect and monitor the health issues or failure of different components of the aircraft, ultimately enabling proactive maintenance to prevent aircraft subsystems/components from breakdowns.

References

Annamdas V, Soh CK (2006) Embedded piezoelectric ceramic transducers in sandwiched beams. Smart Mater Struct 15:538–549

Barr M (2009) Real men program in C. In: Embedded systems design, p 2. TechInsights (United Business Media). Retrieved 23 Dec 2009

Barr M, Massa AJ (2006) Introduction. Programming embedded systems: with C and GNU development tools. O'Reilly, pp 1–2. ISBN 978-0-596-00983-0
Chang C, James JS (1918) Metal-coated optical fiber damage sensors. Proc SPIE Int Soc Opt Eng 1993:138–144
Claus RO, Gunther MF, Wang A, Murphy KA (1992) Extrinsic fabry-perot sensor for strain and crack opening displacement measurements from −200 to 900 °C. Smart Mater Struct 1. https://doi.org/10.1088/0964-1726/1/3/008
Collinson RPG (2002) Introduction to avionics systems. Springer, Berlin
di Scalea FL et al (2002) Guided wave ultrasonics for NDE of aging aircraft components. Proc SPIE 4704:123–132
Du H, Klamecki BE (1993) Characterization of force sensors embedded in surfaces for manufacturing process monitoring. In: American society of mechanical engineers, production engineering division (Publication), vol 64, pp 207–216
Du H, Klamecki BE (1999) Force sensors embedded in surfaces for manufacturing and other tribological process monitoring. J Manuf Sci Eng 121(4):739–748
Francioso L, De Pascali C, Casino F, Siciliano P, De Giorgi MG, Campilongo S, Ficarella A (2015) Embedded sensor/actuator system for aircraft active flow separation control. In: AISEM annual conference, vol XVIII, pp 1–4. IEEE
Friswell MI, Inman DJ (2000) Sensor validation for smart structures. J Intell Mater Syst Struct 10 (12):973–982
Gao H, Shi Y, Rose JL (2005) Guided wave tomography on an aircraft wing with leave in place sensors. Rev Prog QNDE AIP Proc 760:1788–1794
Giurgiutiu V (2005) Tuned Lamb wave excitation and detection with piezoelectric wafer active sensors for structural health monitoring. J Intell Mater Syst Struct 16:291–305
Giurgiutiu V, Bao J, Zhao W (2003) Piezoelectric wafer active sensor embedded ultrasonics in beams and plates. Exp Mech 43:428–449
Grattan KTV, Sun T (2000) Fiber optic sensor technology: an overview. Sens Actuators A Phys 82:40–61
Güemes A (2013) SHM technologies, and applications in aircraft structures. In: Proceedings of the 5th international symposium on NDT in aerospace, 13–15 Nov 2013, Singapore
Güemes A, Fernandez-Lopez A, Fernandez P Damage detection in composite structures from fibre optic distributed strain measurements. In: Proceedings of the 7th European workshop on structural health monitoring, Nantes, France, 8–11 July 2014
Hautamaki C, Zurn S, Mantell SC, Polla DL (2000) Embedded microelectromechanical systems (MEMS) for measuring strain in composites. J Reinf Plast Compos 19(4):268–277
Hummel K, Tiedemann C, Peitsch D (2013) Determination of the pressure distribution and fluctuations of the transonic flow in a highly loaded compressor cascade. In: Proceedings of Deutscher Luft- und Raumfahrtkongress. Deutsche Gesellschaft f¨ur Luft- und Raumfahrt - Lilienthal-Oberth e.V.
Ihn JB, Chang FK (2004a) Detction and monitoring of hidden fatigue crack growth using a built-in piezoelectric sensor/actuator network. I Diagn Smart Mater Struct 13:609–620
Ihn JB, Chang FK (2004b) Detction and monitoring of hidden fatigue crack growth using a built-in piezoelectric sensor/actuator network: II. validation using tiveted joints and repair patches. Smart Mater Struct 13:621–630
Jin XD, Sirkis JS, Chung JK, Venkat VS (1998) Embedded in-line fiber etalon/bragg grating hybrid sensor to measure strain and temperature in a composite beam. J Intell Mater Syst Struct 9(3):171–181
Keilers CH, Chang FK (1995) Identifying delamination in composite beams using built-in piezoelectrics. J Intell Mater Syst Struct 6:649–672
Krantz DG, Belk JH (1997) Remotely queried wireless embedded microsensors in composites. Proc SPIE Int Soc Opt Eng 3044:219–226
Krantz D, Belk J, Biermann PJ, Dubow J, Gause LW, Harjani R, Mantell S, Polla D, Troyk P (1999) Project update: applied research on remotely-queried embedded microsensors. Proc SPIE Int Soc Opt Eng 3673:157–164

Lamb H (1917) On waves in an elastic plate. Proc R Soc A 93:114–120
Lin X, Yuan FG (2001) Dianostic Lamb waves in an integrated piezoelectric sensor/actuator plate: analytical and experimental studies. Smart Mater Struct 10:907–913
Lin M, Qing X, Kumar A, Beard S (2001) SMART layers and SMART suitcase for structural health monitoring applications. Proc SPIE 4332:98–106
Mal A, Chang Z (1998) NDE of rivet holes in aging aircraft components using lamb waves. Proc SPIE 3397:68–75
Matt H, Bartoli I, di Scalea FL (2005) Ultrasonic guided wave monitoring of composite wing skin-to-spar bonded joints in aerospace structures. J Acoust Soc Am 118:2240–2252
Mrad N, Xiao GZ Multiplexed fiber Bragg gratings for potential aerospace applications. In: Proceedings of the international conference on MEMS, NANO and smart systems. Banff, AL, Canada, 24–27 July 2005
Murukeshan VM, Chan PY, Ong LS, Seah LK (2000) Cure monitoring of smart composites using fiber bragg grating based embedded sensors. Sens Actuators A 79(2):153–161
Murukeshan VM, Chan PY, Ong LS (2001) Intracore fiber Bragg gratings for strain measurement in embedded composite structures. Appl Opt 40(1):145–149
Ogisu T, Nomura M, Andou N, Takaki J, Song D, Takeda N (2000) Development of damage suppression system using embedded SMA foil sensor and actuator. Proc SPIE Int Soc Opt Eng 3991:62–73
Pereira CM, Mattice MS, Testa R (2000) Intelligent sensing and wireless communications in harsh environments. Proc SPIE Int Soc Opt Eng 3990:194–203
Rose JL (1999) Ultrasonic waves in solid media. Cambridge University Press, Cambridge
Rose JL, Soley L (2000) Ultrasonic guided waves for the detection of anomalies in aircraft components. Mater Eval 50:1080–1086
Sirkis JS, Chang C, Smith BT (1994) Low velocity impact of optical fiber embedded laminated graphite/epoxy panels. J Comp Mat 28(14):1347–1370
Udd E (1995) Fiber optic smart structures. Wiley (Interscience), New York
Viktorov IA (1970) Rayleigh and Lamb waves. Plenum, New York
Xiao Y, White RG, Aglietti GS (2005) Comparison of structural response and fatigue endurance of aircraft flap-like box structures subjected to acoustic loading. J Acoust Soc Am 117:2820–2834
Yang J, Chang FK (2006a) Detection of bolt loosening in C-C composite thermal protection panels. I Diagn Principle Smart Mater Struct 15:581–590
Yang J, Chang FK (2006b) Detection of bolt loosening in C-C composite thermal protection panels. II Exp Verif Smart Mater Struct 15:591–599
Yang M, Qiao P (2005) Modeling and experimental detection of damage in various materials using the pulse-echo method and piezoelectric sensor/actuators. Smart Mater Struct 14:1083–1100
Zhao X, Gao H, Zhang G, Ayhan B, Yan F, Kwan C, Rose JL (2007) Active health monitoring of an aircraft wing with embedded piezoelectric sensor/actuator network: I. Defect detection, localization, and growth monitoring. Smart Mater Struct 16(4):1208
Zhou G, Sim LM (2002) Damage detection, and assessment in fiber reinforced composite structures with embedded fibre optic sensors—a review. Smart Mater Struct 11:925–939

Chapter 6
Sensors Used in Flying: A Comprehensive Study

Kapil Manoharan and Shantanu Bhattacharya

Abstract The aerospace industry is an ever-growing area where the possibility to explore and achieve next-generation technology has always been a challenge. Passenger safety with power conservation has always been areas of great interest in the global aviation sector. The induction of MEMS (micro-electro-mechanical systems) and other micro- and nanotechnologies have revolutionized the area of sensors. It has helped in drastic improvements in the performance characteristics of the systems such as sensitivity, stability, service life, etc., while reducing the overall size, weight, and power requirements. These systems are now being used aboard aircraft for various specific applications such as autopilot systems, navigation systems, etc., which were traditionally controlled manually using mechanical and electrical systems. This chapter aims at discussing traditional flight instruments and the recent advancements that have occurred due to use of MEMS. Also, we have discussed AVS (air vehicle systems) which are miniature forms of aircraft used for several specific purposes including military operations. The requirement for these vehicles in miniaturized navigation, power and communication systems can only be achieved using MEMS and other micro- and nanotechnologies.

Keywords Aerospace engineering · MEMS · Flight instruments
Sensors · AVS

K. Manoharan (✉) · S. Bhattacharya
Microsystems Fabrication Laboratory, Department of Mechanical Engineering,
Indian Institute of Technology Kanpur, Kanpur 208016, UP, India
e-mail: kapman@iitk.ac.in

S. Bhattacharya
e-mail: bhattacs@iitk.ac.in

S. Bhattacharya et al. (eds.), *Sensors for Automotive and Aerospace Applications*, Energy, Environment, and Sustainability,
https://doi.org/10.1007/978-981-13-3290-6_6

6.1 Introduction

It has been recognized since the starting of manned flight that supplying information to the pilot regarding the operations of the aircraft is very useful and leads to safer journey. The Wright Flyer flown by the Wright brothers had very few instruments such as an engine tachometer, a stopwatch and an anemometer (wind meter) as they were concerned about the engine of the aircraft and the flight progress. Since then a lot of researches have been done to develop various instruments which help the crew during a flight. Various information like condition of the aircraft and the engine, speed of the flight, its altitude, other navigation data, communication system, and weather and cabin environment are provided by the instrument systems developed to help the flight crew.

The range of desired information with the ability to capture and convey the data more accurately and to understand easily is ever growing and is a big challenge for the aviation industry. With this growing need the size and complexity of aircraft systems have grown which further demands for smaller sensory systems to avoid overloading and over-cluttering of the cockpit. One such development is the use of computer-controlled digital interface with flat-panel screens for displaying and prioritizing messages.

Instruments on an aircraft can be basically classified into three groups according to the job they perform as given below:

- ***Flight Instruments***: These instruments are used to display the aircraft altitude, airspeed, attitude, and direction to the pilot. Instruments such as altimeter for altitude, airspeed and vertical speed indicators for flight speed, and the magnetic compass for direction are used as shown in Fig. 6.1. These are usually positioned in top center position in front of the pilot and the copilot. The analog instruments used originally were solely dependent on electricity which could be dangerous for the flight during power failures. Advancement in sensing and display techniques due to development of MEMS devices has helped in developing more reliable primary and secondary instrument systems.
- ***Engine Instruments***: These are the instruments that are used to measure the aircraft engine parameters such as pressure, temperature, fuel quantity, engine speed, etc. and display it to the crew. Pressure gauges, tachometers, and temperature gauges are some of the commonly used instruments and are usually placed in the center of the cockpit for better visibility. A particular engine data or parameter is displayed on a single gauge with multiple pointers to show the information of all engines in case of a multiengine aircraft.
- ***Navigation Instruments***: These instruments are used by the pilot to navigate an aircraft along a fixed route and also direct them while approaching for landing at an airport. Global positioning systems (GPS) are used to locate aircraft nowadays and help in providing specific information to pilot from ground control regarding takeoff, flying, and landing safely in a specific region.

Fig. 6.1 Screenshot of an aircraft flight instrument panel. Six pack instruments containing an airspeed indicator, an artificial horizon or attitude indicator, an altimeter, a turn coordinator, a heading indicator, and a vertical speed indicator (from top left, marked in red outline) comprises the flight instruments panel. In addition, there is also a magnetic compass added to the panel (Reprinted with permission from (Gray et al. 2016). Copyright (2016) Elsevier.)

Most of these modern instruments and information systems not only compute but also sense their environmental conditions and create specific responses. MEMS-based devices are now being used for development of these systems due to their miniature size, lighter weight, localized distributed signal conditioning, higher operating frequencies and wider bandwidth, short thermal constants, and higher resonant frequencies obtained. Also, development of processes such as photolithography has helped in bringing down per unit cost of the devices for batch production.

Another area of interest in recent years is in the field of air vehicle systems (AVS). It includes development of both micro air vehicle (MAV) and nano air vehicle (NAV) and defining the basic characteristics and challenges to development of these unmanned air vehicles (UAVs) (Pines and Bohorquez 2006). DARPA in 1997 started the MAV project and extended it to NAV (Hylton et al. 2012), wherein they defined a certain set of requirements for these systems.

In this chapter, we will discuss the basic working of flight instruments and their advancements due to development of MEMS. Also, we will discuss AVS and its requirements and how MEMS help in its development and progress.

6.2 Six Pack Flight Instruments: Basic Working

The most important and basic instruments of an aircraft have remained same since very long time and are called as the six pack (Fig. 6.1).

Basically, three of these instruments are connected to the pitot-static system and work on principle of pressure difference while the other three work on principle of gyro (FAA Handbook 2016) as shown in Figs. 6.2 and 6.3. Mechanical systems are used to sense either the change in pressure or as the system for gyroscopic effect. Major disadvantage to these systems is the weight and the sensitivity. Recalibration is also tedious and errors due to friction, wearing out, and eddies are very high.

The three instruments connected to pitot-static system are as follows:

- ***Airspeed Indicator (ASI)***: It shows the indicated airspeed using the pitot and static source. The difference in pressures of the pitot tube and the pressure in the static system gives the indicated speed. When the aircraft is on the ground both pressures are equal while in motion the difference in pressure causes the expansion of the aneroid capsule inside the indicator which in turn moves the needle to show the readings on the instrument. Figure 6.4 shows a depiction of the airspeed indicator system. The data shown can be erroneous as air density depending on pressure and temperature and positional error caused due to

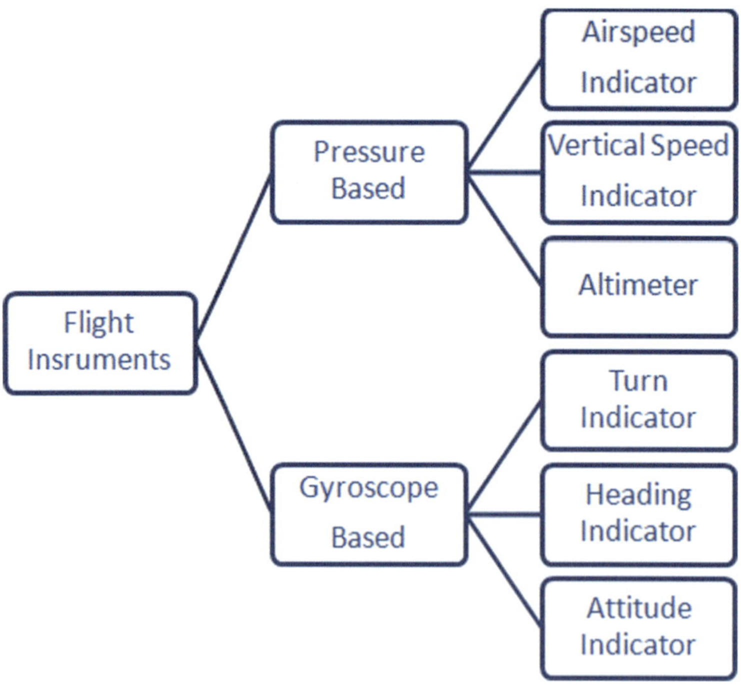

Fig. 6.2 Flight instruments and their classification

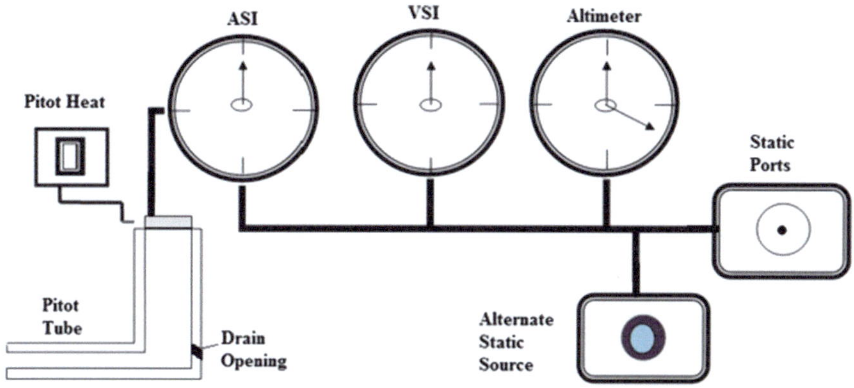

Fig. 6.3 Schematic of the three instruments based on pressure sensing and connected to the pitot-static line

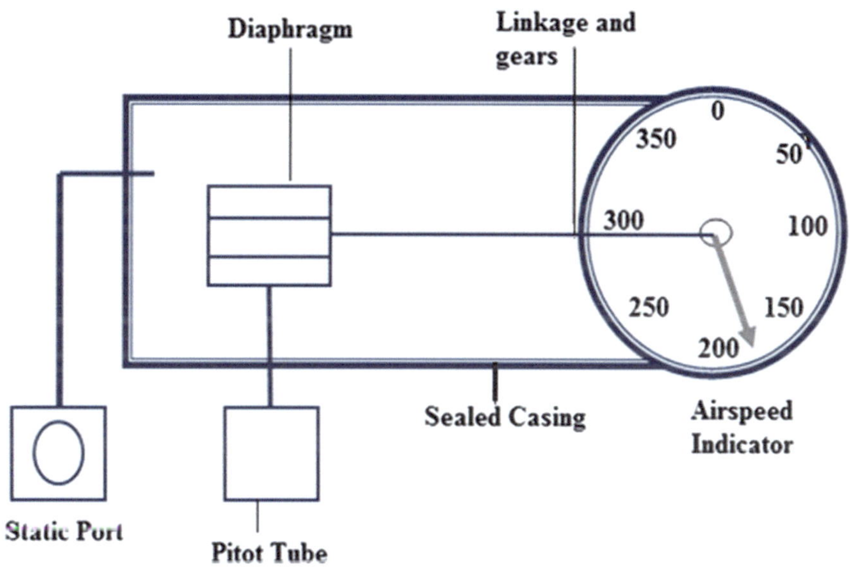

Fig. 6.4 Schematic of an airspeed indicating system

eddies, add up. Addition of these errors gives the actual or the true speed of the aircraft.

- ***Vertical Speed Indicator (VSI)***: It indicates whether the aircraft is in level or climbing, or descending and is usually indicated in feet per minute (fpm). A properly calibrated VSI should show zero in level flight or when on the ground. It displays two types of information: trend information which indicates the increase or decrease of a flight rate of climb or descent and the rate

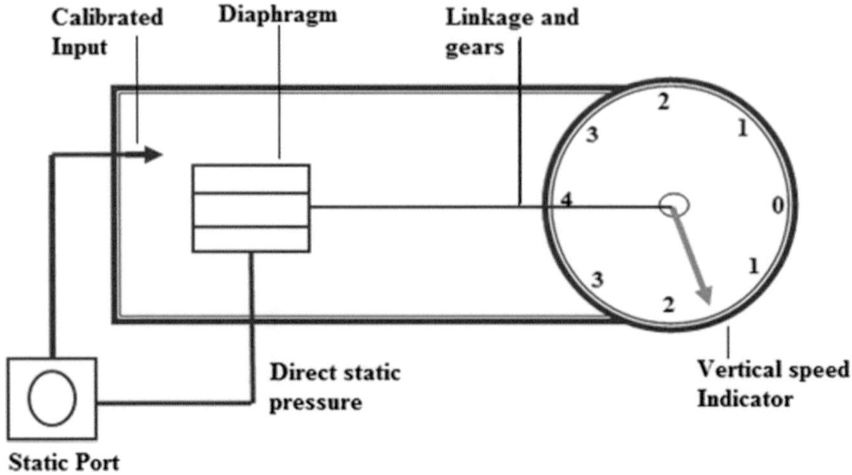

Fig. 6.5 Schematic of a vertical speed indicating system

information which shows rate of change of altitude once the pitch attitude is made constant. Although there is a lag (6–9 s) in showing of the data which further becomes erratic due to turbulence and rough controlling techniques.

- It consists of an airtight casing with metered static pressure from static line. The casing consists of a diaphragm with unrestricted air from the static line and linkages and gear system. As flight climbs or descends, the pressure inside diaphragm changes instantly while the change is slower in the outer casing due to restrictions. A schematic of the system is shown in Fig. 6.5. This causes pressure difference which is shown by deflection in needle.
- ***Altimeter***: It is an instrument that measures the height of an aircraft above a given pressure level and is the only instrument capable of showing altitude. The pressure altimeter uses an aneroid barometer which measures the pressure at different levels to indicate the altitude in feet. As altitude changes the atmospheric pressure decreases with increase in altitude and the difference between this pressure and that at the sea level indicates the altitude in the altimeter. An altimeter consists of a sealed aneroid wafer with linkages and gear mechanism inside a sealed casing. The wafer is evacuated to an internal pressure of 29.92 in. of mercury and is free to expand and contract with changes in the static pressure applied to the sealed casing. Compression of wafers indicates decrease in altitude and expansion indicates an increase. Multiple pointers indicate multiple ranges of altitudes starting from hundreds of feet to tens of thousands of feet (Fig. 6.6).

Since most altimeter readings are affected by air density due to changes in pressure and temperature, each altimeter is equipped with a knob to adjust the barometric pressure setting in the window provided in altimeter. Altimeters are adjusted as the flight progresses from one station to other and the data is provided

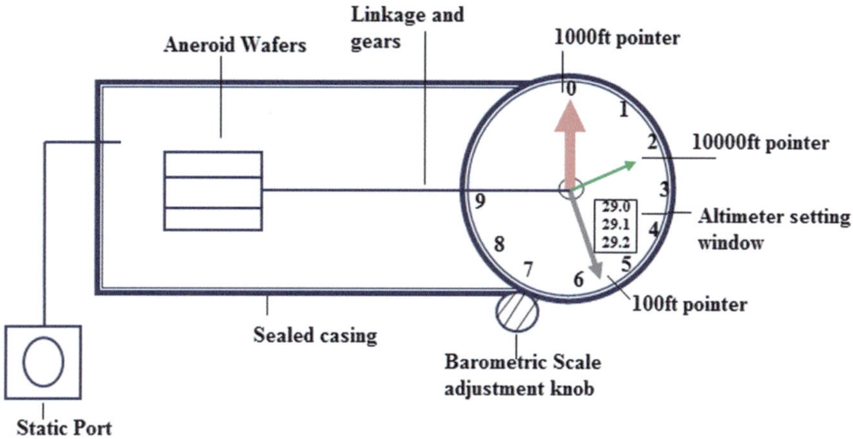

Fig. 6.6 Schematic of an altimeter system

by air traffic control (ATC) of that region against the conditions (pressure and temperature) of that area. Although there are still errors since the data provided is for ground level of that area and may not consider for irregularities during a flight at a certain height above ground.

The other three instruments work on the principle of gyroscope. Most of the aircraft use two power sources to ensure at least that one of the sources is available if other fails. The two sources of power used are electrical system which is mostly used for turn indicators and vacuum or pressure systems which are used for attitude and direction indicators. The three instruments are as follows:

- ***Turn indicators***: Two types of turn indicators are used in an aircraft: a turn and slip indicator which provides only the rate of turn in degrees per second and a turn coordinator which provides the roll rate initially and then the rate of turn. The gyro in the turn and slip indicator corresponding to the longitudinal axis of the aircraft rotates in the vertical plane. Restraining springs avoid tumbling and maintain a center position while a single gimbal restricts the plane in which the gyro tilts. For a turn coordinator, the gimbal is canted disk to which the gyro can sense both roll rate and turn rate. A turn coordinator does not display a specific angle of bank and only indicates the rate and direction of turn.
- ***Attitude Indicator***: It provides the relationship of the aircraft to the actual horizon with its miniature aircraft and horizon bar display. The horizon bar is fixed to the gyro which is mounted in a horizontal plane and remains so even if the aircraft is pitched or banked about its longitudinal and lateral axis. Adjustment knob is provided to adjust the miniature aircraft so that its wings overlap the horizon bar when the aircraft is in level and straight flight. The gyro rotates in horizontal plane and resists deflection of path of rotation and since it relies on rigidity in space, the aircraft actually rotates around the spinning gyro.

The maximum limit of banking plane is 100°–110° while that for pitch is 60°–70° depending upon the make and model of the instrument.

- ***Heading Indicator***: It is basically an instrument developed to reduce errors obtained from a magnetic compass to receive the direction of the flight movement. A magnetic compass shows erroneous data especially in turbulent environments and other forces while no such errors exist in the heading indicator due to use of gyro system. The rotor remains rigid in space and turns in a vertical plane and the instrument case and the whole aircraft rotates around the vertical axis of the rotating gyro. A compass card is fixed to the rotor and provides the heading information accurately and clearly. Friction, improper lubrication, and wear out parts can cause drifting or creeping. Since gyro is oriented in space and earth rotates in space by 15° per hour so that such an error has to be also compensated. Some horizontal situation indicators (HSI) use magnetometers to help negate this error.

6.3 Recent Advancements in Flight Instrument Systems

Electronic flight displays are now replacing free-spinning gyros and pitot tube systems with advancement in field of MEMS. Some of the advancements are as follows:

- ***MEMS accelerometers***: The basic principle of operation of these devices is defined as a simplified second-order damper mass–spring system (Beeby 2004). A MEMS sensing element with very high stability, electronics of very high quality, and best of technologies for assembly and packaging are the key to making a stable and ultrahigh precision MEMS accelerometer. Although capacitive-based MEMS accelerometers are widely used due to the simplicity in fabrication and operation, newer optical MEMS inertial sensors have better sensitivity and stability (Waters and Jones 2007). A Fabry–Perot interferometer and a photodiode on a silicon substrate are integrated using bulk or surface micromachining on a silicon substrate resulting in a compact and minimal-error-based system.
- ***MEMS gyroscopes***: MEMS vibratory gyroscopes work on the principle of Coriolis effect as is experienced by an object due to rotation and inertia effects (Acar and Shkel 2008). These vibratory gyroscopes are classified as angle gyroscopes which measure orientation angles directly and rate gyroscopes which measure the angular velocity or rate of rotation of the object (Shkel 2006). Rate gyroscopes are more widely used. MEMS gyroscopes developed usually use the capacitive drive and sense electrodes but are a major limitation due to mechanical and electrical crosscoupling which impacts the angular resolution and stability. Navigation grade performance has been reached by optical MEMS gyroscopes which optically sense the Coriolis force. The capacitive cross talk is eliminated by an electro-optical MEMS design (Waters et al. 2010).

- ***MEMS pressure sensors***: Micromachined pressure sensors are being inducted in aerospace applications due to demands of decreasing size, weight, cost, and improving the reliability and processing of signals with good stability. Most of the pressure sensors are designed on piezoresistive sensing principle on bulk and surface micromachined thin silicon diaphragms. Surface micromachined sensors are being developed since bulk micromachining requires anisotropic etching of silicon and requires postprocessing for bonding of glass and the silicon together. These are not required for surface micromachined sensors which can be directly connected with ICs for signal processing and other device functionalities (Lin and Yun 1998).

Any one or combination of any of these sensors can be used for specific aircraft applications on a single chip. Altimeters and airspeed sensors make use of MEMS pressure sensors. Additionally, accelerometers are also connected to circuit for better airspeed sensing. Inertial navigation systems (INS) make use of combination of accelerometers, gyros, magnetometers (IMU), and pressure sensors with addition of a global positioning system (GPS) to provide almost 10 DOF inertial data of the aircraft (Ananda et al. 2014). An attitude and heading reference system (AHRS) also makes use of combination of accelerometers, gyroscopes, and magnetometers to provide the attitude information of the aircraft with the yaw, pitch, and roll information (Guerrero-Castellanos et al. 2011).

6.4 Micro and Nano Air Vehicles

NAVs are usually defined as ultrasmall and ultralight weight air vehicles having wingspan less than 15 cm and overall weight of 25 g with operating range of 1–2 km. Also, the flight altitude is maximum up to 100 m. MAVs, on the other hand, can weigh from 100 g to up to 5 kg depending on the payload with a wingspan more than 15 cm and operating ranges of around 10 km for over an hour operations (UAS 2010).

An AVS keeps features of a large aircraft in very small volumes which increases the complexity and also presents challenges for its development. Physical challenges include high reduction in certain physical parameters like Reynolds number and the complexities included in integration of the system at such small levels. Similar to a large aircraft an AVS has to be also fully operational with respect to the acceleration, stability, altitude, and speed data. Figure 6.7 shows different sensors, signal processing units, and other systems which are an integral part of an AVS (Petricca et al. 2011). All these systems have to be integrated with limited weight and minimal power consumption requirements.

Figure 6.8 shows the weight budget allocation of the various parts in an AVS in percentage. It shows that battery and the airframe with motor components have the major share of weight in any AVS (Petricca et al. 2011).

Major parts in any AVS are as follows:

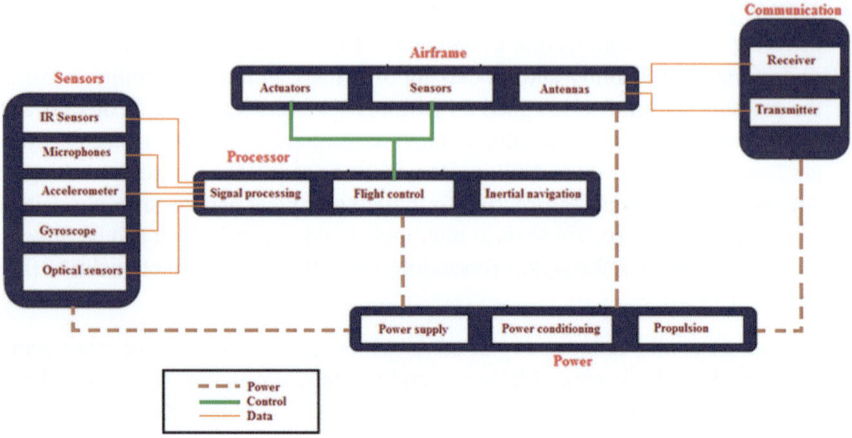

Fig. 6.7 System integration of an AVS

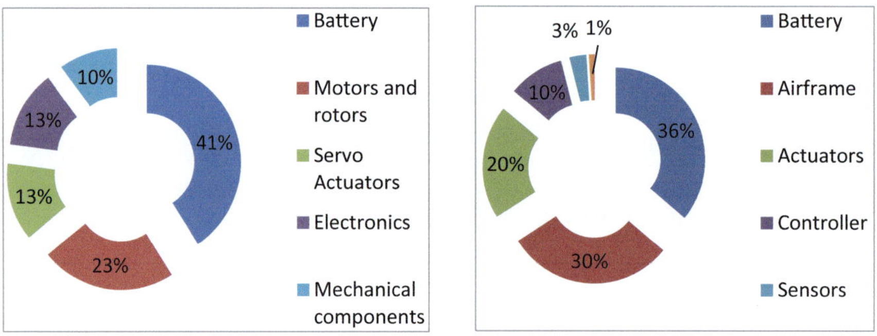

Fig. 6.8 AVS weight budget allocation: for a 15 g AVS and a 197 g AVS

- ***Airframe***: Drop-based shape gives the best aerodynamic performance at subsonic speeds. It reduces the drag force during forward flight and thereby reducing power consumption (Cavendish 2006). Designing is done to optimize space and positioning of the center of mass and uses a generic algorithm keeping in account all physical constraints (Ng and Leng 2006). Carbon fiber composites are usually preferred as material for airframes as they have high strength-to-weight ratio and are also easily accessible.
- ***Propulsion system***: All AVS require to generate motion and upward lift. Most common way is the use of electric motors which have high reliability, efficiency, smaller in size, lighter, and easily controllable. Since the systems are coreless there are no iron losses which increases the efficiency as well as lighter rotor implies small inertia implying faster accelerations and decelerations. Major disadvantage is overheating and thermal problems due to lack of iron core which restricts size of the motor.

Since more than half of the energy generated is used to generate lift, it is more advantageous to replace these motors with gas turbines or internal combustion engines (ICE) (Morris and Lane 1997). The size and weight with the noise generated is the major disadvantage of ICEs for application in NAVs. Although recent advances have helped in miniaturizing combustion engine motors (Sher et al. 2009), still noise produced is very high, and hence such systems cannot be used for tactical and stealth missions. Although lot of work is being carried out for miniaturized gas turbines using MEMS, very less success is seen till now. Another possibility is a hybrid system of electric motor and combustion engines (Glassock et al. 2009) but such a system is only suitable for larger air vehicles due to the overall size.

- ***Energy storage systems***: Electric energy is the primary requirement for all AVSs for supplying to electric motors for propulsion and feeding electronic circuits, communication devices, actuators, and sensors. Development in consumer electronics such as smartphones and notebooks have helped in tremendous improvements in the energy densities of the batteries. All Ni–Cd batteries have now been replaced by Lithium–ion batteries which are less toxic and more energy dense. Also, development of advanced batteries with circuitry to optimize the discharge curves with respect to the load of the cells (Mandal et al. 2008). Although many of the most advanced batteries provide very low energy densities than gasoline or methanol sources. Gas turbines or IC engines may be a better option for energy supply which may altogether negate the use of any batteries.

  Fuel cell systems are another alternatives which transform the fuel present in the system into electric current by some electrochemical process. Proton exchange membrane (PEM) fuel cells (*FuelCellStore*) which use hydrogen as fuel or direct methanol fuel cell (DMFC) which uses methanol as fuel is the most promising cells for AVS use. Although PEM cells are environment friendly as the residual waste is only water, the hydrogen for fuel is not available freely and requires chemical processes which in itself use energy. Also, it is difficult to store hydrogen since it has very low volumetric density (10,700 J/l) but very large mass–energy density (143,000 J/g). Alternate fuel cell like DMFC uses pure methanol but overall efficiency is lower as compared to PEM. The Hornet is an example of MAV which uses fuel cells (DARPA 2003). It has a wingspan of 38 cm and total weight of 170 g including that of the fuel.

  Since the weight of NAVs is very limited ultracapacitors are being increasingly used (Kroo et al. 2000). Fast charging and virtually unlimited charge and discharge cycles are major advantages of ultracapacitors while major disadvantages are lower energy density compared to other sources and dependence of the output voltage on the charge status (Spectrum 2001). Solar or photovoltaic cells can also be used but their usage gets limited due to size and weight constraints and reduced efficiency in indoor and low-light applications.

- ***Transmissions***: Two kinds of signals are basically transmitted in a AVS: data signals and control signals. Control signals are signals which are transmitted from the ground station or user to the vehicle to control the piloting of the vehicle as well as controlling other sensors on the AVS. Data signals mainly transmit the data or information from onboard sensors, cameras, microphones, etc. to the user for taking appropriate actions. To reduce the dimensions of an AVS, the major challenge is to reduce the size and weight of the antennas, filters, and resonators which strongly depend on the operating frequencies. Micromechanical or MEMS resonators have been developed which require low power and weigh very less as compared to the conventional quartz resonators (Discera, no date; Kim et al. 2005).
- ***Sensors and Actuators***: Sensors on an AVS can be broadly classified as sensors that are necessary for flight control and sensors which help in receiving mission-specific information. Although a 3-D accelerometer and a 3-D gyroscope can help in navigating an AVS but they suffer from some drifts and eddies additional sensors like magnetometers (Palaniappan et al. 2010) or some compensation circuitry has to be used (*The Paparazzi Project*). Data collecting sensors such as gas sensors, microphones, cameras, radiation biological sensors, etc. provide useful information to the user and their use depends on the nature of application. For example, a camera is useful for piloting a vehicle when the user cannot directly see the AVS. Development in MEMS and micro- and nanotechnologies has helped in improving the efficiency and reducing the overall size of these sensors. One such example is improvements seen in case of gas sensor whose sensitivity is dependent on the specific surface of sensing (Ding et al. 2009). Building nanostructures with large sensing area are possible and can be used to develop sensors either reducing the size with constant sensitivity or increasing the sensitivity by keeping the size constant. Also, development in materials technology has helped in improving the gas sensing capabilities (Palzer et al. 2008).

Actuators are needed in an AVS for flight control such as turning of vehicle or for sensor movement such as moving camera or for building useful tools such as micropliers to pick up samples. Table 6.1 shows the comparisons for some of the linear and rotary actuators in use (Conn et al. 2006). Both linear and rotary actuators are used in AVS according to the specific applications.

It can be seen that the size, weight, and power requirements are the primary criteria's on which a part is used for an AVS. The requirements of smaller sizes and weights with better performance characteristics can only be achieved by use of MEMS and other micro- and nanotechnologies. The challenges increase in case of NAVs where the requirements are much harsher. Power requirements and powering technologies and sources is a major challenge due to limitation in flight times. Although lithium–ion batteries are a major source but much more research has to be carried out in fuel cell and ultracapacitor fields to make systems much more lightweight for the requirements of an NAV.

Table 6.1 Comparison of linear and rotary actuators

Actuators	Advantages	Disadvantages
Linear actuators		
Piezoelectric ceramic	Excellent performance except for strain output which can be magnified using bender arrays	Requires high activation voltage
Shape memory alloy	Excellent performance except for frequency range	Poor fatigue life
Magnetostrictor	Excellent performance except for strain output	Requires high activation voltage
Solenoid	High strain	Low energy density
Electroactive polymers (EAP)	Dielectric elastomers outperform muscle in both stress and strain outputs	Only ionic EAPs operate on low voltage and novel technology not widely available
Rotary actuators		
Electric motors	Efficiency, reliability, versatility	Weight, dimensions

6.5 Conclusion

This chapter gives the idea about the basic flight instruments used in an aircraft and the recent advancements that have happened due to induction of MEMS in sensing and actuation systems for aircraft. MEMS-based accelerometers, gyroscopes, and pressure sensors are now being used individually or in combination according to the use for sensing, controlling, and navigation purposes in an aircraft. Also, a brief overview of the AVS is given which undermines the need for miniaturization and is being helped with newer developments in field of MEMS and other micro- and nanotechnologies.

References

Acar C, Shkel A (2008) MEMS vibratory gyroscopes: structural approaches to improve robustness (MEMS Reference Shelf), 2nd edn. Springer Publishing Company, Incorporated

Ananda CM, Akula P, Prasad S (2014) MEMS sensor suites for micro air vehicle (MAV) autopilot, pp 21–22

Beeby SP (2004) MEMS mechanical sensors. Sensor Rev Emerald 24(3):319–320. https://doi.org/10.1108/sr.2004.24.3.319.2

Cavendish M (2006) Growing up with science. Marshall Cavendish. https://books.google.co.in/books?id=acomopX55gwC

Conn A et al (2006) From natural flyers to the mechanical realization of a flapping wing micro air vehicle. In: 2006 IEEE international conference on robotics and biomimetics, pp 439–444. https://doi.org/10.1109/robio.2006.340232

DARPA (2003) Micro air vehicle powered entirely by fuel cell makes debut flight, 28 April. http://www.defense-aerospace.com/articles-view/release/3/17450/first-flight-of-fuel-cell-micro_uav.html

Ding B et al (2009) Gas sensors based on electrospun nanofibers, Sensors (Basel, Switzerland). Mol Div Preserv Int (MDPI) 9(3):1609–1624. https://doi.org/10.3390/s90301609
Discera (no date) Mems based oscillators. http://www.discera.com
FAA Handbook (2016) Flight instruments. https://www.faa.gov/regulations_policies/handbooks_manuals/aviation/phak/
FuelCellStore (no date) http://www.fuelcellstore.com
Glassock R et al (2009) Multimodal hybrid powerplant for unmanned aerial systems (UAS) robotics. In: Twenty-fourth bristol international unmanned air vehicle systems conference, pp 1–14
Gray R, Gaska J, Winterbottom M (2016) Relationship between sustained, orientated, divided, and selective attention and simulated aviation performance: training and pressure effects. J Appl Res Memory Cognit Soc Appl Res Memory Cognit 5(1):34–42. https://doi.org/10.1016/j.jarmac.2015.11.005
Guerrero-Castellanos JF et al (2011) Design and implementation of an attitude and heading reference system (AHRS). In: 2011 8th international conference on electrical engineering, computing science and automatic control, pp 1–5. https://doi.org/10.1109/iceee.2011.6106610
Hylton T et al (2012) The DARPA nano air vehicle program. https://doi.org/10.2514/6.2012-583
Kim B et al (2005) Frequency stability of wafer-scale encapsulated MEMS resonators. In: The 13th international conference on solid-state sensors, actuators and microsystems, 2005. Digest of technical papers. TRANSDUCERS '05, vol 2, pp 1965–1968. https://doi.org/10.1109/sensor.2005.1497485
Kroo I et al (2000) The mesicopter: a miniature rotorcraft concept phase II interim report
Lin L, Yun W (1998) MEMS pressure sensors for aerospace applications. https://doi.org/10.1109/aero.1998.686941
Mandal SK et al (2008) IntellBatt: towards smarter battery design. In: 2008 45th ACM/IEEE design automation conference, pp 872–877. https://doi.org/10.1145/1391469.1391690
Morris SJ, Lane L (1997) Design and flight test results for micro-sized fixed-wing and Vtol aircraft
Ng TTH, Leng GSB (2006) Design optimization of rotary-wing micro air vehicles. Proc Inst Mech Eng Part C-J Mech Eng Sci 220(6):865–873. https://doi.org/10.1243/09544062jmes104
Palaniappan M et al (2010) Unmanned vehicle group semester report
Palzer S et al (2008) Nano- and microsized metal oxide thin film gas sensors. Microsystem technologies-micro-and nanosystems-information storage and processing systems. https://doi.org/10.1007/s00542-007-0473-3
Petricca L, Ohlckers P, Grinde C (2011) Micro- and nano-air vehicles: state of the art. Int J Aerosp Eng. https://doi.org/10.1155/2011/214549
Pines DJ, Bohorquez F (2006) Challenges facing future micro-air-vehicle development. J Aircraft Amer Inst Aeronaut Astronaut 43(2):290–305. https://doi.org/10.2514/1.4922
Sher I, Levinzon-Sher D, Sher E (2009) Miniaturization limitations of HCCI internal combustion engines. Appl Thermal Eng https://doi.org/10.1016/j.applthermaleng.2008.03.020
Shkel AM (2006) Type I and type II micromachined vibratory gyroscopes. In: 2006 IEEE/ION position, location, and navigation symposium, pp 586–593. https://doi.org/10.1109/plans.2006.1650648
Spectrum I (2001) Circuit could swap ultracapacitors for batteries. http://spectrum.ieee.org/semiconductors/%0Adesign/circuit-could-swap-ultracapacitors-for-batteries
The Paparazzi Project (no date). http://paparazzi.enac.fr/wiki/MainPage
UAS (2010) UAS: the global perspective, 7th edn
Waters R et al (2010) Design and analysis of a novel electro-optical MEMS gyroscope for navigation applications. In: 2010 IEEE sensors, pp 1690–1695. https://doi.org/10.1109/icsens.2010.5689922
Waters RL, Jones TE (2007) MEMS navigation-grade electro-optical accelerometer, pp 1–16

Chapter 7
Overview of Electric Vehicles (EVs) and EV Sensors

Aviru Kumar Basu, Shreyansh Tatiya and Shantanu Bhattacharya

Abstract Electric vehicles, consisting of hybrid electric vehicle (HEV), fuel cell electric vehicle (FCEV), and battery electric vehicle (BEV), are day by day becoming more common in the automobile companies and automotive sectors. As the recent trend suggests, this kind of transportation is going to substitute internal combustion engine (ICE) vehicles in the future scenario. The technologies present in EVs are getting prominent and attractive, for the industry as well as for the customers. EVs are making significant impact on the power system and environment with lesser number of sensors, miniaturized components, and reduction of greenhouse gases. In this chapter, we have given a small overview of different types of electric vehicles, then we have discussed in details about the different kinds of sensors and its design for electric vehicles as well as for automotive vehicles. Finally, we have reported about the present research on MEMS-based miniaturization going on for sensors and devices for various applications in EVs.

Keywords Electric vehicles · Sensors · MEMS

A. K. Basu (✉) · S. Tatiya · S. Bhattacharya
Design Programme, Indian Institute of Technology Kanpur,
Kanpur 208016, UP, India
e-mail: aviru@iitk.ac.in

S. Tatiya
e-mail: tshrey@iitk.ac.in

S. Bhattacharya
e-mail: bhattacs@iitk.ac.in

A. K. Basu · S. Tatiya · S. Bhattacharya
Microsystems Fabrication Laboratory, Department of Mechanical Engineering,
Indian Institute of Technology Kanpur, Kanpur 208016, UP, India

S. Bhattacharya et al. (eds.), *Sensors for Automotive and Aerospace Applications*, Energy, Environment, and Sustainability,
https://doi.org/10.1007/978-981-13-3290-6_7

7.1 Introduction

Increase in the greenhouse gas (GHG) emissions has created a shift in vehicles from a conventional internal combustion engine (ICE) to an electric motor powered by a battery. An electric car thus derives some or all of its power from electricity. Electric vehicles (EVs) are easy to operate, quite energy efficient, and also cost-effective than their gasoline or LPG-powered counterparts. EVs have been around since very long, and the first small electric car model was built by Professor Stratingh in Groningen during the year 1835, but the lack of batteries during that period prevented it upscaling. In recent years with renewable sources gaining rapid popularity, there has been a revival of such clean modes of transport. Table 7.1 shows a comparison chart of existing conventional vehicles and upcoming electric vehicles.

EVs are generally considered to be consisting of the different subsystems. Each of this subsystem forms a coordination among themselves to make an EV work. There are multiple technologies which are applied in EVs to make all the subsystems work together. The details of this subsystem are shown in Table 7.2.

Different types of electric vehicles and their comparison are described in Sect. 7.2. In Sect. 7.3, an overview of various kinds of sensors related to the position and battery monitoring is mentioned. Sections 7.4 and 7.5 detail about different kinds of MEMS-based sensors for automotive application, which are common in EVs and about recently developed MEMS-based sensors specifically for EVs mentioned.

Table 7.1 Comparison of conventional and electric vehicles

Conventional vehicles	Electric vehicles
Runs on nonrenewable sources	Runs on renewable sources
No tailpipe emission	Greenhouse gases emission
Requires more maintenance	Requires less maintenance
Produces noise	Does not produce noise
Costly refueling	Cheaper refueling
At present range is around 400 miles	Range varies between 60 and 200 in a single charge

Table 7.2 Major subsystems for electric vehicles

Body design	Energy source	Electric propulsion	Auxiliaries
Body structure	Batteries	Electronic controller	Brakes
Frame	Battery charger	Power converter	Steering
Bumper		Electric motor	Auxiliary supply
Suspension		Transmission wheels	Temperature control
			Energy management

7.2 Types of Electric Vehicles

An EV can be broadly categorized into two types: Battery electric vehicle (BEV) one that runs solely on electricity and the hybrid electric vehicle (HEV) that combines the electric energy with any other source as shown in Fig. 7.1.

Hybrid electric vehicles (HEVs) have an inherent advantage and it can stretch the fuel economy further by combining the best of both battery and an engine. Thus while in a populated/urban area the vehicle could be operated on battery and then could switch to the engine when outside the city. Hybrids can further be subdivided into plug-in hybrid electric vehicle (PHEV) and fuel cell electric vehicle (FCEV)

EVs could, therefore, be categorized into

1. Battery electric vehicles (BEVs),
2. Hybrid electric vehicles (HEVs),
3. Plug-in hybrid electric vehicles (PHEVs), and
4. Fuel cell electric vehicles (FCEVs).

1. Battery Electric Vehicles (BEVs)

Electric vehicles that rely only on batteries for power are called BEV; it is without the traditional ICE and must be plugged into an external energy source to recharge its battery. The capacity of the battery directly influences the range of the BEV. Typically, a BEV can cover 100–250 km in a single charge (Grunditz and Thiringer 2016). High-range BEV models are also available but the trade-off is the price, as these are considered to be luxury models and usually have a hefty price tag. BEV like all other electric vehicles can recharge its battery via regenerative braking that slows down the vehicle using the motor and in turn recover some energy that is converted to heat (Carley 2014). Advantages of BEV include simple construction, ease in operation, and is completely noise free. It is also environment-friendly due

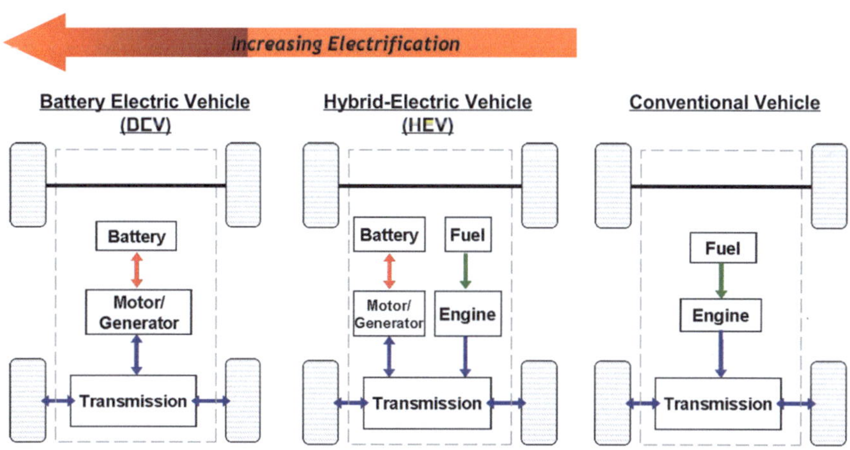

Fig. 7.1 Basic working of different types of vehicles (Irene Berry 2009)

Table 7.3 Comparing various characteristics between types of electric vehicles (Un-Noor et al. 2017)

EV type	Driving component	Energy source and infrastructure	Key features	Drawbacks
BEV	Electric motor	Battery and ultracapacitor	• Zero emissions • Short range • Crude oil independent • Commercially available	• Range • Battery capacity • Charging facilities
HEV	Electric motor and ICE	Battery, ultracapacitor, and ICE	• Very low emissions • Long driving range • Oil-dependent • Complex structure of drivetrains • Commercially available	• Management of energy sources • Battery sizing
FCEV	Electric motor	Fuel cell	• Ultra-low emission • High energy efficiency • Crude oil independent • Currently at a high cost • Independent to the supply of electricity • Under development	• The high cost of fuel cells • Feasible production of hydrogen • Lack of fueling systems

to the absence of any greenhouse gas emission. Only disadvantage is shorter range per charge which coupled with its advantages makes it a perfect choice for the urban household.

2. Hybrid Electric Vehicles (HEVs)

The International Technical Committee 69 (Electric Road Vehicles) defined HEV as "Vehicles utilizing two or more energy sources or storage such that at least one provides electrical energy" (Chan 2002). HEV commonly has an engine with a fuel tank and an electric motor with a battery. Energy is solely derived from gasoline and from regenerative braking (Carley 2014). HEVs could be further be classified into four types as shown in Table 7.3 based on the drivetrain structure:

- Series hybrids,
- Parallel hybrids,
- Series–parallel hybrids, and
- Complex hybrids.

The range of HEV is higher than BEV but it has few disadvantages of being expensive to operate the BEV and also cannot be conveniently charged at home.

3. Plug-in Hybrid Electric Vehicles (PHEVs)

Plug-in hybrids were developed to increase the range of HEVs (Gao and Ehsani 2010). PHEVs utilize an electric motor and a battery that could be charged via the power grid. The battery is also supported by an ICE that can recharge or replace when the vehicle is running on a low battery. Fuel saving is higher in comparison to HEVs as PHEVs utilize electricity directly from the power grid. However, like any hybrid, they too have an inherent limitation in finding a mechanic having the expertise in both types of engines.

4. Fuel Cell Electric Vehicles (FCEVs)

A fuel cell has properties of both a battery and an ICE; it generates electricity from an electrochemical reaction like a battery and it can run indefinitely if it is supplied with a source fuel (hydrogen) similar to an ICE (Matthey 2013). The automotive industry uses proton exchange membrane fuel cell (PEMFC) as its fuel type. FCEVs only produce water as a by-product which is ejected out of the tailpipes. Underdeveloped refueling infrastructure and lack of a cheap and sustainable method to produce hydrogen make fuel cell costly. Hydrogen also ignites faster than petrol and its flames are very hard to recognize. Nevertheless, researchers are addressing these issues and maybe in near future, they are eradicated.

7.3 An Overview of Sensors for Electric Vehicles

In a battery, electric vehicle basic engine and most of the sensors get vanished. An electric vehicle needs only one or two positions and speed sensors, while a hybrid electric vehicle needs 14 different position, speed, temperature, and pressure sensors. For example, BMW i3 presently provides a combination of EV and a 6000 cc motorcycle engine. The battery is charged by the engine which actually acts as a generator and not as a propulsion unit. As a result, the requirement of a number of sensors is low, and the loading on the engine is essentially constant. Hence, the exhaust sensors to measure pressure, temperature, and oxygen are no longer required

Generally, in an EV, it may be HEV or BEV, three sensors are a common one for temperature, second for voltage measurement, and third for current. A complex architectural diagram of an electric car shown Fig. 7.2 generally incorporates 15–20 current sensors and thermistor temperature sensors around 20–30.

Resolver, an expensive mechanical device is commonly used as a position sensing device for the main traction motor supplied by companies like Tamagawa Seiki. Other than this, 2–3 current sensors are needed for brushless (BLDC) electric motor control. The EVs of companies like BMW, Volkswagen, etc. use this kind of motor-control-current devices. In battery electric vehicles, the number of gears needed for transmission gets reduced in a large number and as a result, the number of position sensors needed is only one or two. This substantially reduces the additional sensing requirement compared to other EVs in BEVs. In the next section, we go into further details about the different sensors presently used.

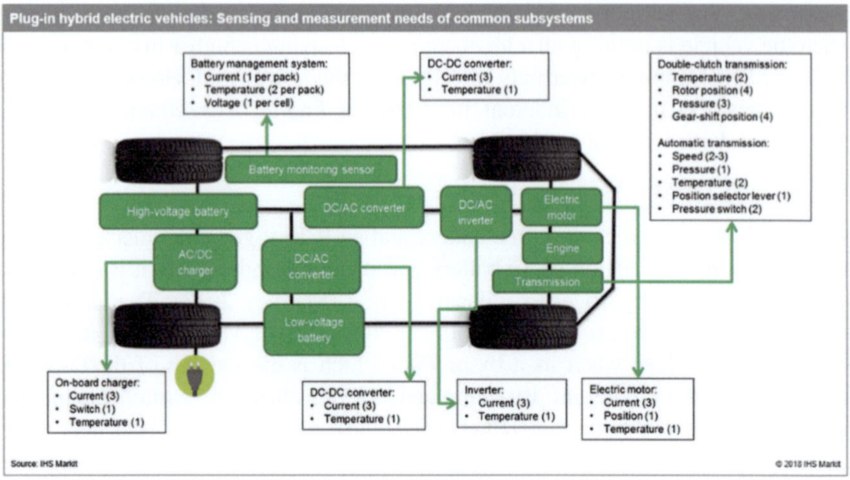

Fig. 7.2 Schematic of sensing and measurement needs for an electric vehicle (IHS Markit)

7.3.1 *Position Sensors*

Our main goal in this section is to understand about types of position sensors that are mostly best suited for tough environmental conditions which exist in automotive systems. Rotary position sensors, or angle position sensors, are mostly used during the transformation of angular mechanical motion data into an electrical signal accurately. Resolver and encoder the mostly used two sensors in motor control will be discussed here further (Verma 2015).

7.3.1.1 Resolvers

The resolver is an analog position sensor, which looks like a small transformer. The coil of resolver solver has total of three windings, one primary (located on the shaft of resolver) and two secondary (oriented at 90° to each other) shown in Fig. 7.3. The sensing mechanism is done by calculating the effective turns ratio and polarity between the two sets of coils. It acts just like a rotary transformer; the data obtained through this mechanism reports the variation in the angle of the shaft as it rotates.

The excitation of resolver coils is done through an AC reference input voltage, which is set at a constant frequency and gets induced in the primary winding. The stator and rotor give the same frequency output with a difference in phase of 90° due to the alignment of coil windings. Due to the 90° phase difference, a signal obtained is sine and cosine. The voltage induced on the windings is equivalent to the reference voltage multiplied by the value of the angle ($\sin\theta$ or $\cos\theta$) between the input shaft and the secondary coils. Here, θ denotes the shaft angle.

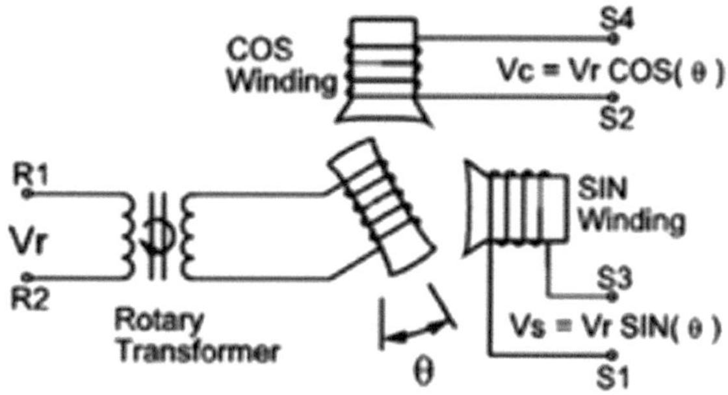

Fig. 7.3 Components of the resolver. Reprinted with permission from Texas Instruments

7.3.1.2 Encoders

Other than the resolvers, the encoders are used as a motor position sensor. The three mostly used encoders are an incremental sinusoidal signal, absolute position value, and incremental with “1” and “0” output.

The absolute position value rotary encoder consists of a disk split in a specific pattern shown in Fig. 7.4. For example, encoder splits into eight different patterns in binary digit for a 3-bit digital output. It finds significant importance when the equipment runs in an unusual manner, and power gets turned on/off in between uses.

Incremental encoder with a reference mark as shown in Fig. 7.5 is used for determining the motor current rotary position; it shows digital output in binary digit high or low. Pulses produced are directly proportional to the distance rotation, with a reference mark considered as angle zero. The rotary position can be determined by counting the number of pulses.

The third one, incremental sinusoidal encoder shown, generates an output pattern which resembles like a sine wave. The output pattern generally consists of a sinusoidal wave and cosine wave, along with a reference mark. The outputs generally require an electronic circuitry to further process these output signals. The most important thing is that the output signals vary significantly from the sine and

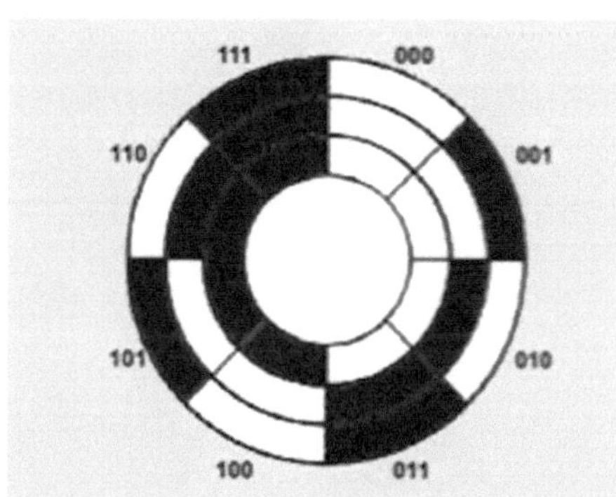

Fig. 7.4 Absolute position value rotary encoder. Reprinted with permission from Texas Instruments

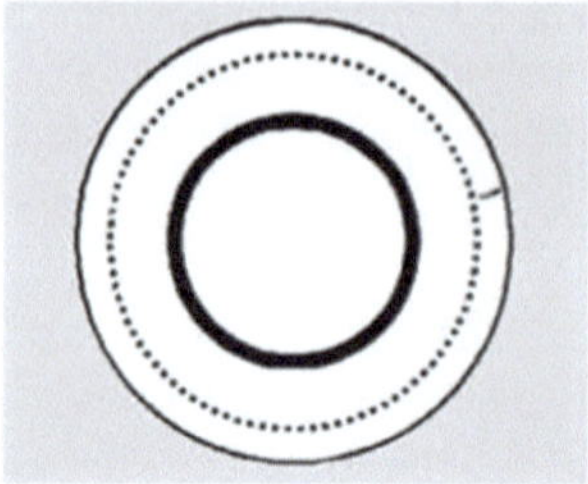

Fig. 7.5 Incremental encoder. Reprinted with permission from Texas Instruments

cosine of the resolver output. The output signals obtained from the resolver are amplitude-modulated sine and cosine waves; however, the output from the encoder does not have any "carrier" frequency and consists of higher frequency signals.

7.3.2 *Battery Monitoring Sensors*

Other than the position sensors, the most important sensor module in hybrid electric vehicles and electric vehicles is the battery monitoring sensor (BMS) shown in Figs. 7.6 and 7.7. The module generally consists of analog devices along with shunt resistors. The battery monitoring sensors are generally attached above 12 V lead acid batteries for its health monitoring. The health monitoring (aging) and state of charge of the battery become important for systems and motors operating at a higher voltage. The three main measurements of the sensor in this regard are voltage, temperature, and current. Voltage measurements among them are most critical, as batteries consist of a group of cells, with each cell attached with a voltage measurement chip. For current measurements, shunt resistors or open-loop hall sensors are generally used. For temperature measurements, NTC thermistors are generally used.

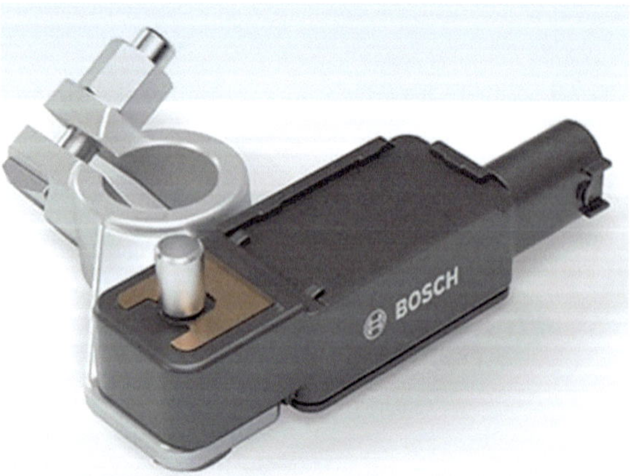

Fig. 7.6 Electronic battery sensor (EBS2). Reprinted with permission from BOSCH Co

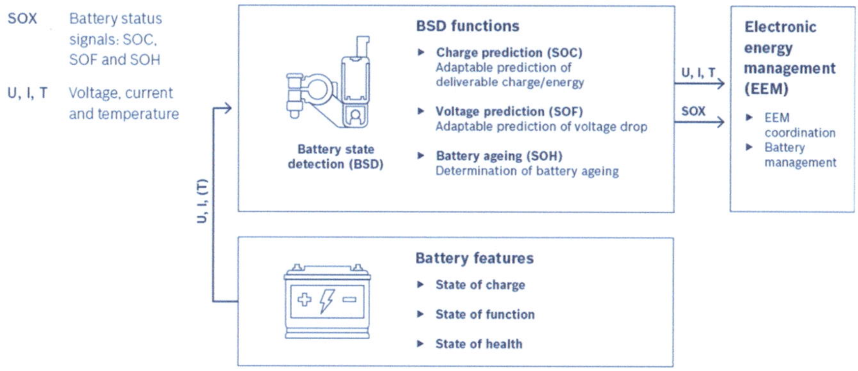

Fig. 7.7 Principle mechanism of battery monitoring sensor. Reprinted with permission from BOSCH Co

7.4 MEMS Technology in Automotive Vehicles

MEMS sensors extensively used in modern vehicles. The most commonly used systems are airbags and vehicle stability control.

MEMS-based technologies are generally classified into four divisions: accelerometers, gyroscopes, inclinometers, and flow and pressure sensors. The most emerging applications are like MEMS-based energy harvesters, oscillators, IR sensors, etc. Some of the thrust areas in which MEMS-based sensors are mostly used are summarized in Fig. 7.8 (Dixon 2007).

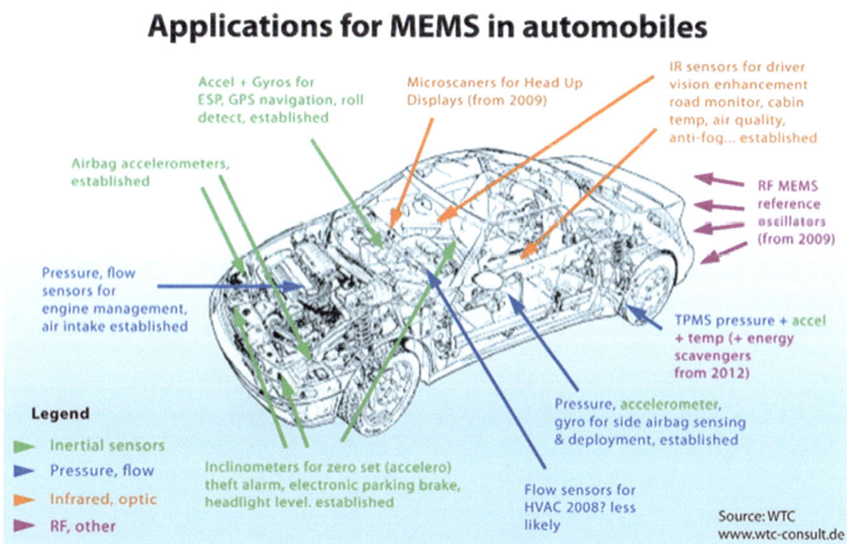

Fig. 7.8 MEMS-based sensors in automotive applications (Dixon 2007)

7.4.1 MEMS-Based Sensors for Passenger Safety

Prevention of collision of cars by opening the airbags in response to rapid change of accelerations of car finds an extensive use in all kinds of vehicle. The airbags employed in vehicles mostly contain MEMS-based fabricated chips of the accelerometer, which eliminates the further need for expensive g-switches in the airbag (Ernest 2010). The accelerometer continuously monitors the vehicle acceleration. This microfabricated element shown in Fig. 7.9 comes into play when there is a sudden deceleration in the vehicle, which causes a change in capacitance which is quickly detected by the chip, and sends necessary signals to the airbag.

7.4.2 MEMS-Based Sensors for Skidding and Rollover Detection

MEMS-based gyroscopes shown in Fig. 7.10 are most used in vehicles for measurement of angular velocity. Gyroscope measures and maintains the orientation on the principles of conservation of momentum. This is mostly used in trucks for rollover detection (Ernest 2010).

7.4.3 Tire Pressure Sensors

Tire pressure sensor basically informs the drivers about the leaks, under inflation, and also about the air pressure decrease which automatically happens with time. Previously, battery lifetime for the sensors is used to require replacement of the whole package which was quite expensive, but nowadays MEMS-based sensors have drawn the attention of vehicle manufacturers to the MEMS sensors fabricating companies. With the progress in technology, tire pressure sensor modules are

Fig. 7.9 Sensor used in the airbag. Reprinted with permission from BOSCH Co

Fig. 7.10 MEMS-based gyroscope for rollover detection. Reprinted with permission from BOSCH Co

becoming more compact and battery less. MEMS-based energy harvesting system is presently incorporated in the tires for making the modules minute in size.

7.4.4 Electronic Stability Control of Vehicles

Electronic stability control is presently regarded as one of the most crucial electronic chips for the safety of vehicles. It actually helps the drivers to maintain stability and control over the vehicles at sudden jerk or movement. A continuous growth in the electronic stability control system has created a need for the use of gyroscopes and accelerometers (Ernest 2010). This MEMS-based device has increased the efficiency and reduced the size. These compact devices are generally highly economical.

7.4.5 MEMS Sensors for Engine Management

MEMS sensors are also used in pressure sensing and as an air mass flow sensor. Figure 7.11 shows a barometric pressure sensor used for fuel combustion process adjustment with the changes in atmospheric pressure.

7.4.6 The Electronic Parking Brake System

In the present vehicles to apply accurate brake pressure, a low g-range accelerometer is used for measuring the continuous acceleration and inclination

Fig. 7.11 Barometric air pressure sensor. Reprinted with permission from BOSCH Co

of the car. The electronic parking brake generally consists of three modules, electronic brake module, actuator, and electric switch in cabin. A command is sent by the driver to the electronic brake module; this module further sends command to the actuator. Finally, the brake pads create force on the wheel to stop the motion of the wheels.

7.4.7 *Sensors for Antitheft*

Car manufacturers and industries are nowadays highly concerned with the theft of expensive vehicles. Previous sensors were not capable of detecting minor tilt in the vehicle when pulled by a rope or chain. MEMS-based three directional axis accelerometers are presently installed in the car, which measures the inclination angle with respect to the earth. A device for antitheft is shown in Fig. 7.12.

Fig. 7.12 Sensor for antitheft protection and navigation. Reprinted with permission from BOSCH Co

7.4.8 Vehicle Navigation Based on MEMS

The vehicle navigation is one of the most common setups installed in most of the cars, especially for commercial vehicles. GPS plays a major key role in tracking; however, it is not possible for the GPS, alone to provide all the information required. Elevated buildings/walls generally block this signals. Along with, GPS MEMS-based gyroscopes and low g-accelerometers are needed for information about the position and forward path.

Other than the above applications, MEMS-based sensors are used for many other applications like headlight leveling, motor damping, fuel line, fuel evaporation, wheel speed, etc.

7.5 Recent Trends in MEMS-Based Sensors in Electric Vehicles

Microelectromechanical systems (MEMS) comprise of miniscule mechanical and electromechanical elements produced by a technique of microfabrication. MEMS technology has an integration of 4 Ms: Microsensors (Basu et al. 2016, 2018), microactuators, microelectronics, and microstructures. In the era of automation, MEMS technology has been embraced by researchers with open arms as it has significant benefits like low cost, reliability, and increased performance from other microscopic electromechanical sensors.

MEMS-based sensors convert a mechanical signal to an electrical signal. Electric vehicles (EVs) and hybrid electric vehicles (HEVs) due to their complex nature make sensor integration tricky, and thus most interesting applications are found only in high-end models. However, in the coming era, many of these will be integrated as a standard equipment.

7.5.1 Battery Characterization

Electrical and mechanical changes that occur in a battery due to its charging and discharging are essential to monitor in an EV. Battery failures may occur in an electric vehicle due to many reasons like loss of function in associated electronics, heat runaway on a charge, and connector failures. Precise detection of these failures was reported recently using a UV-based MEMS sensor that can detect electric arcs as they develop (Us et al. 2018).

7.5.2 *Longitudinal Velocity and Road Slope Estimation*

Fuel economy optimization and improvement in traction control have been another major area of research in the HEVs. Direct measurement of the longitudinal velocity is complex and a costly affair, and thus it is estimated from sensors like accelerometers, wheel speed sensors, wheel torque, etc. In addition to this, slope angle estimation in a turn is also vital for accurate dead reckoning from vehicle acceleration. Information from the electric motor on wheel torque can be used for early detection of excessive wheel slip with high accuracy using MEMS technology (Klomp et al. 2014).

7.5.3 *Piezoresistive MEMS Sensor*

Measurement of pressure using a MEMS sensor is not new to the automotive industry; it has been in mass production for measuring air and oil pressure. However, MEMS piezoresistive sensor usage in a fuel cell electric vehicle (FCEV) is an area of ongoing research; one such application is to check the leakage among unit cells. This is done by converting the piezoresistor value from a diaphragm deformed due to pressure difference into an output voltage (Lee et al. 2014).

7.5.4 *Rotor Flux Sensing*

Nonlinear reluctance torque control due to its complex nature has prevented the usage of hybrid reluctance permanent magnet (HRPM) motors in an electric vehicle propulsion. MEMS scnsors havc also bccn dcmonstratcd for thc dctcction of magnetic flux levels, thus overcoming the hurdle in the adoption of HRPM commercially (McCann and Damugatla 2005).

7.5.5 *Two-Wheel Self-balanced Electric Vehicle*

Evolution of vehicles has witnessed many different types and designs in automotive industry suitable for various terrains. Places like campuses, theme parks, museums, and foot streets often have huge areas that cannot be accessed by a car and can be covered easily now by using small compact vehicles like a two-wheel self-balancing electric vehicle. This system shown utilizes two parallel wheels for maintaining balance and employs micro-silicon accelerometer and gyroscopic sensor as position detection sensors which can be fabricated using MEMS (Sun et al. 2009).

7.5.6 State of the Charge Sensing

Accurate estimation of the state of charge (SOC) of a lithium–ion battery is not easy as voltage remains almost constant for a considerable level of SOC. MEMS-based sensors are being studied for identifying other parameters that vary with SOC so that accurate estimation of battery level can be possible (SIMS lab, University of Waterloo).

7.6 Negative Impacts of Electric Vehicles

Electric vehicles create carbon emissions that are not visible from the tailpipe. Majority of the electricity generated is using coal-fired power plant and thus indirectly emit more or less same proportion of greenhouse gases. EVs also consume material resources such as nickel in lithium–ion batteries and other rare metals used in electric motor magnets and in some battery chemistries. These resources often are mined extensively, and their extraction, processing refining utilizes heavy machinery and equipment, producing high CO_2 emissions. Nickel mining specifically is notoriously associated with plumes of SO_2 choking the skies, blanket of cancerous dust, and rivers running blood red. Currently, due to low quantum of electric vehicles sales, rare materials used in batteries are not feasible to recycle. Still used lithium–ion batteries are not dumped into landfills and battery recycler plants are used to recover a lot of materials from these batteries.

7.7 Conclusion

EVs are having huge potential for the future transport communication, by replacing the present conventional vehicles. EVs will be becoming much more eco-friendly by saving the planet from global warming, by reducing the greenhouse gases emitted from present vehicles. The sensor-based technologies for different configurations of EVs are discussed in detail in this chapter. Sensors for automotive applications, some of them which are also common in EVs, are discussed in details. Finally, we have reported about the various kinds of microfabricated sensors which are recently coming into play through MEMS-based researches and can be used for applications like motion sensing, battery sensing, energy harvesting, etc. This miniaturized sensor will help to reduce the cost, space, and give better sensing capability for the upcoming vehicles. Further, research needs to be done by the laboratories in collaboration with the automobile industries on EVs and its sensors to give all of us a better future with pollution-free environment.

References

Basu AK, Sah AN, Pradhan A, Bhattacharya S (2018) BSA Detection on polymeric nanocantilever, pp 1–6

Basu AK, Sarkar H, Bhattacharya S (2016) Fabrication and resilience measurement of thin aluminum cantilevers using scanning probe microscopy. In: Proceedings of the 3rd international conference C2E2 in foundations and frontiers in computer, communication and electrical engineering, Taylor and Francis, Mankundu, pp 457–460

Carley (2014) The beginners guide to electric vehicles (EV)

Chan CC (2002) The state of the art of electric and hybrid vehicles [Prolog]. Proc IEEE 90:245–246. https://doi.org/10.1109/JPROC.2002.989872

Dixon (2007) Prospects for MEMS in the automotive industry

Ernest P (2010) MEMS@Bosch: Automotive Application and beyond

Gao Y, Ehsani M (2010) Design and control methodology of plug-in hybrid electric vehicles. IEEE Trans Indust Electron pp 633–640

Grunditz EA, Thiringer T (2016) Performance analysis of current BEVs based on a comprehensive review of specifications. IEEE Trans Transp Electrif 2:270–289. https://doi.org/10.1109/TTE.2016.2571783

Irene Berry MK (2009) What's the deal with hybrid and electric cars?

Klomp M, Gao Y, Bruzelius F (2014) Longitudinal velocity and road slope estimation in hybrid electric vehicles employing early detection of excessive wheel slip. Veh Syst Dyn 52:172–188. https://doi.org/10.1080/00423114.2014.887737

Lee M, Nam K, Lee S, Kim H, Kim C, Park Y, Ahn B (2014) Design and fabrication of low-pressure Piezoresistive MEMS sensor for fuel cell electric vehicles. 912:1616–1621

Matthey (2013) Fuel cell electric vehicles: the road ahead

McCann RA, Damugatla SR (2005) Analysis of MEMS based rotor flux sensing in a hybrid reluctance permanent magnet motor. In: IEEE conference on 2005 vehicle power and propulsion, pp 194–199. https://doi.org/10.1109/vppc.2005.1554555

Sun H, Zhou H, Li X, Wei Y, Li X (2009) Design of two-wheel self-balanced electric vehicle based on MEMS. In: 4th IEEE international conference Nano/Micro Engineered and Molecular Systems NEMS, pp 143–146. https://doi.org/10.1109/nems.2009.5068545

Un-Noor F, Padmanaban S, Mihet-Popa L, Mollah MN, Hossain E (2017) A comprehensive study of key electric vehicle (EV) components, technologies, challenges, impacts, and future direction of development. Energies 10

Us A, Us C, Feel E, Difference T, Limited G (2018) Ultrasonic sensing MEMS predicts electric vehicle battery failure, pp 7–10

Verma A (2015) Rotary position sensing for electric vehicles

Chapter 8
Fabrication Processes for Sensors for Automotive Applications: A Review

Aviru Kumar Basu, Shreyansh Tatiya, Geeta Bhatt and Shantanu Bhattacharya

Abstract MEMS technology has revolutionized the industries of twenty-first century by combining the micromachining technology with the silicon microelectronics. The miniaturized devices produced from MEMS-based fabrication process have good adaptability and flexibility due to its low power consumption and compactness. MEMS devices are hugely deployed for automobiles sectors like accelerometer, gyroscopes, pressure sensors, etc. These devices are fabricated through batch fabrication in industries, using various types of fabrication technologies. In this chapter, first we have given an overview about MEMS. In the second section, we have given details about various kinds of materials which are utilized in the formation of sensors. In the third, fourth, and fifth sections, in details, different lithographic, bulk, and surface micromachining techniques, and thin film deposition techniques are discussed. At the last section, we have mentioned recent progress in MEMS-based processes, which are presently used mostly in atomic level for the fabrication of sensor and devices for automotive application.

Keywords MEMS · Sensors · Actuators · Microfabrication

A. K. Basu (✉) · S. Tatiya · G. Bhatt
Design Programme, Indian Institute of Technology Kanpur,
Kanpur 208016, UP, India
e-mail: aviru@iitk.ac.in

S. Tatiya
e-mail: tshrey@iitk.ac.in

G. Bhatt
e-mail: geetabht@iitk.ac.in

A. K. Basu · S. Tatiya · G. Bhatt · S. Bhattacharya
Microsystems Fabrication Laboratory, Department of Mechanical Engineering,
Indian Institute of Technology Kanpur, Kanpur 208016, UP, India
e-mail: bhattacs@iitk.ac.in

S. Bhattacharya et al. (eds.), *Sensors for Automotive and Aerospace Applications*, Energy, Environment, and Sustainability,
https://doi.org/10.1007/978-981-13-3290-6_8

8.1 Introduction

This chapter deals in details about the various processes involved in the MEMS technology to create tiny miniaturized devices, a combination of electrical and mechanical components. These devices are fabricated using various batch processing methods. The devices' size ranges from few micrometers to few millimeters. These devices generally have the capability for actuation and sensing mechanism.

The interdisciplinary MEMS area utilizes electrical engineering, mechanical engineering, material science engineering, chemical engineering, fluid mechanics, optics, and packaging. MEMS can be easily found in systems and machines ranging across medical, automotive, communication, medical, electronics, and in various sectors of defense. The present MEMS devices include accelerometers, gyroscopes, inkjet printer heads, microvalves, blood pressure sensors, and various biosensors.

MEMS is considered one of the most emerging technologies for this century which is by combining micromachining technology and various silicon-based microelectronics. Fabrication of semiconductors at past was considered one of the greatest revolutions. Presently, MEMS has brought another revolution in our lives.

This book chapter is divided into different sections: the first section deals with the general introduction in MEMS. Second section is on the MEMS fabrication processes mostly used. Third section is on the recently developed fabrication processes developed for MEMS devices.

MEMS is originated in the United States, which is referred to as micromachines in Japan and microsystem technology in Europe. The device chips generally utilize IC fabrication technology, but for fabricating the micromechanical components, micromachining processes are applied upon highly sophisticated substrates of silicon and other substrates. Two most commonly used processes which are used are

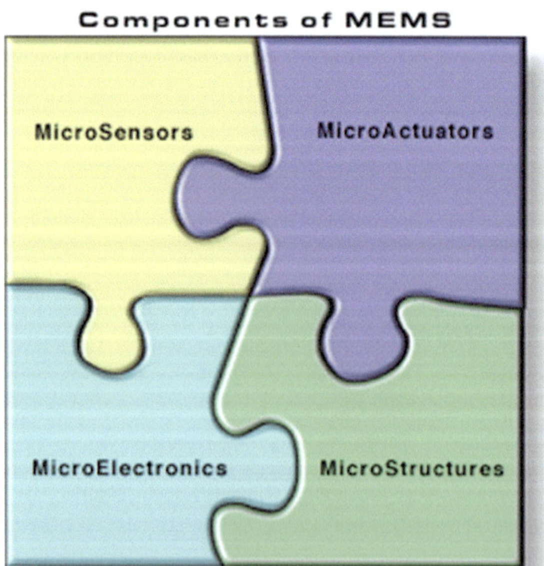

Fig. 8.1 Schematic of components of MEMS (Picture Courtesy: MNX: MEMS and Nanotechnology Exchange.)

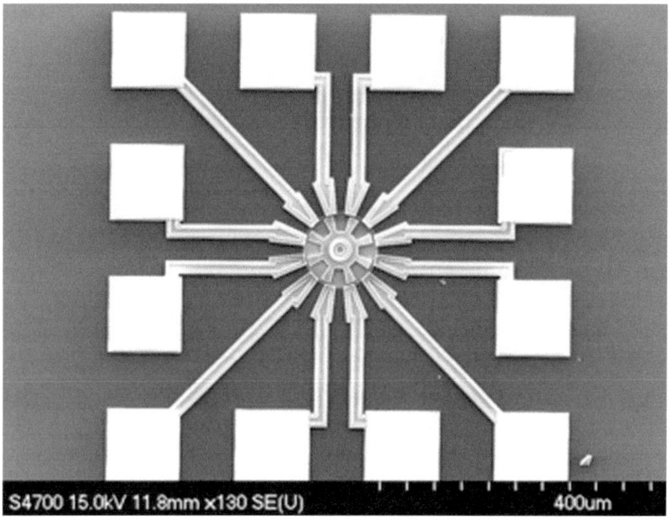

Fig. 8.2 MEMS-based microactuator (Picture Courtesy: MNX: MEMS and Nanotechnology Exchange.)

bulk micromachining and surface micromachining which selectively remove parts of silicon or deposit on it to form mechanical and electromechanical components. MEMS takes advantages of mechanical properties of the substrates (e.g., silicon) as well as its electrical properties. An integration of MEMS-based microstructures, sensors, and actuators as well as microelectronics makes the MEMS device complete shown in Fig. 8.1.

Microsensors generally check the sudden changes in the environment by measuring thermal, mechanical, chemical, and magnetic field-related phenomena or information. The microscopic MEMS devices like micromotors and microgears are already fabricated. But MEMS is not only a miniaturization technique but also actually a manufacturing technology platform for fabricating complicated mechanical system and devices with integrated electronics.

MEMS nowadays is not only restricted in research laboratories but also hugely deployed in industry-based products, for example, in automobile industries for airbags based on accelerometer, gyroscopes, and pressure sensors. MEMS finds a lot of application in data storage and micropositioning. Other than this, it is widely used in biomedical and telecommunication sectors.

Compared to the existing manufacturing technology, several advantages are associated with MEMS. The micromachining techniques involved with MEMS have resulted in a wide range of applications. Second, the batch fabrication techniques involved in MEMS enables the device to be fabricated with much reliability, lower size, weight, and cost. Figure 8.2 shows a microactuator developed by MNX.

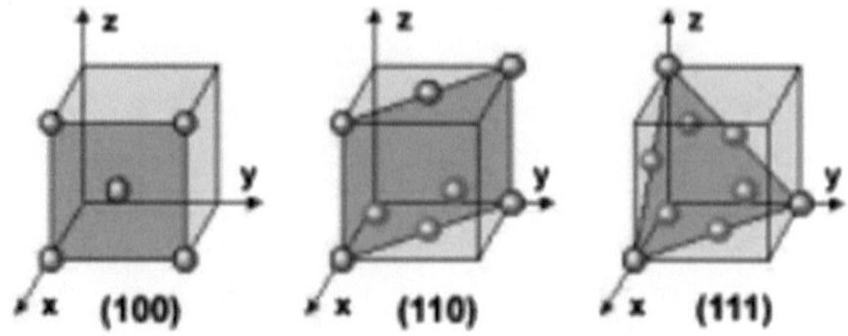

Fig. 8.3 Crystallographic index planes of silicon (Picture Courtesy: An Introduction to MEMS, Prime Faraday Partnership.)

8.2 Materials for Micromachining

The most commonly used material as the substrate is silicon. It is a well-established material for MEMS devices in microelectronics industry and for miniaturization.

Some of the reasons for silicon being the most commonly used substrate:

(i) Mostly abundant and inexpensive material.
(ii) Can be easily deposited on thin films, and
(iii) Bath fabrication of circuits is possible.

Germanium (Ge) and gallium arsenide (GaAs) are other materials, which are used as substrate. But one of the important features of silicon is that it can be easily oxidized to become an insulator called SiO_2 which is one of the most attractive features of Si, for batch fabrication.

The homogenous crystallographic structures provide the required electrical properties required in circuits. Other than the electrical property, the mechanical properties of Si are much better than steel with much higher elastic limit in both compression and tension.

The crystal orientation of silicon is the most important part during fabricating devices. Because of some of the etchants, both wet and dry attack the crystal structure in different directions and different rates (Fig. 8.3).

Other than the substrate additive films and materials which are most commonly used MEMS devices are as follows:

- Silicon: Amorphous, single crystal and polycrystalline,
- Silicon compounds (SiO_2, Si_xN_y, SiC, etc.),
- Metallic compounds and metals (Au, Al, Cu, GaAs, ZnO),
- Ceramics (Al_2O_3, etc.), and
- Organics (polymers, diamond, antibodies, DNA, etc.).

8.3 Lithography Techniques

8.3.1 Photolithography

A photomasking is a technique to transfer patterns from the mask, onto the surface of the silicon wafer (substrate) (Basu et al. 2016, 2018). A mask is generally printed through a mask writer, on a glass with coatings of chromium. The substrate is generally covered with a thin film of photoresist (epoxy/organic polymer) positive or negative in nature, patterns created upon exposure with light. The resist which is coated on the substrate is very much light sensitive. When the wafer is exposed to UV-light radiation, chemical reactions (crosslinking) occur in the polymeric films. Films are generally washed away or stay back depending upon the nature of the photoresist, while it is developed in developer solution. The resulting image is either original pattern of the mask copied on the substrate or just its negative image.

The lithography techniques can be further subdivided into three categories depending upon the distance between the mask and the substrate, and upon the optical setup.

8.3.1.1 Projection Lithography

Mask image is generally projected on the mask at particular distance and generally demagnified into a smaller image at 1:4–1:10 magnification.

Advantages: It can be of extremely high resolution ($\sim$0.007 μm or better than that), and distance between the mask and the substrate prevents the damage in the mask and increases its life.

Disadvantages: The equipment are much more expensive and complicated. Diffraction effects are present on the obtained pattern on the substrate.

8.3.1.2 Proximity Lithography

Resist is much closer but not in contact with mask: 1:1 magnification.

Advantages: Equipments are not expensive and low resolution up to ($\sim$1–2 μm or better).

Disadvantages: Accurate patterns are not obtained on the substrate due to diffraction patterns. The method is much lesser repeatable than contact methods.

8.3.1.3 Contact Lithography

The photoresist is in direct contact with the mask: 1:1 magnification.

Advantages: The equipment are not much expensive. Patterns with high resolution are obtained.

Disadvantages: Direct contact with mask creates scratches, damages, and attachment of dirt particles on the mask. The nonuniform resolution is obtained on the substrate due to the variation in gap between the mask and the wafer (Fig. 8.4).

8.3.2 Electron Beam Lithography (EBL)

Electron beam lithography is an extremely powerful technique for fabrication of nanostructures that are difficult to form with existing conventional lithographic

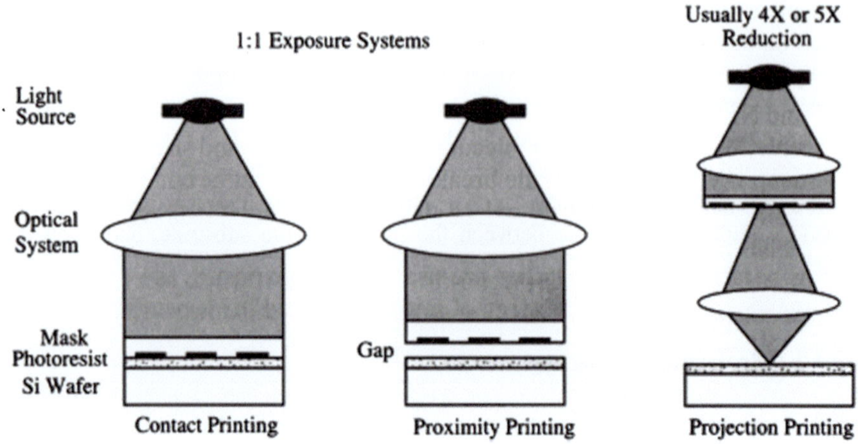

Fig. 8.4 Mask exposure techniques (Lithography and Pattern Transfer, Georgia Tech, ECE 6450 Course)

techniques. Resolutions up to nanometer can be easily achieved using EBL systems.

A focused beam of electrons is utilized for patterning a sample, with patterns designed in standard CAD tools.

An electron-sensitive film is coated on a substrate, through spin coating. The beam of electrons induces change on the film or resist upon exposure, by changing its molecular structure and nature of solubility. After the film is exposed, the film is developed in a particular solvent which selectively dissolves the unexposed areas or exposed areas of interest.

After beam exposure and development of the film, the obtained pattern can be further used as mask to transfer this pattern to other samples as shown in Fig. 8.5.

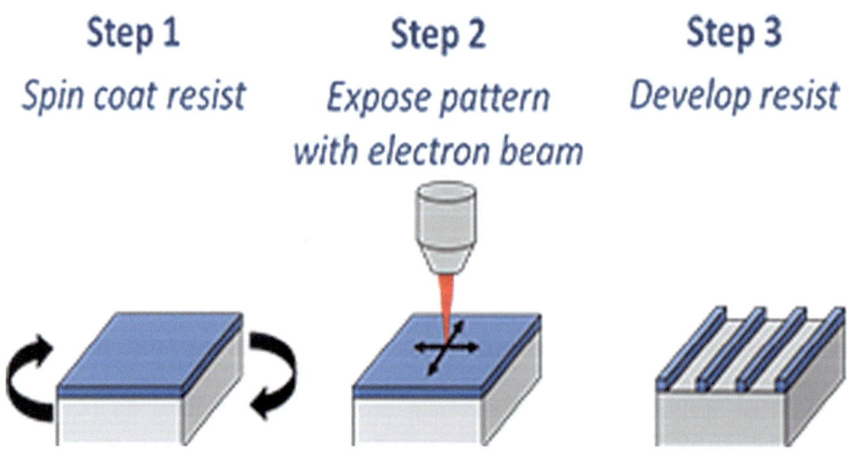

Fig. 8.5 Resist pattern schematic (EBL Overview, University of Sheffield)

8.3.3 *Maskless Lithography*

The most commonly used lithography in semiconductor industry and in MEMS field is optical projection lithography (OPL). In this process, at first a pattern is created on a mask at 4 times magnification compared to the desired size. The image of the created mask is projected on to silicon wafer substrate by using costly reductive lenses. Despite much advances in OPL technique, lot of disadvantages and crisis also associated with this technique. The cost and long amount of duration of time are associated with making the masks for OPC. The cost can generally exceed above $2 million but still only few masks get produced. The cost associated with designing and repairing the masks created a huge problem for the scientists. Hence, a new lithography technique is replacing the existing conventional lithography technique.

This technique enables to perform high-resolution lithography of thin film resists spin-coated on a glass or silicon substrate, avoiding the need of hard mask. The cost and delays associated with making the hard masks can be completely avoided in this technique. This process meets the requirement of industries with high resolution and throughput (Rammohan et al. 2011).

The various forms of maskless lithography techniques are scanning probe lithography, dip-pen lithography, scanning electron beam lithography, and FIB lithography. A schematic of maskless lithography setup is shown in Fig. 8.6.

8.3.4 *Grayscale Lithography*

Grayscale lithography is a simple, inexpensive, one-step lithography process to create three-dimensional micro/nanostructures in a thin photoresist layer using

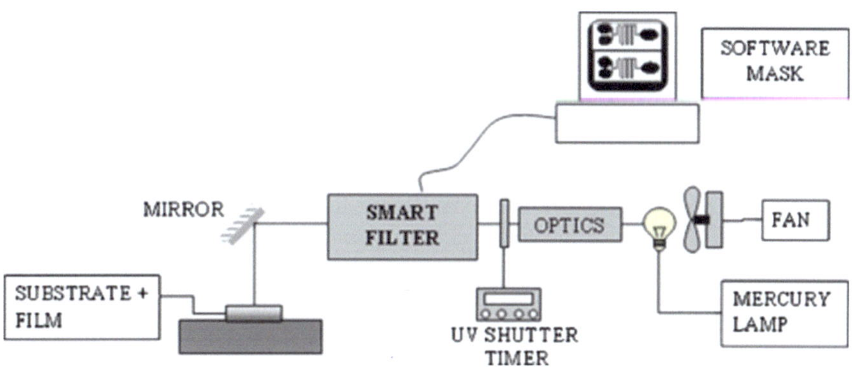

Fig. 8.6 Schematic of an SF-100 maskless lithography system [Reprinted with permission from Elsevier (Rammohan et al. 2011).]

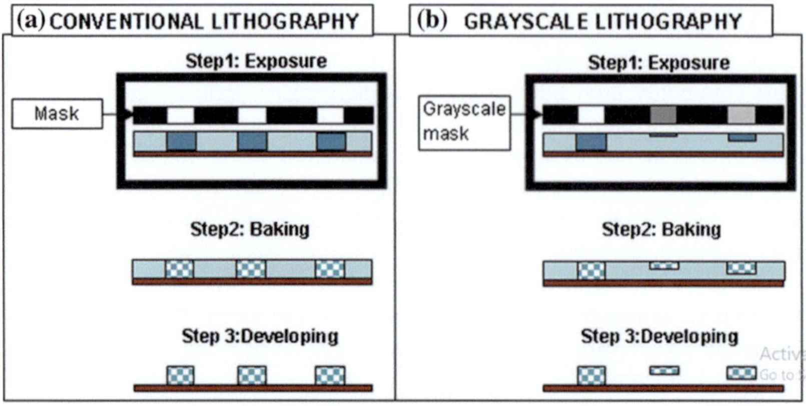

Fig. 8.7 Fabrication flowchart of conventional lithography and grayscale lithography [Reprinted with permission from Elsevier (Rammohan et al. 2011).]

optical gray mask. It uses a two-dimensional binary optical mask with sub-resolution, as well as a stepper motor system which modifies the intensity of UV light. The ultraviolet light is passed through the grayscale mask for exposure on the thin photoresist film. The modulated intensity of UV light exposes the photoresistive layer to particular depths. A gradient intensity pattern can be easily created by changing color combination in each cell or pixel.

In a recent paper (Rammohan et al. 2011), it is reported using a combination of micro-mirrors and software mask (gray tone masks); cantilevers and microfluidic channels were fabricated in a single step as shown in Fig. 8.7. This polymeric cantilever fabrication process can be used for batch fabrication of accelerometers and gyroscopes in industries.

8.4 MEMS-Based Etching Techniques

8.4.1 Bulk Micromachining

Functional MEMS structure on a surface is only possible after deposited substrate is etched out. The oldest known micromachining technology is the bulk micromachining. It involves selective removal of material/substrate to form microscaled mechanical components. Bulk micromachining can be of either physical or chemical type. Most common type of technique is the chemical wet etching. It involves immersion of the masked wafer in liquid comprising of a chemical etchant. It can further be of two types: anisotropic wet etching and isotropic wet etching.

Isotropic wet etching forms uniform structures by etching in all directions at more or less the same rate. Here, etch rate is determined by the concentration of the etchant and not the orientation of the mask edge. But more widely used method for silicon micromachining is the anisotropic wet etching. It is characterized by the formation of complicated patterns due to different etch rates in different directions. This property is utilized for the formation of micromechanical and microelectrical structures. Etch rate here varies with crystallographic orientation of the substrate. Reason for this is attributed to different bond configurations and densities that are exposed to different planes in the etchant solution. Silicon can be etched by both isotropic etching using an etchant like HNA (combination of HF, nitric acid and acetic acid) and by anisotropic etching using substances like potassium hydroxide (KOH) and ethylene diamine pyrocatechol (EDP) (Ashby). Figure 8.8 shows differences in etching patterns obtained from all the discussed etching techniques.

Modern very large-scale integration (VLSI) avoids wet etching and uses plasma or dry etching instead. It helps in achieving very deep etching range (<100 nm) and also high vertical aspect ratio (50 to 1). Plasma used in the process of dry etching produces energetic neutral free radicals that attack the wafers from all directions. Dry etching only affects the surface and thus is part of surface micromachining. Table 8.1 gives a summary of comparison between dry and wet etching.

8.4.2 *Surface Micromachining*

Surface micromachining builds microstructures by first deposition and then later etching different layers from the surface. Generally, polysilicon is employed as one of the layers and silicon dioxide layer is etched out to create necessary void. Surface micromachining is more popular due to reasons like

i. Precise dimensional control in both horizontal and vertical dimensions;
ii. More combinations possible in variety of structures, etchant, and sacrificial layer;
iii. Relatively simple usage as it processes wafer from single side; and
iv. High integration density and lower per die cost.

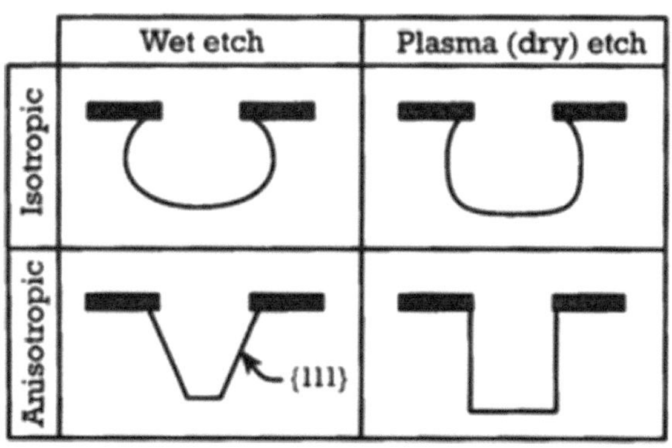

Fig. 8.8 Comparison of different types of etching (Picture courtesy: Srinivasan T: Process Flows and Bulk Micromachining CMOS Processing.)

Table 8.1 Comparison of dry etching and wet etching (Jency et al. 2014)

	Wet etching	Dry etching
Method	Chemical etchant	Plasma/reactive ion
Conditions	Atmosphere, Bath	Vacuum chamber
Advantages	• Low cost • Easy implementation • Good selectivity for most material	• Deep etch in the range of <100 nm • Higher etch rates • Can be easy automated
Disadvantages	• Feature size of <1 μm cannot be defined • Wafer contamination issues • Difficult to maintain a constant etch rate	• Poor selectivity • Usage of toxic and corrosive gases • High cost and hard to implement
Directionality	Isotropic (except crystalline materials)	Anisotropic

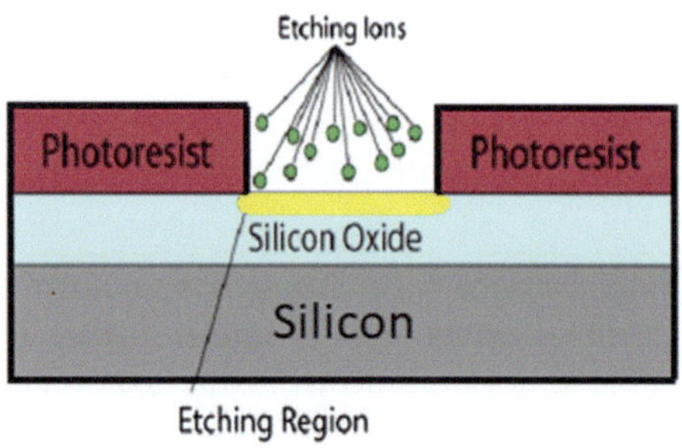

Fig. 8.9 Reactive ion etching technique

Dry etching can be divided into three classes called reactive ion etching (RIE), vapor phase etching, and sputter etching. In RIE, external energy in the form of radio frequency (RF) is used to break the gas molecules into ions. Generated ions are accelerated toward the substrate that is etched in the direction of movement of ions. Figure 8.9 shows a schematic of the ion etching technique. Involvement of both reactive neutrals and ions together helps to achieve ten times greater etch rates than both of these separately (Rangelow 2001).

Vapor phase etching involves wafer to be kept in a chamber with the introduction of one or more gas. The chemical reaction of the gas molecule with surface causes the material to be dissolved and etched out. Figure 8.10 shows parallel plate type of vapor etching technique. It has been extensively used in the manufacturing of both RF and capacitive switches, and piezoelectric MEMS devices like pumps, bolometers, and microphones (Hammond 2015).

Sputter etching involves the creation of plasma by discharging neutral gases like helium and argon. Potential gradient causes bombardment of the target with the ions causing it to form vapors that are then condensed and deposited on the substrate.

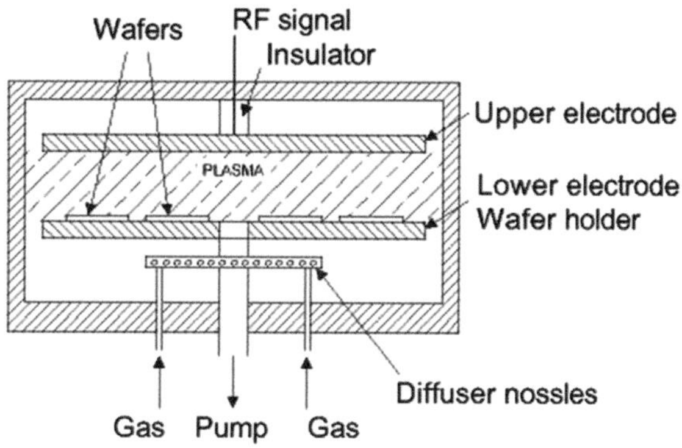

Fig. 8.10 Parallel plate vapor phase etching technique (Picture courtesy memsnet: portal for MEMS and Nanotechnology community.)

Deep reactive ion etching (DRIE) is a special subclass of RIE that is relatively new in the fabrication technology. It can achieve very high aspect ratio (up to 40:1) of etching and also deep etch depths with almost vertical sidewalls. A combination of two different gases is introduced into the reactor. The reaction of the first gas deposits a polymer on the surface of the substrate. Etching of substrate is then carried out by the addition of the second gas in the reactor. The slow dissolution of the polymer in the chemical part of etching causes it to deposit only on the sidewalls and not the horizontal surface. Figure 8.11 shows the major steps involved in the DRIE process.

8.4.3 LIGA

Another technique to get high aspect ratio in micromachining is called LIGA. It is a German acronym for *lithography* (particularly X-ray lithography), *G*alvanoformung (electrodeposition or electroplating), and *A*dformung (plastic molding). The letters of the acronym also indicate the process sequence. LIGA is a versatile process that can be utilized to produce parts of heights ranging from micrometers to centimeters. But due to the requirement of a special mask and X-ray as the radiation source, the process becomes expensive and is risky to human health. Application of deep microstructures exists in many sectors of the industry (Malek and Saile 2004) like production of high-precision actuators, moving elements like shafts, connecting

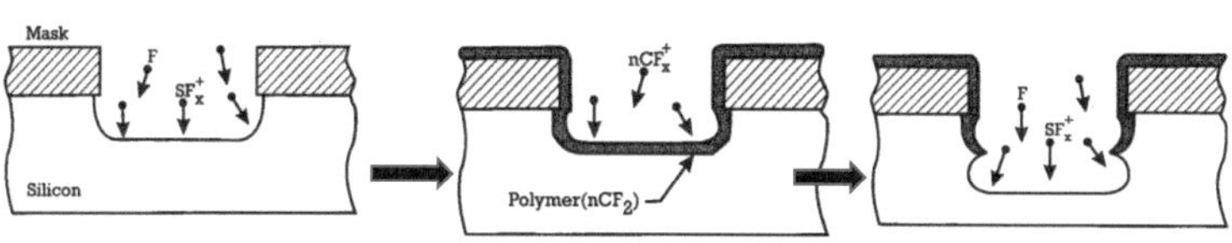

Fig. 8.11 Steps involved in DRIE

Fig. 8.12 Application of LIGA in industry [Reprinted with permission from Elsevier (Hirata 2003).]

parts, miniaturized gear system, robotics and assembly, the formation of piezoelectric composite like lead zirconate titanate (as shown in Fig. 8.12c) (Hirata 2003). Sensors based on electrostatic or electromagnetic-based processes such as micro-gyroscope and capacitive acceleration are also fabricated using LIGA. Figure 8.12 shows one of the applications of LIGA in industries.

8.5 MEMS-Based Thin Film Deposition Processes

Ability to deposit a thin layer of materials is the other most vital aspect of MEMS fabrication. The thickness of the film can be in the range of a few nanometers to about 100 μm. Material to be added can either be structural layers or can act as a spacer to be etched out later. MEMS-based deposition techniques are broadly categorized as either physical or chemical (Madou 1997).

Depositions involve a chemical reaction:

I. Chemical vapor deposition (CVD): In the process, the substrate is placed in a chamber with a supply of a number of gases. Gaseous phase is formed by the chemical reaction and/or thermal decomposition. CVD technologies that are used widely in MEMS are low-pressure chemical vapor deposition (LPCVD) and plasma-enhanced chemical vapor deposition (PECVD). As LPCVD operates in low pressure, better uniformity and homogeneity of the deposited layer are obtained due to slower mass transfer (Curley et al. 2012). In PECVD, plasma is added in the chamber with gases to create desired solid surfaces. This causes fabrication possible with higher film density and operation at low temperatures. Figure 8.13 shows the fabrication of accelerometer using LPCVD technique.
II. Electrodeposition (ED): It is a versatile deposition method that can yield thin layers of metals and its alloys. ED employs electricity to reduce cations of a material from the solution and then coats it on a conductive substrate surface. It has been used in industry for the formation of corrosion and wears resistant coating, battery grids, electronic connectors, transformer core material, etc. (Sobha Jayakrishnan 2012).
III. Epitaxy: Epi(top)- Taxis (ordered arrangement) is a process of growing a single crystalline film on a substrate. It is used for coating of thin film on a

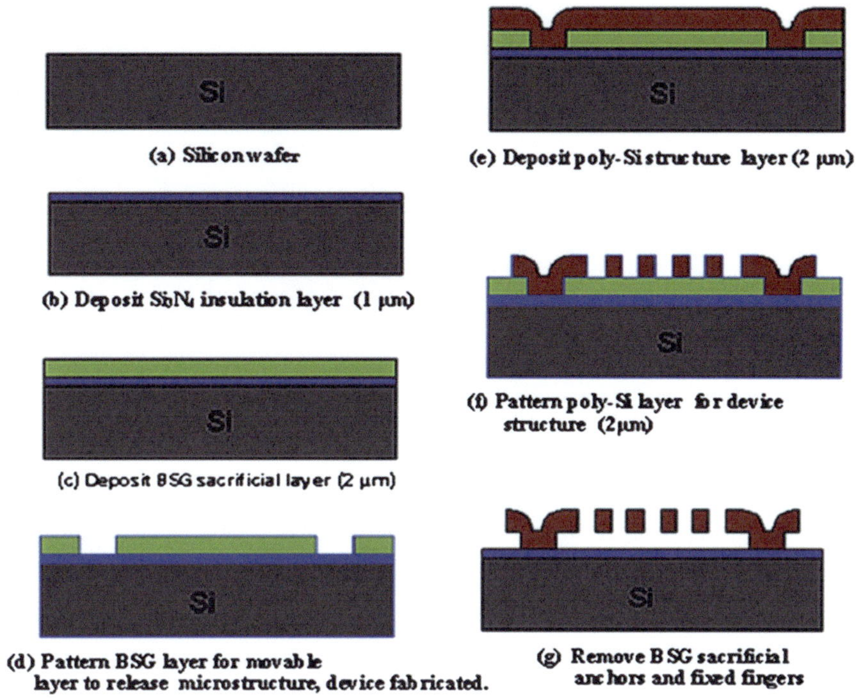

Fig. 8.13 Fabrication process flow involving deposition by LPCVD for accelerometer (Kavitha et al. 2016) (Reprinted with permission from Elsevier.)

semiconductor surface. It can be either homoepitaxy (crystal grown epitaxially on a substrate of same material, e.g., silicon film grown on a silicon substrate) or it could be heteroepitaxy (when a crystal is grown on a foreign substrate, e.g., silicon film grown on sapphire). This method is used for industries in the manufacturing of compound semiconductor lasers in opto-electronic devices and high-speed electronic devices (Ucer 2010).

IV. Thermal oxidation: It is a chemical oxidation process where silicon dioxide (SiO_{2-}) is grown generally at high temperature (800–1100 °C) in the presence of O_2. It is used to form films for electrical insulation and etch stops or masks for future processing steps. Figure 8.14 shows a schematic of a thermal oxidation furnace.

The deposition involves a physical reaction:

I. Physical vapor deposition (PVD): It comprises all the techniques in which the material is physically moved from source to the substrate such that thickness can be controlled too. Most important among them are sputtering and evaporation. PVD is preferred over CVD in metals as it involves a lower process risk and is relatively cheaper. It is used successfully in the fabrication of nanobelts

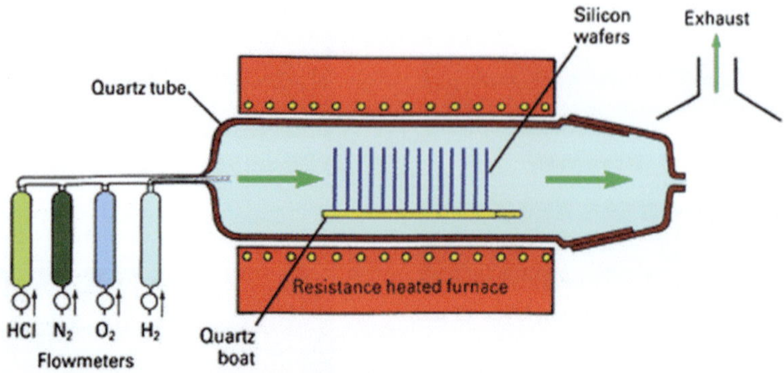

Fig. 8.14 Thermal oxidation furnace (Picture Courtesy: Microelectronics processing course J. Salzman 2002.)

and nanowires with different characteristics. Figure 8.15 shows the fabrication of silicon-on-insulator (SOI) wafer for MEMS-based devices like resonators, accelerometers, gyroscope, etc. (Xie 2015).

8.6 Recent Developments in MEMS-Based Fabrication Process

MEMS-based on-chip devices are being attracted more and more due to increased demand for electronic-based instruments in automotive as well as in industries. These fabricated devices provide a large range of benefits with respect to low cost, occupancy of lesser space, lesser power consumption, high-speed response, and many more. So because of these attractive benefits, this MEMS fabrication field is explored more and more to opt for new rapid, faster, and effective fabrication techniques. In this context, this section discusses several recent developments in the area of MEMS-based fabrication. It majorly focusses on development related to methods like deposition, etching, and printing.

8.6.1 Deposition

Deposition is a very broad area to discuss as it ranges from chemical vapor deposition to sputtering for deposition of metals/nonmetals or oxides. These techniques themselves help well in microfabrication but recent advancements have introduced some more new techniques; one of them is atomic layer deposition (ALD), broadly considered to be a subclass of chemical vapor deposition. ALD is a thin film deposition technique using gas-phase chemical processes sequentially. It

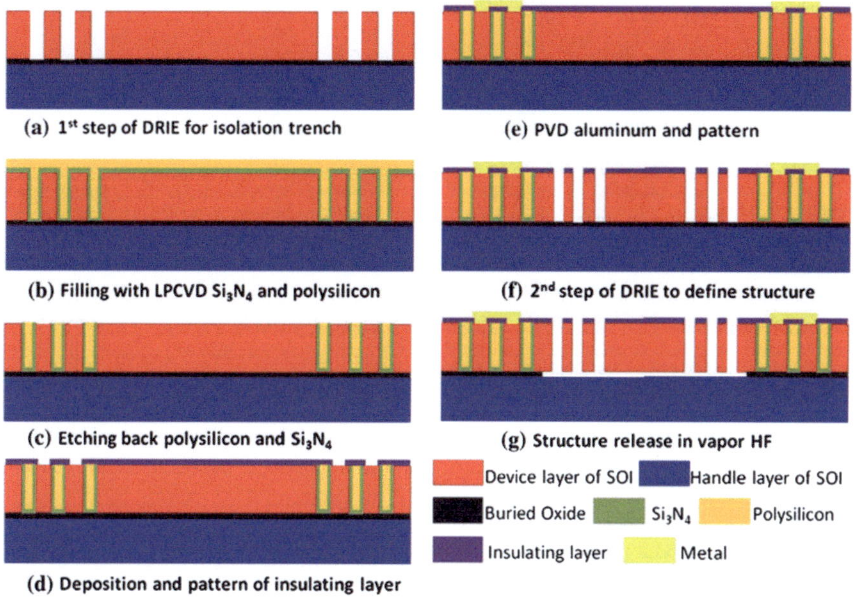

Fig. 8.15 SOI devices fabrication process with refilled isolation trenches [Reprinted with permission from Springer (Xie 2015).]

uses different chemicals as precursors in a sequential self-limiting manner to deposit thin films slowly on the substrate.

Figure 8.16 shows the schematic of thin layer deposition through ALD. ALD is a low-temperature deposition technique which provides excellent conformity and atomic layer thickness/composition control. ALD has proved to be efficient in many micro/nanodevice/feature fabrications like nanoCMOS, nanodot memory, non-volatile memory, display devices like TFT, LED, nanolaminates, nanotubes, nanoalloys, etc. (Kim et al. 2009). Various micro/nanodevices are also fabricated through ALD which are used as the sensing element in the automotive as well. ALD is used for fabricating different nanodevices like high mobility FETs and nonvolatile memory devices and 1D FET which are used for sensing and detection in electronic devices.

ALD has also shown its application in enhancing the oxygen reduction reaction activity in highly active but unstable solid oxide fuel cell cathode ($La_{0.6}Sr_{0.4}CoO_{3-\delta}$) by coating ZrO_2 films on its surface using ALD (Gong et al. 2013). The ALD-coated ZrO_2 has shown a remarkable reduction in degradation rate and area-specific resistance as compared to the original cathode surface. The thermal accelerometer has also been fabricated through plasma-enhanced ALD, which allows high-density ultrathin platinum films with higher accuracy and greater stability (Everhart et al. 2018). The fabricated accelerometer has achieved a 100X

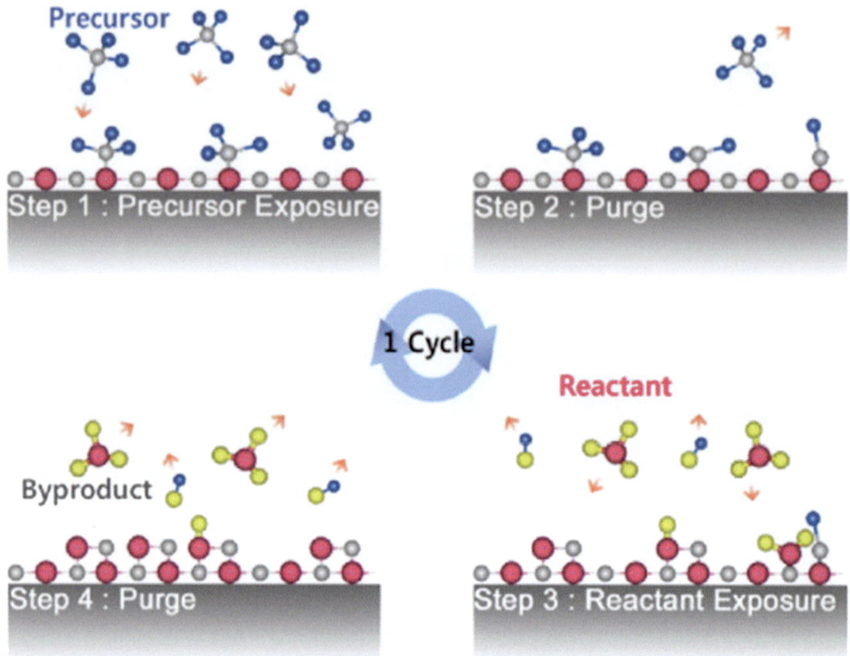

Fig. 8.16 Schematic showing thin layer growth process in ALD (Reprinted with permission from Kim et al. (2009). Copyright (2009) Elsevier.)

reduction in cross section to enhance the efficiency of heating and also help in reducing thermal time constants.

8.6.2 *Etching*

Etching is a very old technique to remove the material from a substrate by selecting a proper etchant corresponding to the selected substrate. Several substrate–etchant combinations are available in the literature to facilitate etching of various substrates. In the current era, depending on the evolved requirements, the advancement in the ordinary etching technique is done and these are atomic layer etching (ALE) and maskless etching technique.

ALE is a better controlled modified etching technique as compared to the reactive ion etching technique. There are two basic steps in this process, first is self-limiting modification step which modifies only the top atomic layer of the substrate while the next step is etching, which selectively removes the top layer which has been chemically modified initially. This technique has a potential to resolve various challenges related to continuous or pulsed plasma processes like

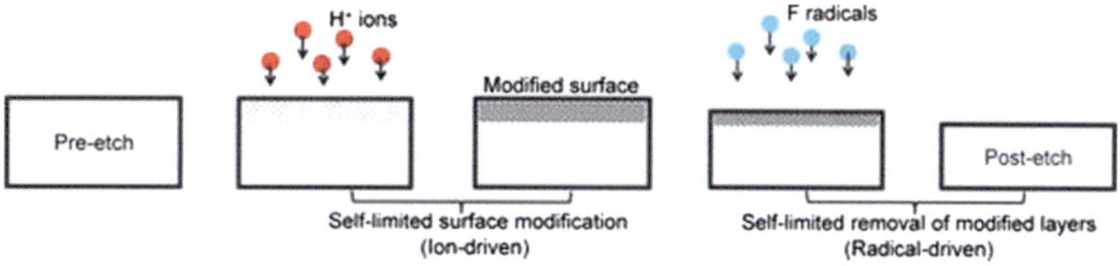

Fig. 8.17 Schematic representation of a single cycle of quasi-ALE of silicon nitride (Reprinted with permission from Sherpa and Ranjan (2017). Copyright (2017) AIP publishing.)

selectivity, aspect ratio, and profile dependency. As silicon is the most used substrate for devising MEMS devices, its ALE has always been an interesting aspect. Etching of silicon is reported using various gas plasmas combining argon (Ar) with other gases like Ar/C_4F_8, Ar/CHF_3 (Metzler et al. 2016) and Ar/Cl_2 plasma (Huard et al. 2017). 2D trenches, as well as 3D features, have been fabricated through Ar/Cl_2 plasma. It has been observed that controlled pulse plasma in every cycle corresponds to a very efficient etching at the atomic level. Ar/Cl_2 plasma combination can also be used to etch SiO_2 substrate. Self-limited etching of silicon nitride, which is another common semiconductor substrate, has also been performed in the plasma etching chamber (Sherpa and Ranjan 2017) by self-limiting surface modification with hydrogen plasma with the etching step completed with fluorinated plasma. Figure 8.17 represents the schematic of the quasi-ALE of silicon nitride. It has also been observed that this process is anisotropic in nature and is selective to oxide (>100).

Molybdenum disulfides (MoS_2) are one of the highly explored semiconductors that form a stable 2D structure with bandgap varying from 1.29 to 1.9 eV depending upon the number of layers (bandgap increases with the decrease in a number of layers). Hence to control the MoS_2 layers in a very fine nature, ALE is used (Kim et al. 2017). ALE is a clean adsorption and reacted-compound-desorption process, which restrains the inducing damage and contamination that occurs due to other processes while reducing MoS_2 thickness. The combination of chlorine radicals (Cl) and low energy Ar radicals is used as adsorption and desorption reagent, and it has been observed that the electrical characteristics of the etched surface remain similar to that of the original surface.

Maskless etching, the other modification of etching, is also used for creating 3D features in the silicon substrate in KOH. The etching which was initially restricted to (100) plane has attained different planes by using this technique. It has been observed that the cutting planes for convex edges are {311} while for that of convex corners is {411} (Li et al. 1996). Maskless etching of 3D MEMS structures in SOI (SiO_2 and SiN_2) dielectric membrane has been utilized for CMOS (Mansoor et al. 2018). This process eliminates the need of dedicated mask as the patterns are transferred through CMOS metallization layer-based microstructures.

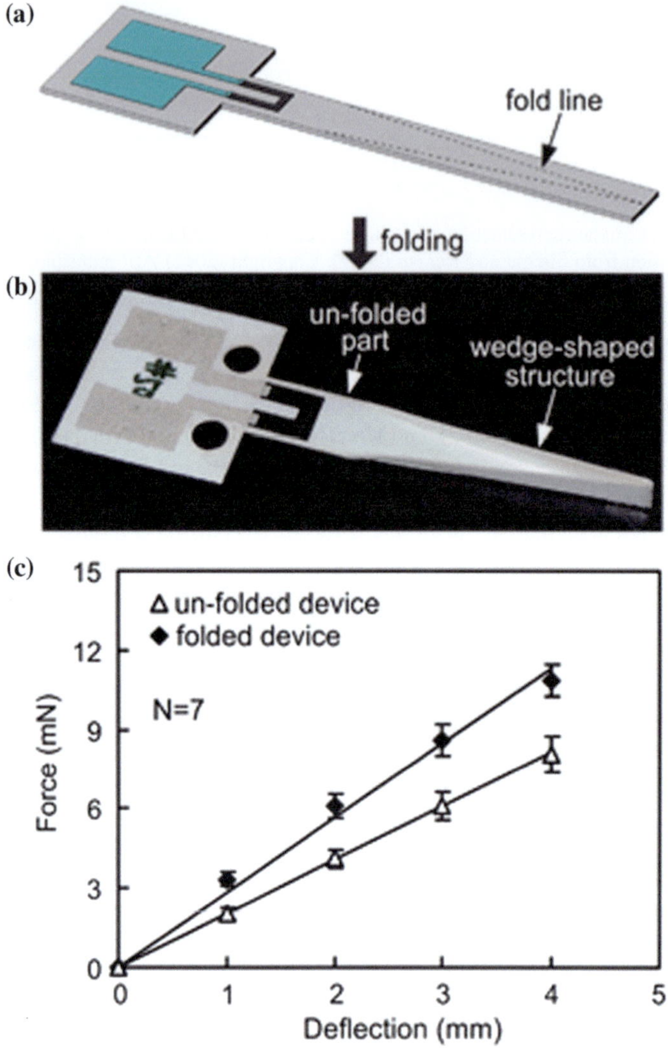

Fig. 8.18 Paper-based piezoresistive force sensor **a** Schematic view of a paper-based force sensor using carbon resistor (shown in black) as the sensing component, contact pads with silver ink (shown in blue color) with perforation of fold lines made by laser cutter; **b** Folded sensor; **c** Force–deflection curve of unfolded and folded cantilever (Reprinted with permission from (Liu et al. 2011). Copyright (2011) Royal Society of Chemistry.)

8.6.3 Printing

The simple screen printing technique is also used to develop MEMS-based force sensors (Liu et al. 2011) on a paper substrate only. These sensors work on the piezoresistive effect produced by the conductive material (the material to be printed is in the form of ink) printed on the paper substrate. It used Whatman chromatography paper to fabricate cantilevers with screen-printed carbon resistors (through graphite ink) connected via silver ink-based contact pads. The force sensors developed through this technique can be used in the 2D environment, by directly installing the printed sensors for an application or in the 3D environment by folding these paper printed sensors. The folded paper cantilevers shown in Fig. 8.18 has further shown higher stiffness as well as higher sensitivity.

8.7 Conclusion

MEMS devices and sensors with miniaturization are nowadays heavily getting deployed in both the automotive and aerospace sectors. This miniaturized low power consumption device, with a combination of electronics and mechanics, gives the enormous potential for applications. The existing processes used for fabricating MEMS devices have been much improved and advanced nowadays, due to scaling down of devices from MEMS to NEMS scale. In this chapter, a total review of different fabrication approaches for MEMS-based devices is explained thoroughly. As the size of the sensors is also reduced with the introduction of NEMS-based devices, upcoming technologies with capabilities of etching and deposition at the atomic level are mentioned at the last section of the chapter.

References

Ashby MF for MEMS designers. 1–12

Basu AK, Sah AN, Pradhan A, Bhattacharya S (2018) BSA Detection on polymeric nanocantilever, pp 1–6

Basu AK, Sarkar H, Bhattacharya S (2016) Fabrication and resilience measurement of thin aluminium cantilevers using scanning probe microscopy. In: Proceedings of the 3rd international conference c2e2 in foundations and frontiers in computer, communication and electrical engineering, Taylor and Francis, Mankundu, pp 457–460

Curley R, Mccormack T, Phipps M (2012) Low-pressure CVD and Plasma- Enhanced CVD. 1–5

Everhart CLM, Kaplan KE, Winterkorn MM, Kwon H, Provine J, Asheghi M, Goodson KE, Prinz FB, Kenny TW (2018) High stability thermal accelerometer based on ultrathin platinum ALD nanostructures. In: MEMS, pp 976–979

Gong Y, Palacio D, Song X, Patel RL, Liang X, Zhao X, Goodenough JB, Huang K (2013) Stabilizing nanostructured solid oxide fuel cell cathode with atomic layer deposition. Nano Lett 13:4340–4345. https://doi.org/10.1021/nl402138w

Hammond P (2015) Vapor phase etch processes for silicon MEMS. Elsevier Inc.

Hirata Y (2003) LIGA process—micromachining technique using synchrotron radiation lithography—and some industrial applications. Nucl Instrum Methods Phys Res Sect B Beam Interact Mater Atoms 208:21–26. https://doi.org/10.1016/S0168-583X(03)00632-3

Huard CM, Zhang Y, Sriraman S, Paterson A, Kanarik KJ, Kushner MJ (2017) Atomic layer etching of 3D structures in silicon: Self-limiting and nonideal reactions. J Vac Sci Technol A Vacuum, Surfaces, Film. https://doi.org/10.1116/1.4979661

Benjamin J, Jency G, Vijila G (2014) A Review of Different Etching Methodologies and Impact of various etchants in Wet Etching in Micro Fabrication. Int J Innov Res Sci Eng Technol 3:558–564

Kavitha S, Joseph Daniel R, Sumangala K (2016) Design and analysis of MEMS comb drive capacitive accelerometer for SHM and seismic applications. Meas J Int Meas Confed 93:327–339. https://doi.org/10.1016/j.measurement.2016.07.029

Kim H, Lee HBR, Maeng WJ (2009) Applications of atomic layer deposition to nanofabrication and emerging nanodevices. Thin Solid Films 517:2563–2580

Kim KS, Kim KH, Nam Y, Jeon J, Yim S, Singh E, Lee JY, Lee SJ, Jung YS, Yeom GY, Kim DW (2017) Atomic layer etching mechanism of MoS2 for nanodevices. ACS Appl Mater Interfaces 9:11967–11976. https://doi.org/10.1021/acsami.6b15886

Li X, Bao M, Shen S (1996) Maskless etching of three-dimensional silicon structures in KOH. Sensors Actuators, A Phys. https://doi.org/10.1016/s0924-4247(97)80094-5

Liu X, Mwangi M, Li X, O'Brien M, Whitesides GM (2011) Paper-based piezoresistive MEMS sensors. Lab Chip 11:2189–2196. https://doi.org/10.1039/c1lc20161a

Madou J. M (1997) Fundamentals of microfabrication

Malek CK, Saile V (2004) Applications of LIGA technology to precision manufacturing of high-aspect-ratio micro-components and -systems: a review. Microelectron J 35:131–143. https://doi.org/10.1016/j.mejo.2003.10.003

Mansoor M, Haneef I, Luca A De, Coull J, Udrea F (2018) A maskless etching technique for fabrication of 3D MEMS structures in SOI CMOS devices. J Micromech Microeng. https://doi.org/10.1088/1361-6439/aabe0d

Metzler D, Li C, Engelmann S, Bruce RL, Joseph EA, Oehrlein GS (2016) Fluorocarbon assisted atomic layer etching of SiO 2 and Si using cyclic Ar/C 4 F 8 and Ar/CHF 3 plasma. J Vac Sci Technol A Vacuum, Surfaces, Film. https://doi.org/10.1116/1.4935462

Millimeter-wave IT bulk micromachining for GaAs MEMS

Rammohan A, Dwivedi PK, Martinez-Duarte R, Katepalli H, Madou MJ, Sharma A (2011) One-step maskless grayscale lithography for the fabrication of 3-dimensional structures in SU-8. Sensors Actuators, B Chem. https://doi.org/10.1016/j.snb.2010.10.021

Rangelow IW (2001) Dry etching-based silicon micro-machining for MEMS. Vacuum 62:279–291. https://doi.org/10.1016/S0042-207X00442-5

Sherpa SD, Ranjan A (2017) Quasi-atomic layer etching of silicon nitride. J Vac Sci Technol A Vacuum, Surfaces, Film 35:01A102. https://doi.org/10.1116/1.4967236

Sobha Jayakrishnan D (2012) Electrodeposition: the versatile technique for nanomaterials. Woodhead Publishing Limited

Srinivasan T MEMS fabrication i : process flows and bulk micromachining CMOS processing

Ucer KB (2010) NAN242—Thin film fabrication (Epitaxy). Wake for University

Xie J (2015) Fabrication challenges and test structures for high-aspect-ratio SOI MEMS devices with refilled electrical isolation trenches. Microsyst Technol 21:1719–1727. https://doi.org/10.1007/s00542-014-2357-7

Chapter 9
Wireless Sensing Systems: A Review

Sudarsana Jena, Ankur Gupta, Rohit Kumar Pippara, Pramod Pal and Adit

Abstract Wireless sensing is a state-of-the-art technology of acquiring information about a remote object and its characteristics without bringing into physical contact with it. Humans have various sensory systems, viz., vision, hear, and smell to observe the object from the remote in day-to-day life. The data from the surrounding can be acquired by means of acoustic waves, distribution of forces, and electromagnetic energy. The collected data can be analyzed to obtain useful information about the object and its characteristics. Swift progress in the microelectronics has resulted in faster response and accurate mode of signal detection. Keeping the technological progress in mind, this article discusses the fundamentals aspects and reviews the state of the artwork on wireless detection technology.

Keywords Wireless sensing technology · Radar systems · RF communication
RFID tag · Bluetooth · NFC

9.1 Introduction

Wireless sensing modules are widely used in many fields such as military applications, threat detection, land mapping, survey purpose, intelligence, gas sensing, commercial purpose, etc. The object is detected and continued to track based on the signal strength received from the reflected object. The amount of energy reflected from the object is dependent on the property of the object's material, surface condition, incident angle, power density, and wavelength parameter of the radiated energy (Campbell 2011; Parks et al. 2013). Figure 9.1 represents the overview of daily-life situations with wireless detection.

Wireless sensing (WS) involves multidisciplinary sciences such as electrical, RF, mechanical, chemical, materials, telecommunications, spectroscopy, photography,

S. Jena · A. Gupta (✉) · R. K. Pippara · P. Pal · Adit
School of Mechanical Sciences, Indian Institute of Technology Bhubaneswar, Bhubaneswar 752050, Odisha, India
e-mail: ankurgupta@iitbbs.ac.in; ldankurgupta.k10@gmail.com

S. Bhattacharya et al. (eds.), *Sensors for Automotive and Aerospace Applications*, Energy, Environment, and Sustainability,
https://doi.org/10.1007/978-981-13-3290-6_9

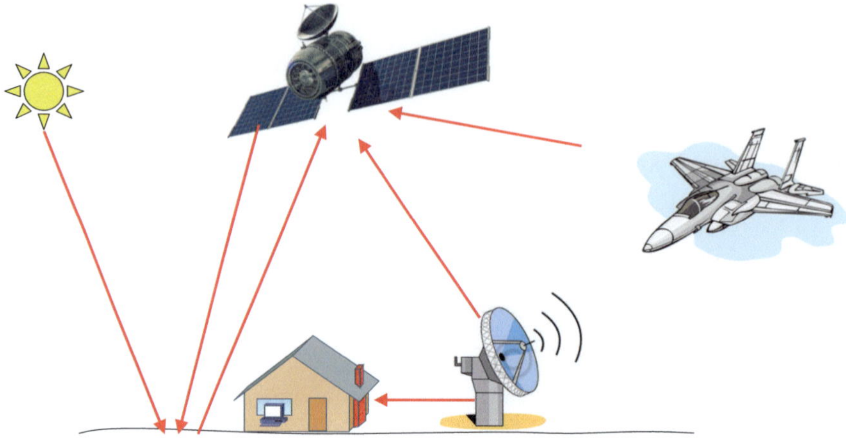

Fig. 9.1 Wireless sensing systems

etc. When all these technologies come together to produce a useful system, then the system is called wireless sensing system (Rappaport 1996).

There are many wireless sensing devices which are used in practical fields which not only make life easier but also give a useful information about the object that is remotely located, so that proper precaution and action plan can be executed. In this chapter, we will discuss major wireless sensing systems including radar system.

One can define a sensor as follows: A device which detects the presence of a particular gas and measures a physical, chemical, or biological change in property and records, indicates, or responds to it, and transmits the information or power between two or more points that are not connected by an electrical conductor is a sensor. These sensors work on the variation in the physical or chemical composition of the detecting element, which when occurs gives the presence of the desired gas. The change in these properties is manipulated into some presentable result. This is usually referred to in the scientific term as the data acquisition system. This system takes into account the change in the property either in the form of the intensity of light in optical sensor type or the change in the voltage reading in case of metal oxide-based electrical sensor (Capone et al. 2003). This leads us to the study of the elements of a gas sensor. So, there are basically three elements of a gas sensor namely,

(a) Sensing element
(b) Instrumentation
(c) Output.

Sensing element is the heart of the gas sensor. They are responsible for detecting gas and respond accordingly through a change in the physical, chemical, or biological change in the property. The target gases are reactive to the chemical present in the sensing element. Initially, use of the electrochemical gas sensor was done which consisted of the electrode which on reaction with the gases could either get chemically oxidized or reduced. So, in this process, exchange of electrons takes place. Thus, there is a flow of current and quantification of the gas which can be done through this amount of current flow. The place where the electrochemical sensor finds its application involves refineries, chemical plants, gas turbines, underground gas storage facilities, etc. Another set of sensors are very effective in detecting the combustible gases that are present in explosion hazard places. The sensing element is a catalytic bead sensor. Platinum wire coils are used in this as a catalyst. Semiconductor gas sensors detect gases through the chemical reaction when gas comes in direct contact with the sensor. Tin dioxide is used as the sensing element in this, and it is the common material used. The difference in the electrical properties of the sensing element that is the electrical resistance in the sensor is decreased on interaction with the gas to be monitored. This change in the resistance can be quantified to measure alcohol vapor, hydrogen, oxygen, and harmful gases such as carbon monoxide. Other sensing elements consist of the utrasonic gas leak detectors, holographic gas sensors, infrared image sensors, infrared (IR) point sensors, photoionization detectors (PIDs), etc. (Pati 2015).

Instrumentation system is the part of the gas sensing that is required to transfer the changes in the composition of sensing element on interaction with the gas. The system consists of the electrical or other types of connections which interpret the output from sensing element such as the change in resistance, voltage, current, change in wavelength of light, etc., into some readable quantity. The most basic instrumentation system can be understood as a sensing circuit with a microprocessor such as arguing, a data sending and receiving system that is connected to the display monitor. The purpose of this system is not limited and changes from one sensing circuit to the other. Like some sensing circuit, it requires current sensor and voltage sensor in their circuit but some does not require. Some circuits require amplification for the voltage output so that it comes under some readable range. Some circuits even consist of heating system to maintain a suitable temperature at which the sensor operates. So, we can say that any kind of discrepancy in the sensing element is corrected and substituted by the instrumentation system. Most of the time the circuit requires voltage amplification, so an Op Amp circuit is often being used in most of the gas sensing techniques (Stetter et al. 2003).

Generally, the instrumentation system itself has some integrated LCD, LED screen to display the output. Still, many of the gas sensing circuit consist of the display monitor for live display of the variation of gas sensing. This output system is generally shown as a two-dimensional plot of the variation of one property with the time and concentration of the gas. The best way to understand the variation of physical, chemical, or biological properties is displayed by the output system. It is a good mode for comparison in the future and a proof of the presence of the gas in the atmosphere. However, the output system requires some software's which are programmed to display the desired output. Some common software's consist of the Arduino, MATLAB, etc. They are needed to be run the whole time of the experiment to measure even the feeble amount of gas present in the atmosphere. So, they consume a lot of power in this way.

9.2 Existing Detection Techniques

Existing sensing techniques have been broadly divided into two categories. The one technique focuses on the wire-based detection and the other is the wireless-based technique. The wire-based sensors have been in use before but the wireless-based detection system has been more profoundly in use nowadays (Fig. 9.2).

9.2.1 *Wire-Based Detection Technique*

The wire-based detection technique makes use of the physical connection between various components of the Instrumentation system involved in detection. The categories of such detection system have their share of advantages and disadvantages. This technology came into existence earlier due to the ease of availability of this technology in the past and also very small development was done in the field of wireless communication. The wire-based detection technique has been done in two ways. The first one is the electrical based and the other one is the optical-based detection technique.

9.2.1.1 Electrical-Based Detection Scheme

The electrical-based detection technique consists of the electrical wire-based connection. The variation of current, voltage or resistance gives us the required value of the output signal. These signals may be in the form resistance (R) or inductance (L) or capacitance (C) change. Such a system has a network of interconnected wires. This network is connected to each and every component of the system that is involved in it. From the sensing system to the power supply to the signal output system, everything is connected through electrical wires. The detection technique based on this electrical scheme is used for the sensing of various VOC gases such as acetone (C_3H_6O), ethanol (C_2H_5OH), propanol (C_3H_8O), acetylene (C_2H_2),

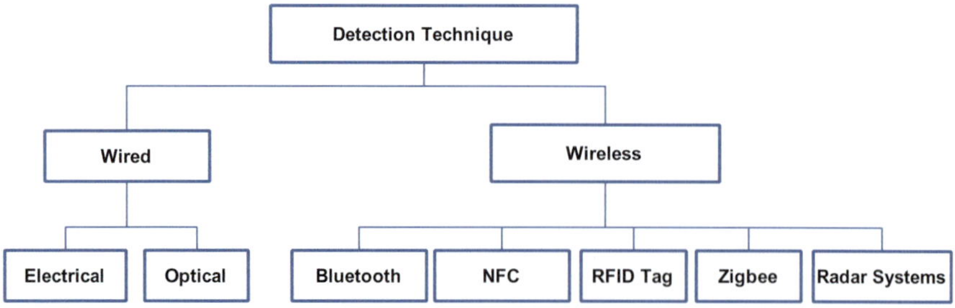

Fig. 9.2 Types of detection techniques

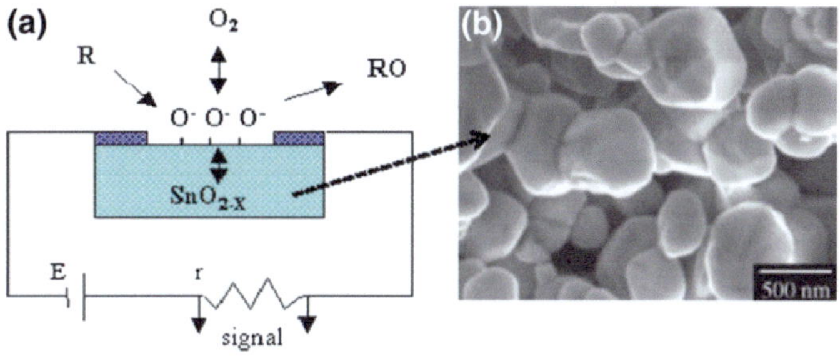

Fig. 9.3 **a** Schematic circuit diagram of a resistive semiconductor gas sensor. **b** A study of structure and size of sensing element based on a metal oxide (reprinted with kind permission from Mirzaei et al. (2016))

benzene (C_6H_6), cyclohexene (C_6H_{10}), formaldehyde (HCHO), n-butane (C_4H_9OH), methanol (CH_3OH), and toluene (C_7H_8). Mirzaei et al. (2016) have reviewed the sensing of various VOC gases. Through the study of the properties of various sensors, they have concluded that conduct metric (resistive) gas sensors mostly based on metal oxides are capable of controlling dangerous volatile organic gases. They also came to a conclusion that smaller the size of a sensor smaller will be the power consumption and better will be the performance (Luo et al. 2007; Sekiguchi et al. 2010).

Figure 9.3a represents the proposed principle and the circuit of a conduct metric sensors consisting of electrical readout. It is based on nonstoichiometric tin dioxide. Figure 9.3b shows the grain size of metal oxide, sensing element. In the sensing mechanism, oxygen species present in the atmosphere occupies the semiconducting metal oxide surface which absorbs the particles in the ionic forms. These ionic forms are as follows: O_2^-(ads), O^-(ads), and O^{2-}(ads) (Ho et al. 2001).

Figure 9.4 shows the choice of the sensor to detect the VOC gases. The sensor used is Figaro TGS2620 which is a metal oxide-based sensor. The sensor exhibit a low detection limit, high sensitivity, fast response, and excellent stability because of its thinner structure and less agglomerated (Dai et al. 2017).

Other researchers, viz., Trouillet et al. (2006) have performed the electrochemical detection of pollutants using the gold nanoparticles. They have used the gold (Aurum) Nanomaterials-based electrode to sense the pollutants. These sensors mainly consist of a physical transducer (electrode) with a recognition element and the interactions between the electrode surface and the recognition element decides its sensing performance (Nayak et al. 2013). Although nanoparticles have a small size and sensing electrodes as well as have a large surface area which results in high sensitivity and rapid response (Li et al. 2015a).

Another work performed in this technique is in the aging management of the instrumentation and control sensors in nuclear power plants. Hashemian (Yadav et al. 2015) has proposed a way to monitor the performance of the various sensors involved in the nuclear power plant such as thermal sensors, pressure sensors, and differential pressure sensors. These sensors along with various other components

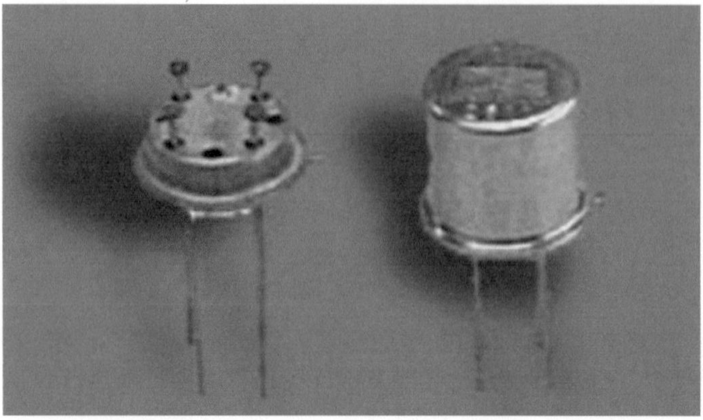

Fig. 9.4 VOCs sensor (Figaro TGS2620) (reprinted with kind permission from Mirzaei et al. (2016))

like wires, transmitters are present at the spots where the sensing needs to be done. They are subjected to high heat, humidity, radiation, temperature cycling, vibration, and dirt. There are high chances of degradation of a component's performance. So, in order to protect them, online monitoring (OLM) encompasses methods for evaluating the performance of sensors and its components. OLM consist of the computer as a monitoring system, communications hardware and software, a data historian, and OLM software package. They perform a test on all the sensors and alert the people in the case of any undesired result. Thus, maintaining and saving instrumentation and control system from aging.

9.2.1.2 Optical-Based Detection Scheme

Optical-based detection technique consists of the use of properties of light to transmit signals from a sensing element to the receiver that interprets these signals in order to get the desired output. The sensing element on interaction with the light changes its properties like intensity, wavelength, etc. The work done involving this technique is in the field of sensing the humidity (Sikarwar and Yadav 2015). The humidity was measured using *n*-type semiconducting metal oxide films and using the property of difference in the refractive index of the light when humidity is present.

Figure 9.5 shows the experimental setup for the optoelectronic method of sensing of the humidity level in the atmosphere. The light from the He–Ne laser was passed through the sensing element where some of the light underwent partial refraction and reflection. The laser light was finally received at the other end at the optical meter where the received intensity of light was measured. The reduction in the intensity of the light gives us the amount of humidity present in the atmosphere.

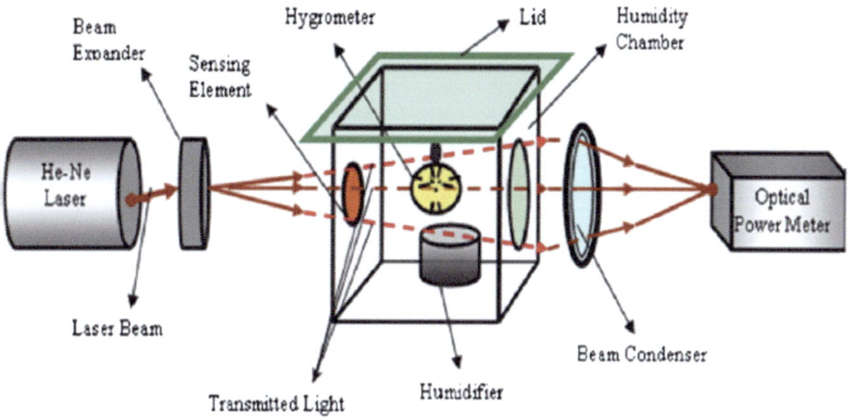

Fig. 9.5 Experimental setup of opt-electronic humidity sensor (reprinted with kind permission from Sikarwar and Yadav (2015))

These humidity sensors find application in a field like construction, food and packaging factory, agriculture industry, medical health purpose, etc.

Another optical-based sensing technique focuses on the use of palladium (Pd) and tungsten oxide (WO_3) thin films to sense the presence of hydrogen gas (Zhang et al. 2017; Gupta et al. 2013). The optical signal is taken in the form of change in the properties of the light such as intensity, wavelength. The Pd film is deposited on the fiber and in the presence of hydrogen, the change in volume, as well as the dielectric constant of Pd, takes place. The change dielectric constant leads to change in the optical signal. The principle on which we sense the hydrogen (Gupta et al. 2014a, 2016) is that the change in concentration of hydrogen leads to change in intensity.

The above Fig. 9.6 shows the hydrogen sensing setup with circular Pd-coated single-mode tapered fiber (Villatoro et al. 2001). There is a coating of the Pd all around this interactive length. It works by absorption change of the evanescent fields in a circular Pd-coated tapered fiber. At room temperature, the detectable limit of hydrogen concentration has a range of 0.018–0.1.

The first optical fiber micro-mirror-based hydrogen sensing was proposed by Butler (1991). So, based on the micro-mirror technique, another improved sensing

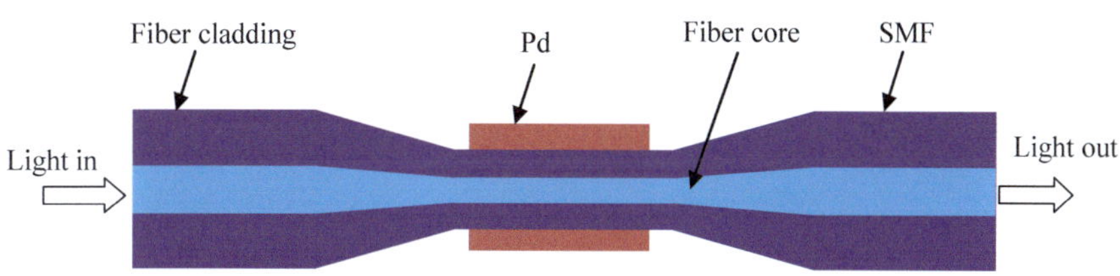

Fig. 9.6 Hydrogen sensing by using a tapered single-mode fiber with Pd coating

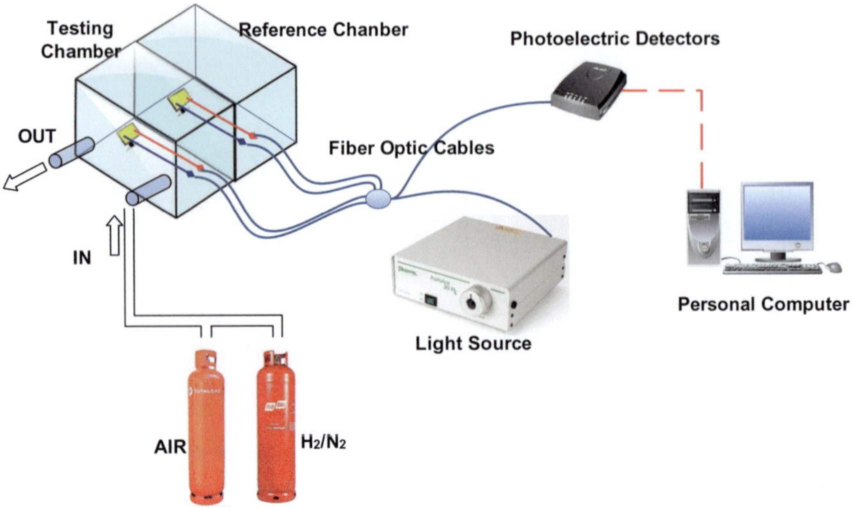

Fig. 9.7 Hydrogen sensor with dual-path micro-mirror configuration

technique setup uses the dual-path micro-mirror to sense the presence of hydrogen gas (Liu et al. 2012a; Gupta et al. 2014b). The Pd/Y alloy film is used to detect the hydrogen gas. The use of dual micro-mirror is to reduce the noises by eliminating noise interference from the fluctuating light source, temperature, and loss by an optical fiber.

Figure 9.7 shows the setup used by Li et al. (2015b), and in this setup, the use of dual-path micro-mirror is done. It is based on the Pt/WO_3 film which senses the presence of the hydrogen gas. The reflective power will reduce with the enhancement of H_2 concentration. The oxide material WO_3 used by sensing film is very stable.

9.2.2 Wireless-Based Detection Technique

The wireless-based detection technique involves the use of wireless-based electronic instrumentation to receive and send signals. The signals are sent in the form of radio signals which do not require any physical connection of wires or a channel to pass the flow of light. This type of instrumentation system has the advantage of not using any heavy connection. The sensing and instrumentation system can be kept at a farther distance depending on the signal range. Due to advancement in microelectromechanical systems (MEMS) technology, digital electronics, and wireless communications, there has been a development of low power, low-cost, and multifunctional sensor nodes smaller in size than the wired ones and communication in short distances (Akyildiz et al. 2002). The wireless-based detection technique has been in use and there are many advancements in the type of electronic instrument which is used to send and receive signals. Initially, the use of Bluetooth was used but later many new technologies came into light such as NFC, RFID tag, and Zigbee (Smith et al. 2006; Ferrer-Vidal et al. 2006; Wheeler 2007).

9.2.2.1 Bluetooth

It is a wireless tool extensively used nowadays. It uses short-wavelength UHF radio waves in ISM band from 2.4 to 2.485 GHz from stationary, portable electronics, and construction of personal area networks for transferring data. It operates at frequencies between 2400 and 2482 MHz, or including guard bands 2 MHz wide at the bottom and 3.3 MHz wide at the top (DeCuir 2014).

It was developed for short range using transceiver microchips and consumes less power. See the Table 9.1.

Indoor conditions are suitable for Bluetooth conditions, and we use a class 2 Bluetooth module for our gas sensing technique. The Fig. 9.8 shows a simple circuit of the MQ7 gas sensor with a bluetooth module.

Roseline et al. (2013) developed a pollution monitoring system using the wireless sensing network. Gas concentration levels of O3, NO2, and CO/VOC are measured by three analog sensors. Recordings of temperature and relative humidity along with GPS data can be stored. An LCD panel displays the immediate visual feedback about the system status. The people can be alerted from, and we can save the environment from the pollution (Fig. 9.9).

Similarly, another researcher Pummakarnchana et al. (2005) developed an urban air pollution monitoring system. The data on pollution level was recorded and disseminated in real time on wireless geographic information systems (GIS). Metal-oxide semiconductors such as ZnO were used for sensing of pollution level of NOx concentration. Solid-state gas sensors with Personal Digital Assistant (PDA) were linked through Bluetooth communication tools and global positioning Ssystem (GPS). The use of the Bluetooth device has also been found in the

Table 9.1 Ranges of Bluetooth devices

Ranges of Bluetooth devices by class			
Type	Maximum permitted power		Typical range (m)
	(mW)	(dBm)	
1	100	20	~100
2	2.5	4	~10
3	1	0	~1
4	0.5	−3	~0.5

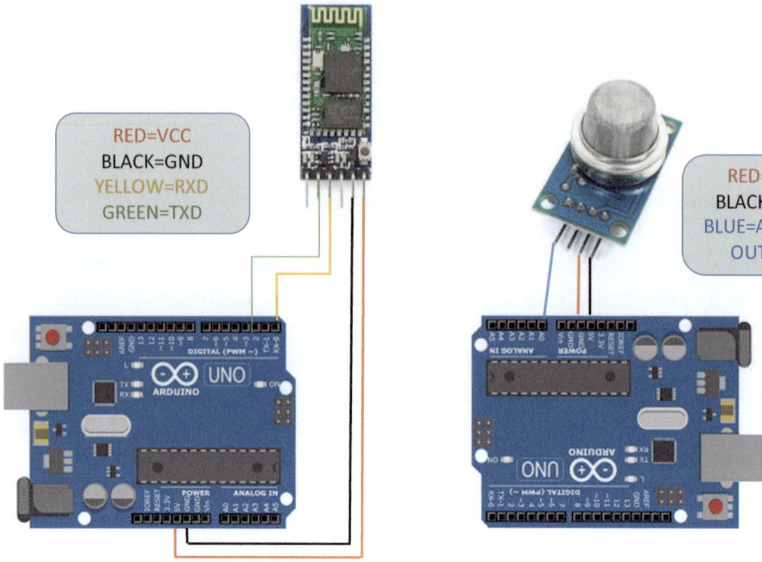

Fig. 9.8 Use of Bluetooth module using Arduino UNO

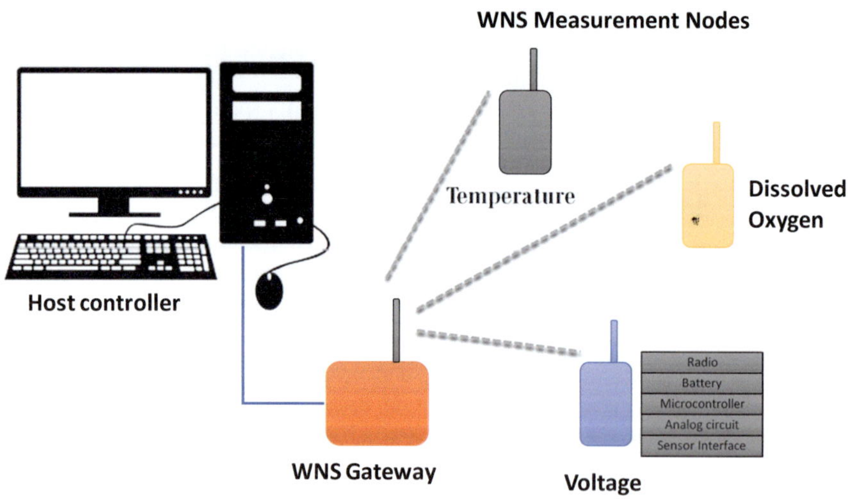

Fig. 9.9 A wireless sensor network for pollution monitoring system

embedded hardware prototype for LPG gas detection (Kurzekar et al. 2017). The monitoring was done on the Android mobile platform. The sensor used to detect the presence of LPG gas is an MQ-6 gas sensor and the levels of gas were monitored through Android applications. This application uses the data received by it from the

Bluetooth module. The Bluetooth module used here was the HC-05 Bluetooth module.

Another gas sensing module focuses on the detection of the NO_2 and CO gases (Moon et al. 2013). The sensor consists of SiO2 and SiNx thin films. The Bluetooth module sends the varying response to the personal digital assistant (PDA). The principal mechanism of this gas sensor operation is to measure the variation in resistance of the metal oxide nanopowder when exposed to gases. These gases can be oxidizing or deoxidizing. The alcoholic gas sensor has been developed which transmits the levels of alcoholic gas concentration wirelessly through Bluetooth to the mobile device (Kim et al. 2017). The sensors used to consist of In_2O_3/Pt nanoparticles and Ag nanowires electrodes. The sensor works similar to the breath analyzer used by the traffic police. The difference is that the data can be transmitted to a farther distance on the smartphone within the range of the Bluetooth.

Further advancement led to hybrid chemical sensing of the volatile organic compound (VOC). The detector consists of tuning fork sensor in the present shown device in Fig. 9.10 which is prearranged with a molecularly imprinted polymer (MIP) that is adjusted for selective sensing of monoaromatic, alkyl hydrocarbon, and also Volatile Organic Carbons (Chen et al. 2013). The Bluetooth chip is connected to a master microcontroller which reads the sensor signal, and is paired with the smartphone for communication and data transmission.

Figure 9.10 shows the key components involved in the sensor. A pre-concentrator unit selectively collects and pre-concentrates analysts from atmospheric air, a separation unit separates the analysts released from the pre-concentrator, and a detection unit comprises a tuning fork arranged with molecularly imprinted polymers which sense the analysts.

There has also been the development of the analyzer of the oxygen in our breath which works on Bluetooth communication to send the data in real-time to more

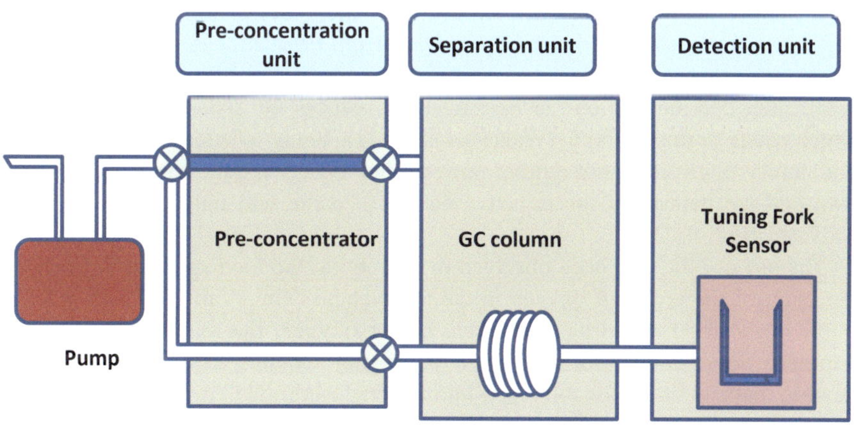

Fig. 9.10 Key component involved in the wireless hybrid gas sensing

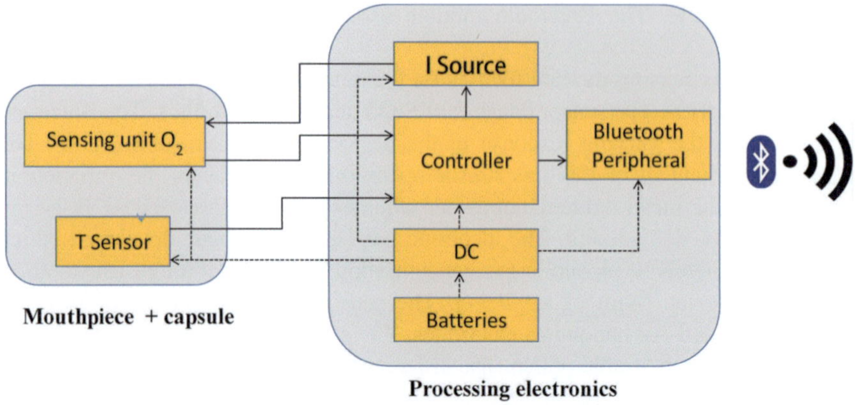

Fig. 9.11 Block diagram of the oxygen levels detection instrument

than one of the portable device such as smartphones or tablets carried out by the observer (Lopez-Ruiz et al. 2015) (Fig. 9.11).

The next wireless gas detection scheme is used at the household level to detect the presence of LPG gas leakage, carbon monoxide, and ammonia gases. This system is used to detect leakage and alert the user along with preventing any further leakage of the gas. The instrumentation system consists of ATmega328 as a microcontroller, Arduino UNO, MQ-6 as gas sensor and HC-05 Bluetooth module. The home appliances are connected to the Arduino board so that they can be easily controlled inside a smart home using any Bluetooth-enabled smartphone (Bhoir et al. 2017).

9.2.2.2 Near-Field Communication (NFC)

Near-field communication known as NFC is a battery-free platform for sensing and computation powered and is read by an Ultra High Frequency (UHF) RFID reader. It has an advantage to link powered devices by using short-range RFID style backscatter communication mechanism. So it reduces the size, as well as battery, which needs to run the NFC (Walker et al. 2017). Being a battery-less device does not signify that it does not require power. They do not require the power of their own and are dependent on an active device to come into range before they are activated (Fig. 9.12).

The use of NFC has been observed in preventing the food spoilage through the detection of the levels of oxygen in the atmosphere (Zhu et al. 2017). In order to improve the food and drug safety sensors for oxygen, the monitoring system is required. So, this monitoring system packaging requires low-cost, battery-free devices that can be easily combined into the enclosure. NFC is the best choice for this. The sensing element consists of a Fe(II)-P4VP-SWCNT composite displays in response to oxygen on the chemiresistor.

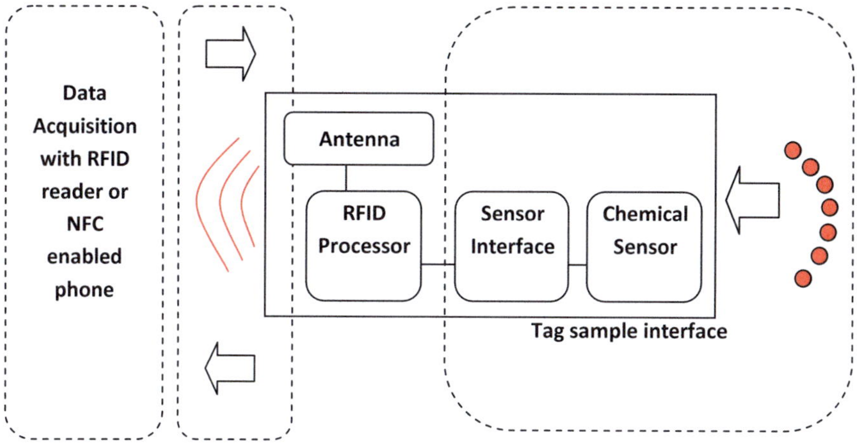

Fig. 9.12 Data acquisition with RFID or NFC enable phone

Figure 9.13 shows the circuit of the NFC converted into a sensing device called as p-CARD. It consists of tuning capacitor C, an inductor (antenna) L, the resistance of integrated circuit RIC, the capacitance of IC CIC, and chemiresistor.

Use of Near-Field Communication tags for sensing biochemicals with electronic gadgets is done to detect the ammonia gas present in the environment. So for this, the modification of the NFC tag into a sensor was done. Gas sensing test on Carbonic Interdigitated electrodes covered with monolayer graphene was performed which acts to be a good sensor (Xu et al. 2017a). As we can see almost, all the NFC-based sensors use the method of modification to of NFC into a sensor that senses as well as sends the signal to the smartphone devices for the output of the concentration levels of the gases. Similarly, a very low-cost sensing device has been

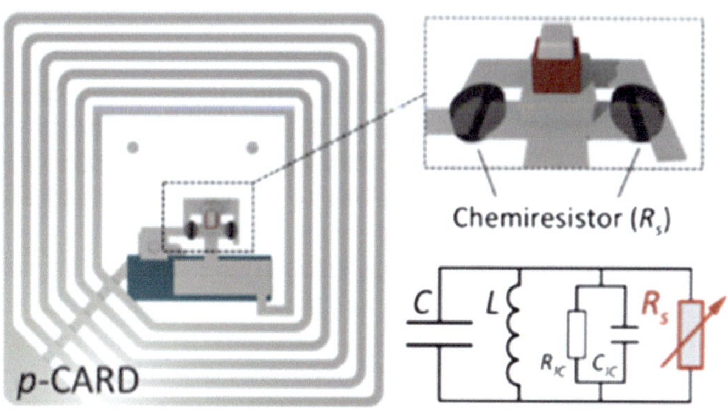

Fig. 9.13 NFC tag converted into the wireless sensor (p-CARD) is attached to the inside of the food packet and the circuit diagram of the p-CARD (reprinted with kind permission from Zhu et al. (2017))

developed to sense the cyclohexanone vapors which are a common constituent of plastic explosives (Azzarelli et al. 2014).

The use of (NFC)-enabled digital wireless sensor is widely used by people with diabetes to measure glucose concentration nowadays. The nonenzymatic abiotic methodology is used to detect glucose. Abis-boronate glucose recognition structure is polymerized onto the surface with steady fluorescent indicator molecule which interacts directly with glucose. High-precision wireless IC sensor allows for measuring fluorescence which can be remotely powered and connected through an NFC front-end (DeHennis et al. 2016).

9.2.2.3 RFID Tag

Radio frequency identification (RFID) tag technology is based on wireless communication systems, and it can be integrated into the Internet cloud system. The use of RFID tag nowadays has been in healthcare, environmental pollution, food safety, public transport, logistics, etc. RFID technology is a wireless communication method in which objects having RFID tags (microchips) on it are identified and are tracked utilizing RF waves. RFID system has two basic components: a memory microchip (tag) stores electronic information and an antenna-connected reader network to receive and analyze signals.

An RFID tag-based sensor was developed having the carbon nanotubes (m-MWCNTs) and polypyrrole (Ppy) coated over two comb-like interdigitated gold electrodes (Chavali et al. 2008). RF module has been integrated with the developed sensing electrode. The gas that has been detected here is Sevoflurane [fluoromethyl 2,2,2,-trifluoro-1-(trifluoromethyl) ethyl ether]. Sevoflurane is a volatile anesthetic agent. Another group of the researcher has used RFID tag to monitor the ripening of vegetables during the food transportation in the refrigerated cargo vehicle (Formisano et al. 2015). They developed a device called “TinyNose” to find the ripening parameter of the vegetable product during transport. They have integrated chemiresistive MOX sensors with the RFID tag to sense the quality of vegetable in the container.

This uniform is useful for the people working in a hazardous gas area. They can be alerted when the toxic gas is present in the atmosphere. It consists of an RF actuation circuit which consists of the following parts: an antenna, a rectifier, a resonance circuit, and a comparator. The sensor is a profitable catalytic gas sensor which is used for measuring combustible gases composition.

The highly sensitive wireless H_2S gas sensors have been developed that uses RFID (Asad and Sheikhi 2016a). The sensor is made up of CuO-single-wall carbon nanotube (SWCNT) nanostructures. They offer a very sensitive detection of H2S gas. For observing the quality of healthcare-developed wireless sensor is demonstrated by placing on the skin of the hand. The RFID tag used in this sensor is

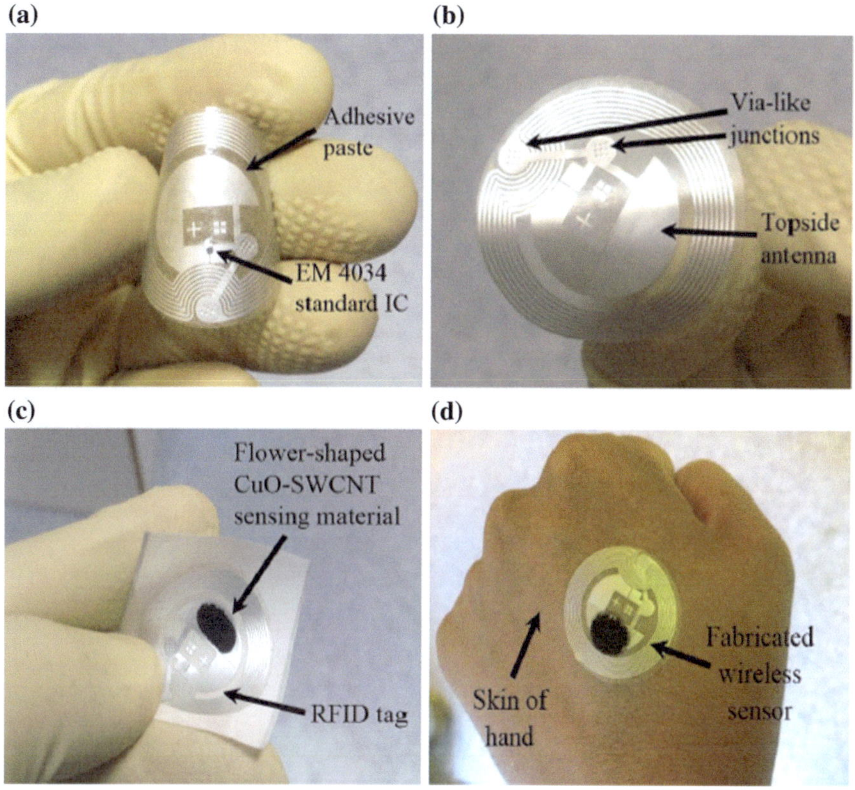

Fig. 9.14 **a** RFID tag on the flexible substrate with EM4034 IC. **b** The antenna is mounted on the topside of the tag to attach it on the body skin. **c** Deposited CuO-SWCNT sensing material in the shape of a flower. **d** Developed sensors are demonstrated by placing it on the skin (reprinted with kind permission from Asad and Sheikhi (2016))

13.56 MHz RFID tag. The flower-shaped CuO-SWCNT is deposited on the RFID tag circuit. Figure 9.14 shown describes how the actual sensor along with the RFID tag looks like.

Another set of developed sensing devices uses the inkjet printing as a way to fabricate the sensor. Inkjet printing is preferred as it is low-cost practices for directly producing electronic materials on large surface areas. The printing of the RFID tags on the flexible substrate in this way is cheaper and fulfills the desired purpose (Singh et al. 2017). The process has been employed in the fabrication process of the PtNP/rGO hybrid-based RFID tag sensor for detecting hydrogen gas (Lee et al. 2015). During communication, Pt-rGO sensor tag can be operated without power during communication which ensures a long lifespan for the sensor tag. The demonstration showing the schematic of the working of the hydrogen gas sensor gas has been shown in Fig. 9.15.

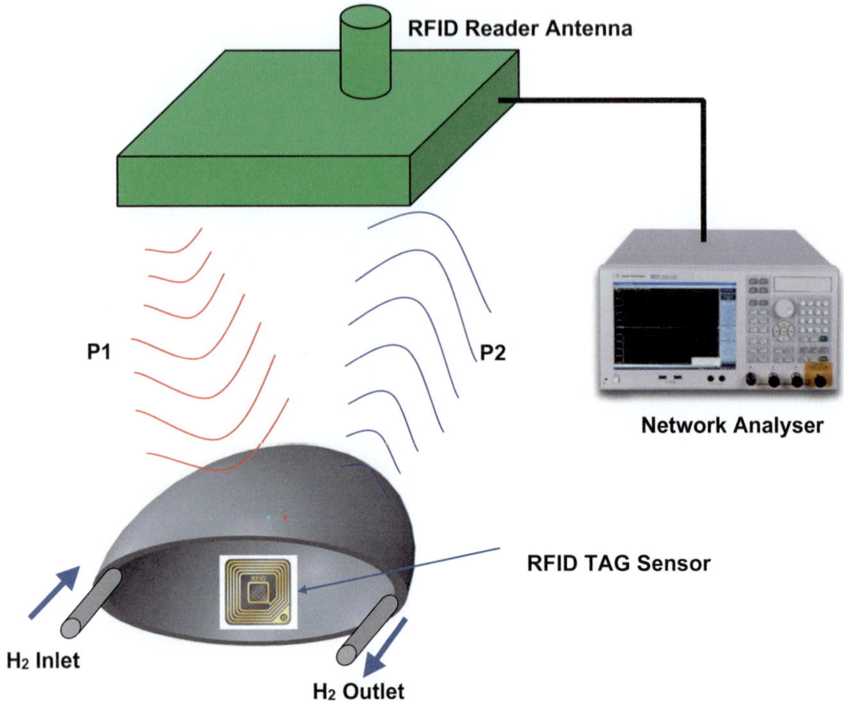

Fig. 9.15 Schematic diagram of hydrogen gas sensing by the UHF-RFID tag sensor system having RFID sensor tag and antenna

As shown in Fig. 9.15, an interrogation signal was emitted by network analyzer at a power level of P1 to activate the sensor tag subject to the condition in the RFID reader antennas electromagnetic field. The signal was processed then redirected back at a power level of P2.

Another inkjet-based sensor involves the printing of the sensor antennas and carbon nanotubes to sense gases like the NH3 and NOx (Yang et al. 2009). An ultrasensitive composite is prepared that can be printed directly on the paper substrate together with the antenna. The RFID tag is printed on that paper. Similarly, the last sensing device with another sensor uses a paper substrate for printing the circuit and sensing element. It considers the following parameters such as surface roughness, thickness, porosity, permittivity, and cost of the paper. The choice of paper they have done is a Kodak photo paper (Quddious et al. 2016). The gas to be sensed here is H_2S. So, in order to sense that, gas copper acetate is used. The chemiresistive ink of copper acetate is deposited on the paper substrate. The ink has a low conductivity which in presence of H_2S increases the conductivity of the circuit.

Another inkjet-printed sensor and the RFID tag have been developed that is used for the sensing of gas such as ethanol, methanol, acetone, and ammonia (Vyas et al.

2011). Another set of researcher developed an inkjet-printed wireless and chipless sensor for detection of the CO_2 and temperature detection (Vena et al. 2015). Also, RFID tags are employed in the food logistics that are tracked while transported in the cargo (Abad et al. 2007).

9.2.2.4 Zigbee Module

This is the latest and currently the most preferred technology for the sending and receiving of the signals. It has wide application in the research in military applications, home automation, building automation, and utility metering. Zigbee was released in the year 2004. Since then, they have replaced the above devices discussed. Several major advantages that could justify and have a leading edge over the abovementioned device are:

- Long-Range connectivity up to 100 m
- Easy to program and use
- End-to-End Encryption.

The experiments on detection of the methane gas have been performed. So, in this experiment, methane has the lower limit of its explosion is 5% vol. (2000) but as the CH_3 concentration is up to 1% volume, the dangerous situation occurs and the system closes the gas controller. The wireless sensing setup consists of two parts, namely the sensing circuit and the wireless actuator. The sensing circuit performs the function of the detection of the gas and then communicates with the actuator. The wireless actuator performs the function of sending the input signal from the sensing circuit to the output device (Somov et al. 2014). The wireless actuator consists of the 2.4 GHz ISM band ZigBee/802.15.4 transceiver. A Wheatstone bridge is used for sensing circuit (Liu et al. 2012b).

The above was used specifically at the hazardous workplace but it is not just limited to that place. Another gas sensing setup uses Zigbee to do environmental monitoring of the levels of the harmful gases (Krishna et al. 2016). The gas sensed here in the environment is the CO_2. The CO_2 gas sensor does the level of detection of CO_2 by observing the amount of infrared radiation absorbed by CO_2 molecules. The sensor works on a low range (0–10,000) parts per million and also on high range (0–100,000) parts per million. Apart from the CO_2 gas detection, the sensing of humidity, temperature and soil moisture is done. Apart from carbon dioxide, the gas that poses more harmful and could even cause death at higher levels of intake is carbon monoxide. A Wi-Fi-based real time for observing the quality of a CO system is proposed for application in the construction locations (Yang et al. 2015a). Node measuring sensor design is carried out by the low-frequency variation method to acquire CO absorption consistently. Sensing module consists of the pyroelectric sensor. This sensor has the advantage of fast response rate as the thermal response time of 150 ms and enables real-time monitoring. However, the wireless communication discusses the various modules and discards the Zigbee on the grounds of being a high price. The system used here is the Wi-Fi.

The advancement in the field of technology and the emergence of nanotechnology led to the compact size of the devices. The gas sensors also have gone through the same phase and have given very impressive results. The use of nanotechnology has been employed in the detection of toxic, inflammable gases. The need for the long-range communication for these nanoscale sensors led to the emergence of the Zigbee module. With this wireless sensor platform, environmental gas can be monitored by analog and digital gas sensors (Spirjakin et al. 2015). For the wireless communication, IEEE 802.15.4 ZigBee-based Wireless Sensor Network has been used which is processed for low power consumption operation.

Another wireless sensing network system uses the smartwatch as a mode to detect smoke and other combustible at the industrial level. Smartwatch is used for industrial parameter monitoring (Sandra et al. 2017), again due to low-cost and low power connectivity, ZigBee is used which is defined by the approved IEEE 802.15.4 standard. ZigBee TarangP20 is used for sending the sensor values to the database and to the Arduino-based wearable smartwatch simultaneously. The gas sensor used here is MQ2. The ppm levels of toxic gases are shown on the OLED display and LCD display too.

Today's state-of-the-art wireless sensing network has benefits of low costs for installation, maintenance, and repair with longer functioning time. The wireless sensor network can be used for stationary sensing or mobile sensing (Doni et al. 2018). Oikonomou et al. have developed capacitive sensors for monitoring of VOCs gases at the industrial workplace (Oikonomou et al. 2016). They have made use of the ZigBee network for the wireless data transmission. The result shown by the sensor was very good sensing performance with long-term stability and high repeatability.

Similarly, Lorwongtragool et al. developed a real-time monitoring of VOCs with wearable electronic nose based on a ZigBee network (Lorwongtragool et al. 2014). Based on inkjet printing technology, they have developed carbon nanotubes (CNTs)/polymer sensor array. This sensor can be used as a wearable e-nose which can sense various complex orders. They are used for the monitoring of the auxiliary door released from the human body. ZigBee technology, which is known as a wireless network communication at low price, was employed in the system to exchange the data between the device and the computer. The low-cost ZigBee module used here is Xbee-PRO RF module based on ZigBee/IEEE 802.15.4. The result is used to indicate the level of skin hygiene compared to the normal stage (Table 9.2).

Table 9.2 A table on recent development in the past year in wireless gas sensors

Year	Author	Wireless module	Target gas	Sensing film	Operating temperature	Ref.
2005	Pummakarnchana et al.	Wireless GIS network	Level of NO_x in air	Metal-oxide semiconductors such as ZnO is used	Typical 200–400 °C	Pummakarnchana et al. (2005)
2008	Chavali et al.	UHF-5 (433 MHz wireless RF sensor	anesthetic agent: sevoflurane	Multiwalled carbon nanotubes (-MWCNTs) and polypyrrole (Ppy)		Chavali et al. (2008)
2015	Yang et al.	Wi-fi	Carbon Monoxide in the industry		28–29 °C with relative humidity 50–60%	Yang et al. (2015a)
2016	Krishna et al.	ZigBee	Carbon dioxide			Krishna et al. (2016)
2016	DeHennis et al.	Near-field communication (NFC)	Implantable Glucose sensor			DeHennis et al. (2016)
2016	Asad and Sheikhi	13.56 MHz radio frequency identification tag	H_2S gas sensor	CuO-SWCNT hybrid nanomaterials	Room temperature	Asad and Sheikhi (2016b)
2017	Xu et al.	Near-field communication (NFC)	Biochemical; ethanol			Xu et al. 2017a
2017	Zhu et al.	Wi-Fi	Oxygen	Fe(II)-polymer wrapped carbon nanotubes	Room temperature	Zhu et al. (2017)
2018	Kim et al.	Bluetooth Module	Ethanol Vapor	An indium oxide and platinum nanoparticles	−40 to 125 °C	Kim et al. (2018)

9.2.2.5 Radar Systems

Radar is an electromagnetic remote sensing system for detecting and tracking the remotely located objects/targets. The word RADAR stands for RAdio Detection And Ranging. Radar detects the objects using electromagnetic radio waves and finds its location called ranging. It transmits the electromagnetic energy into space by means of a designated waveform and detects the echo signal from the objects. The sense of vision of human eyes can be extended by use of radar. The capability of radar is not only to sense the objects in the environment but also it can classify the types of objects in the space where normal human eyes cannot reach its range. In another way, the radar cannot detect the color of the objects that human eyes can do. Radars are designed to detect the objects even in fog, darkness, rain, haze, and snow (Skolnik "Radar handbook").

The basic radar systems consist of a transmitter, receiver, antenna, processors, display, etc. Transmitter produces an electromagnetic signal by an oscillator and is transmitted through an antenna into space. When this transmitting signal strikes any reflected object, the echo signal coming out from the object is received by the radar antenna. This return signal is mixed with various clutters from the atmosphere. If the strength of the signal crosses the threshold limit, then the radar will declare the signal as an object/target. The received signal is processed by the signal processor and further handover to the data processor to derive deferent object-related parameter such as range, azimuth, altitude, heading, speed, RCS, etc. The range of the object is calculated from the travel time of the signal from radar point to target location and back. Radar actually measures the range and Doppler of the object, and all other parameters of the objects are derived by the data processor. Radar can distinguish moving objects from stationary by use of Doppler Effect (Edde 1993).

(i) Radar System Components

The elementary radar system consists of the following major parts as shown in Fig. 9.16 to accomplish the required task. Major parts are:

(a) **A Transmitter**: It can be generated by the required electromagnetic radiating energy using an oscillator. Waveform generator can be a power amplifier like a Klystron, or wave tube or a local oscillator (LO) (Yu et al. 2008) like Magnetron. Initially, the signal is generated by the signal generator and then is amplified to the required band before sending to the antenna. This process is called signal up-conversion.
(b) **Antenna**: A metallic structure that can transmit or receive electromagnetic waves to/from space is called an antenna. It can be many shapes and sizes such as a parabolic reflector, planar arrays, or electronically steered phased array (Hansen 2009) as per the frequency band and requirements.
(c) **Waveguides**: Radar signals are transmitted to the antenna through transmission lines called waveguides.
(d) **Receiver**: The echo or reflected signal is received by the receiver and the signal is sent to signal processor after down-conversion. Local oscillators are used to

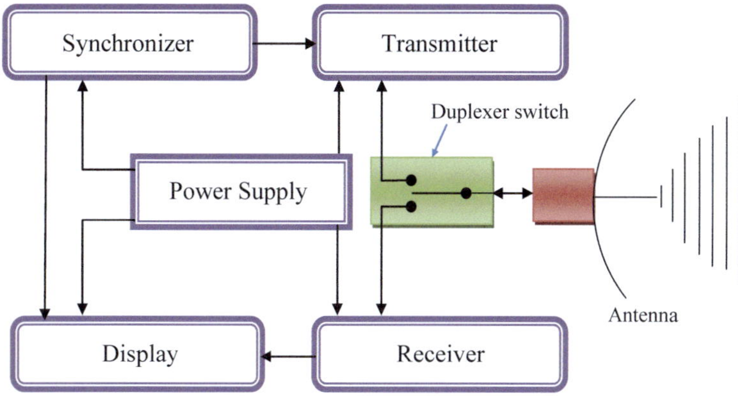

Fig. 9.16 Basic radar components

convert the signal from higher band to lower band that can be processed in signal processor (Krikorian et al. 2007).

(e) **Duplexer**: It operates like a switch. It allows the same antenna as a transmitter during transmission signal and as a receiver during return signal (Meharry et al. 2009).

(f) **Processors**: It extracts the information from the received signal and displays it to a user understanding format. The output signal is compared with the threshold limit set in the radar environment to decide the received signal is a real object or a clutter (Ward 1984).

(g) **Display**: It can be a plan position indicator (PPI) view or range height indicator (RHI) view or 3D view display. This enables the user or operator to understand the air situation of the object in the space.

(h) **Power Supply**: It supplies the power to different subsystems of the radar and monitors the power quality online.

(ii) The Concept of Radar Operations

Electromagnetic energy is generated by a radar signal generator or transmitter and then, it is radiated by the antenna into the atmosphere/space. When this electromagnetic wave hits a reflected object in space, the echo signal is produced from the object and scatter all around it. Some of the reflected energy from the object is collected at radar antenna and further forwarded to the receiver in order to obtain the required data about the reflected object. The range between the object and the radar point is computed from the travel time (τ) taken by the transmit signal to strike the object and back. Since electromagnetic wave travels at the speed of light (c), i.e., 3×10^8 m/s, then object/target range (R) can be expressed as (Curry 2012);

$$R = \frac{c\tau}{2} \tag{9.1}$$

At denominator of the above equation, factor 2 appears because of the round way propagation of the electromagnetic signal.

The object travels with a velocity V. The radial velocity V_R can be given by

$$V_R = V \cos \propto \tag{9.2}$$

In the above equation, α is the angle made by the radar line of sight (LOS) and the object velocity vector is shown in Fig. 9.17. The angle α can also be found out using Doppler shift (f_D) from echo signal as given below

$$V_R = \frac{f_D c}{2f} = \frac{f_D \lambda}{2} \tag{9.3}$$

where

f radar frequency
λ wavelength of the electromagnetic signal.

Object's azimuth and elevation angle are determined from the beam orientation toward the object. Other characteristics of the object are derived from the echo signal received from the reflected object.

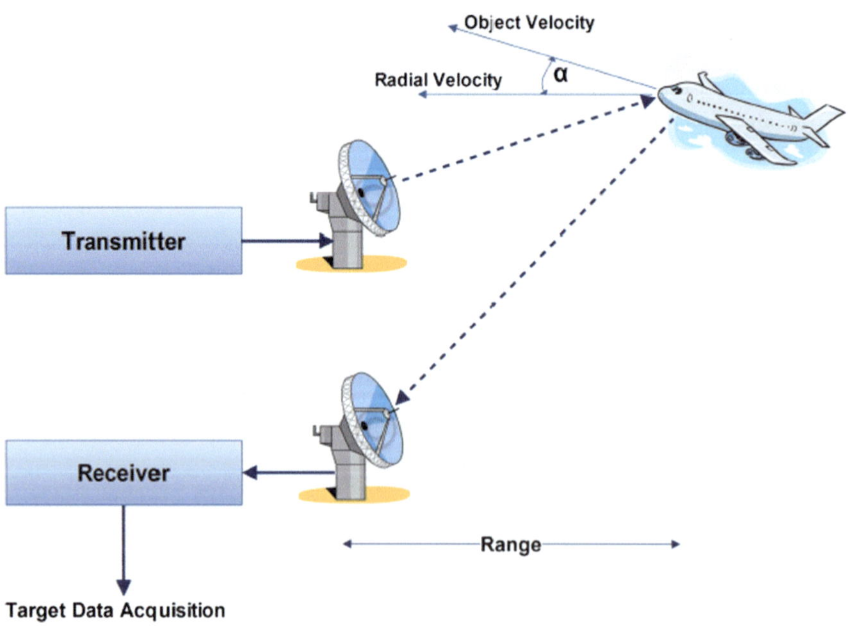

Fig. 9.17 Basic radar concept

(iii) Radar Functions

The moderns radars are multifunction radars, such radars perform more than one function simultaneously. Some of the important functions as mentioned below are commonly applied in a basic radar system:

(a) *Search function or called surveillance*: Volumetric search in space for object detection.
(b) *Detection*: Signal strength crosses the threshold limit indicates the presence of an object in space/atmosphere.
(c) *Measurement*: Determining the range, azimuth, elevation, altitude, size, heading, and radial velocity of the object from radar.
(d) *Tracking*: Generation of object flight path by successive measurements.
(e) *Imaging*: Generation of 2D or 3D image of the object for better visualization and understanding the object/target.
(f) *Classification*: Target/object is classified as target-type-missile (TTM), target-type-aircraft (TTA), slow-moving-target (SMT), etc., based on its characteristics parameter.

When a radar is designed to perform more than two functions at a time, it is said to be multifunction radar. Such radars are capable to handle multiple targets and are effective in a wide variety of functions. Some of the important characteristics of such multifunction radars are mentioned below:

(a) *Phased-array antennas*: Such antennas enable radar beam direction to be changed electronically to a desired angular position with a short span of time and without changing the physical antenna orientation (Jeffrey 2009).
(b) *Multiple waveform types*: This enables various functions to be performed for multiple targets simultaneously (Blackman 1986).
(c) *Digital signal processing (DSP)*: This handles the radar return signal from different waveforms and characteristics parameters are extracted for multiple targets (Mitra 1993).
(d) *Scheduler and radar computer*: To schedule the transmission signal as per user/operator preference and also it enables to configure the radar as per the requirements (Farina and Studer 1986).

(iv) Types of Radars

Radars are classified based on various aspects discussed below:

(a) **Monostatic radars**: In this type of radars, both transmit and receive antennas are placed at single place. Both the antennas are used to search the same volume space. But in modern radars, same antennas are used for transmitting and receiving functions by use of duplexer switch. The time taken for signal travel from radar to object and back is calculated to find out the range and radial velocity of the object is calculated from Doppler shift.

(b) **Bistatic radars**: In this type of radars, both transmit and receive antennas are placed quite separately so that there will be no interference between transmit and receive signal. Such radars enable a single transmitter to operate with a number of passive receivers (Jackson 1986).
(c) **Coherent radars**: The radar's transmitted waveforms are determined from a constant frequency source known as STALO. This can also be used to process the return signal. These type radars have multiple capabilities such as (i) radial velocity can be measured directly, (ii) coherent pulse integration, (iii) background clutter can be removed by moving target indication (MTI), (iv) pulse Doppler technique, and (v) space-time adaptive processing (STAP) methods (Sarkar et al. 2001). All these capabilities are not in noncoherent type radars.
(d) **Pulsed radars**: In such radars, a single pulse is transmitted and wait for the return pulse signal from the object, so there will be no interference between the transmit and return signal. Therefore, monostatic antennas are used to execute both transmit and receive the signal. The target position is found out by computing the time taken from a signal pulse transmission to signal pulse reception (Alabaster 2012).
(e) **Continuous wave (CW) radars**: In this type of radars, a continuous signal is transmitted and reflected signal is received simultaneously. Radial velocity of the object can be straight calculated from Doppler shift. Object range can also be found out in the case of a change in frequency of the continuous wave. Since such radars handle continuous transmit and receive of the signal, there is more chance of interference between the transmitter device and receiver device which may constrain both powers as well as the sensitivity of monostatic CW radars (Geisheimer et al. 2001).
(f) **Over-the-horizon (OTH)**: This type of radars operates in the high-frequency band from 3 to 30 MHz. This utilizes the ionospheric reflection to direct radar waveforms ahead of the usual radar horizon. The selection of frequency is to accomplish both ionospheric condition and object range. Such radars need a bigger antenna size, large transmitter power, and longer processing times. Mainly, this type of radars is used for early warning purpose. It has poor range and angular measurement (Headrick and Skolnik 1974).
(g) **Synthetic aperture radars (SAR)**: Such radars are mounted on satellite or aircraft in order to move along with its path. Such radars transmitted a series of pulses with very small beam width and the received signal is processed. Therefore, such radars have very good resolutions and therefore, 2D images of ground-based targets and terrain can be generated with high resolution (Curlander and McDonough 1991).
(h) **Secondary surveillance radars (SSR)**: The antenna of such radars are in rotating state to transmit the pulses with adequate power to reach the object to be tracked. A transponder is fitted on the object in order to communicate with the base station at a unique frequency (Trim and Mode 1990). The return signal from the object is a coded type pulse that poses object-related information like range, azimuth, altitude, speed etc. Such radars enable to track the object for a longer

range with a one-way transmission that consumes less power. In defense applications, this technique is used to identify the target whether belongs friend or enemy, is popularly known as identification friend or foe (IFF) (Voles et al. 1989).

(v) Frequency Bands

Wavelength (λ) and frequency (f) of the electromagnetic waves radiated from radar antenna are related to each other. The relations are stated below

$$\text{Wavelength } (\lambda) = \frac{c}{f}, \text{ and frequency } (f) = \frac{c}{\lambda} \tag{9.4}$$

(a) **Radar frequency bands**: Frequency spectrum of radar is alienated into different bands which are normally allocated by English letters (as shown in Table 9.3). International Telecommunications Union (ITU) has authorized specific operating frequency ranges for radar use. In most of the radars, the

Table 9.3 Radar frequency bands (2003)

Band designation	Frequency range	Assigned radar frequency ranges	Common radar frequency	Common radar wavelength
HF	3–30 MHz			
VHF	30–300 MHz	138–144 MHz 216–225 MHz	220 MHz	1.36 m
UHF	300–1,000 MHz	420–450 MHz 890–942 MHz	425 MHz	0.71 m
L	1–2 GHz	1.215–1.4 GHz	1.3 GHz	23 cm
S	2–4 GHz	2.3–2.5 GHz 2.7–3.7 GHz	3.3 GHz	9.1 cm
c	4–8 GHz	4.2–4.4 GHz 5.25–5.925 GHz	5.5 GHz	5.5 cm
X	S—12 GHz	8.5–10.68 GHz	9.5 GHz	3.2 cm
Ku	12–18 GHz	13.4–14 GHz 15.7–17.7 GHz	16 GHz	1.9 cm
K	18–27 GHz	24.05–24.25 GHz 24.65–24.75 GHz	24.2 GHz	1.2 cm
Ka	27–40 GHz	33.4–36 GHz	35 GHz	0.86 cm
V	40–75 GHz	59–64 GHz		
W	75–110 GHz	76–81 GHz 92–100 GHz		
Millimeter	110–300 GHz	126–142 GHz		
Wave		144–149 GHz 231–235 GHz 238–248 GHz		

Table 9.4 International telecommunications union frequency bands (2003)

Frequency range	Band designation	Metric designation
3–30 MHz	High frequency (HF)	Dekametric waves
30–300 MHz	Very high frequency (MIT)	Metric waves
0.3–3 GHz	Ultra high frequency (UHF)	Decimetric waves
3–30 GHz	Super high frequency (SHF)	Centimetric waves
30–300 GHz	Extremely high frequency (EHF)	Millimetric waves

operating frequency range is kept around 10% of the center frequency. ITU decides the working range of radar frequency based on bandwidth limitations of radar components.

VHF, UHF, and L bands frequencies are often used in surveillance type radars. It requires large power and bigger antenna size so that these frequencies can be easily obtained. Whereas, in case of tracking radars, they use X, KU, and K bands due to its narrow beam width and high measurement precision. The modern radars like multifunction radars use S & C frequency bands.

(b) **ITU bands**: Frequency bands as designated by the ITU for radar spectrum are listed in Table 9.4.

(c) **Electronic warfare (EW) bands**: The frequency bands used by EW are listed in Table 9.5.

(vi) Radar Equation

Radar range is expressed in term of radar equation by considering the characteristics factor of different components of radar system such as a radar antenna, receiver, transmitter, and also it considers the target type with its environment. Radar equation is not only used to find out the range of the object from the radar but it can

Table 9.5 Electronic warfare frequency bands (Scheer et al. 2010)

EW band	Frequency range
A	30–250 MHz
B	250–500 MHz
C	500–1,000 MHz
E	1–2 GHZ
F	2–3 GHz
G	3–4 GHz
H	4–6 GHz
I	6–8 GHz
J	8–10 GHz
K	10–20 GHz
L	20–40 GHz
M	40–60 GHz
N	60–100 GHz

help to understand the different aspects of radar design based on the target characteristics in space. A simple form of radar equation is derived in this section.

Let us consider P_t as the radar transmitted power. The power density for an antenna which radiates uniformly in all directions called as isotropic antenna can be found at a range R from the radar point as follows (Barton et al. 1991):

$$\text{Power Density for isotropic antenna} = \text{Transmitted power/surface area} = \frac{P_t}{4\pi R} \quad (9.5)$$

In the case of a directive antenna which radiates in a particular direction, antenna gain G plays an important role to determine the power density of a directive antenna. Antenna gain (G) is calculated by dividing the maximum radiation intensity with the radiation intensity from a lossless isotropic source. Mathematically

$$G = \frac{4\pi A}{\lambda^2} \quad (9.6)$$

$$= \frac{4\pi A_A}{\lambda^2 L_O L_E} \quad (9.7)$$

where

$$A = \text{effective aperture area} = \frac{A_A}{L_O L_E} \quad (9.8)$$

A_A Actual aperture area,
L_O Antenna ohmic loss,
L_E Antenna efficiency loss.

Now, the power density for a directive antenna with antenna gain G can be calculated by the following expression:

$$\text{Power density of directive antenna} = \text{Transmitted power } x \text{ Antenna Gain/surface area} = \frac{P_t G}{4\pi R^2} \quad (9.9)$$

When radar incident power strikes on the object, the echo signal from the object scatters in various directions. The measure of this scattered power per unit solid angle in radar direction divided by the amount of incident power per square meter is called the radar cross-section (RCS) (Ruck et al. 1970). RCS is designated by σ, is normally expressed in square meters and also it often expressed in dBsm or dBm^2. The expression for power density in Eq. 9.9 can be rewritten as follows considering the amount of return signal.

$$\text{The power density of the return signal} = \frac{P_t G}{4\pi R^2} \times \frac{\sigma}{4\pi R^2} \tag{9.10}$$

If A is the antenna's effective aperture area that captures the echo signal, then the received power P_r from the echo signal can be expressed as follows:

$$\text{Power Received}(P_r) = \frac{P_t G}{4\pi R^2} \times \frac{\sigma}{4\pi R^2} \times A = \frac{P_t G \sigma A}{(4\pi)^2 R^2} \tag{9.11}$$

The radar maximum range where the object cannot be further detected has occurred when the received power P_r is just equal to the minimum detectable signal (MDS) $S_{\min}$. So, the expression for maximum range $R_{\max}$ can be given as follows:

$$R_{\max} = \left[\frac{P_t G \sigma A}{(4\pi)^2 S_{\min}}\right]^{1/4} \tag{9.12}$$

Substituting the expression for effective aperture area A from Eq. 9.5, the expression in Eq. 9.10 will be

$$R_{\max} = \left[\frac{P_t G^2 \lambda^2 \sigma}{(4\pi)^3 S_{\min}}\right]^{1/4} \tag{9.13}$$

Maximum range expression with effective aperture area A can be written as

$$R_{\max} = \left[\frac{P_t A^2 \sigma}{4\pi \lambda^2 S_{\min}}\right]^{1/4} \tag{9.14}$$

Equations 9.10–9.12 are simplest forms of radar equation. Interpretation of all these three equations conveys some useful design-related information. Equation 9.10 conveys that radar range is independent of wavelength λ, whereas Eq. 9.11 shows that the radar range varies as $\lambda^{1/2}$ and in Eq. 9.12 indicates $\lambda^{-1/2}$. So, the right relationship depends on whether the parameter like antenna gain is considered to be invariable or the effective area is invariable with respect to wavelength (Probert Jones 1962).

Therefore, these simplest forms of radar equations are not sufficient to understand the practical radar performance. Many other factors like line noise, clutters, etc., are also affected by the radar range equation.

(vii) Signal-to-Noise Ratio (SNR)

Signal-to-Noise Ratio (SNR) is an important parameter in radar performance measurement. SNR determines the radar capability to detect, measure, and track the object. As discussed while deriving the radar equation, there are many factors like noise, clutter, jammer, and chaff can interfere with the echo signal which affects the radar to detect, measure and track the real object.

The SNR is estimated at the receiver output. This is the ratio of signal strength to noise strength of the received signal. SNR can be calculated using the radar equation as follows (Johnson 2006):

$$\text{SNR or S/N} = \left[\frac{P_p G \sigma A P_s}{(4\pi)^2 R^4 B k T_s L}\right]^{1/4} \tag{9.15}$$

where

- P_p Peak power transmitted,
- G Antenna gain,
- σ Radar cross-section (RCS),
- A Antenna effective aperture area,
- P_s Signal gain is produced when a pulse compression waveforms is used.
- R Range of the object,
- B Signal bandwidth of radar,
- k Boltzmann's constant (1.38×10^{-23} J/K),
- T_s System noise temperature,
- L System loss.

(viii) Applications of Radar

Radar is widely used for remote sensing for many purposes. Radar has been deployed not only on the ground but also in air, sea, and space. Radar used in the ground is called ground-based radar and is used for detecting, locating, and tracking of aircraft or other targets in space. Radar mounted in the ship is called shipborne radar and is used to locate shorelines, navigation buoys, and other ships for safety purpose. Radar used in aircraft is called airborne radar which is used to detect and track other aircraft, ships, vehicles, etc. Other important applications are these radars can be used for land mapping, for the avoidance of storm and terrain, and for navigation purpose. Radars are also used in space in order to guide the spacecraft and for identifying of land and sea. Although radars are used in civil applications, the major area of applications is in the military. Some of the major applications of radar are mentioned below.

(a) Air Traffic Control (ATC): Radars are widely deployed in the airport across the world to monitor, control, and track the aircraft. Radar is also used to direct the aircraft for a safe landing during bad weather or night.
(b) Aircraft Navigation: Radars are used in aircraft for navigation purpose during bad weather condition, for terrain avoidance and also terrain following. Sometimes, ground mapping radars are also used for navigation purposes.

(c) Ship Safety: Shipborne radars are used in the ship for improving the ship safety while traveling in order to avoid the collision with other ships or objects. It is also very useful to locate buoys, shorelines during poor visibility. The radar used in the ship is small in size as compared to ground-based radar.
(d) Space: Radars are used in spacecraft in order to guide the spacecraft toward orbit and for landing on the moon. There is also some ground-based radar which is bigger in size and having long-range tracking capabilities are used to detect and track the satellites.
(e) Remote Sensing: All the radar systems are used for remote sensing. But particularly, this term is used in the case of detecting geographic objects, or environment like to know weather condition, to probe the moon and the planets. For mapping and measurement of land and another geographic survey purpose, this type of radars is used. Such radars can be mounted on satellites and aircraft.
(f) Law enforcement: Radar is widely employed to monitor the speed of the vehicles in highways by police in order to avoid accident due to high speed.
(g) Military radar: The main application of radar is in defense purpose. Military use involves continuous surveillance of air situation, navigation, and for control and guidance of weapons.

9.3 Various Wireless Circuits: A Review

9.3.1 Solid-State Gas Sensor-Based Used for Wireless GIS Network

This device consists of a solid-state gas sensor incorporated with a personal digital assistant (PDA). The communication between these two is established through Bluetooth and GPS. Nanomaterials have dimensions between 0.1 and 100 nm. It has been implemented in various industries, for example, in space exploration, computing, health, and medicine.

Solid-state gas sensor uses metal-oxide semiconductors based on nanotechnology is very lightweight, exceptionally small in size, low price, robustness, and can be installed anywhere for data acquisition. This data collected through this sensor is uploaded through a wireless GIS network. In this, a metal oxide layer is coated as a sensing element over the substrate with two electrodes. Materials are SnO_2, ZnO, WO_3, and TiO_2 (Choi et al. 2016).

GIS can be accessed through the internet by means of a web browser. Detailed maps can be engendered from databases which are readily available and accessible throughout the world. Using the internet is an economical way to share data (Ma et al. 2016).

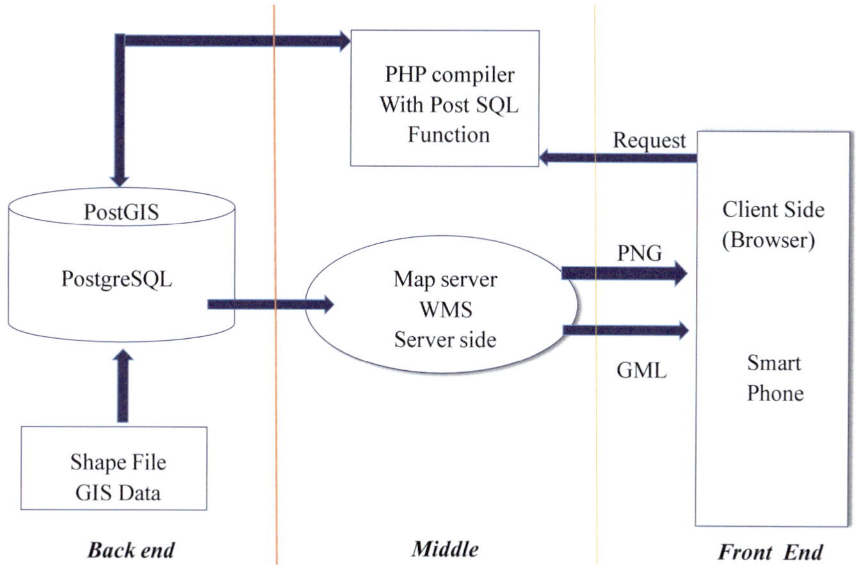

Fig. 9.18 Wireless GIS data login system

As shown in Fig. 9.18, a wireless GIS data login system was developed which has two parts, i.e., software and hardware. On the hardware side, a server gives the back end support. On the front side, we have a user's browser or some pocket PC.

The above Fig. 9.19 gives the detailed view of how Nanosensors along with GPS system are integrated using the internet for real-time monitoring of air pollutant gases.

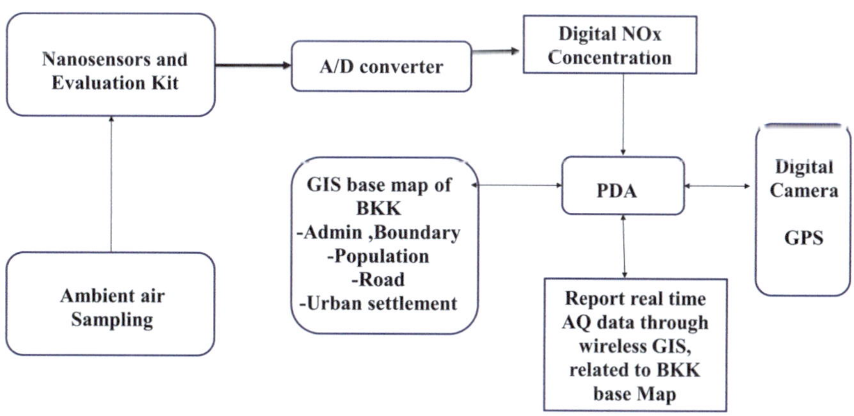

Fig. 9.19 Nanosensors along with GPS system

9.3.2 A Wireless Sensor Network for Wireless Monitoring of Sevoflurane

A wireless sensor network is used for wireless monitoring of sevoflurane. A nanocomposite material of m-MWCNT/polypyrrole sensor is used along with RF-powered chip. A proposed wireless gas sensor network comprises a wireless receiving unit and a wireless sensing unit. Its schematic circuit is shown below. The wireless sensing system has a host computer and an RF module which includes a data acquisition module, the sensor chip, and a microprocessor. The data acquisition module is used to login data from the sensor chip. There are two microprocessors each at sensing and receiving unit. The one at the sensing unit acquires the data from the data acquisition module, and the other one processes the received data to the gateway, connected to the host. All the data of sensing and receiving units are transmitted via wireless RF module (Karmaoui et al. 2015).

The wireless system shown below has one power node with a microprocessor base station and few wireless sensors which are distributed. The symmetry of the sensor communication is exploited by developing network architecture and communication protocol. Mostly, the information flows from the sensor node to the base station, and only few information in the form of command flows from the base station to the sensor nodes.

A schematic diagram of the proposed wireless gas sensor system is given in Fig. 9.20.

The sensing system includes a microprocessor with wireless RF module (MICA2DOT), sensor chip, data acquisition module (MDA500), and gateway (MIB510 Crossbow Technology Inc., CA, USA; http://www.xbow.com/) with a host computer.

9.3.3 A WSN for Real-Time Monitoring of Industry Carbon Monoxide

Carbon monoxide sensor nodes based on NDIR technology has been developed. To obtain minimum jittering, a low-frequency modulation method was suggested. The communication between each router and sensor nodes were created on a Wi-Fi wireless network. This network can locate each sensor node and data is transported to smartphones, a personal computer from the sensor node through routers. The wireless network with sensor nodes and routers constitute a wireless sensor network which is used in industrial areas. CO is a hazardous gas and it is highly flammable, so accurate detection is needed by fast response system. Considering the above measurement module, we use a pyroelectric sensor, which consists of an airway tube and CO concentration measurement module.

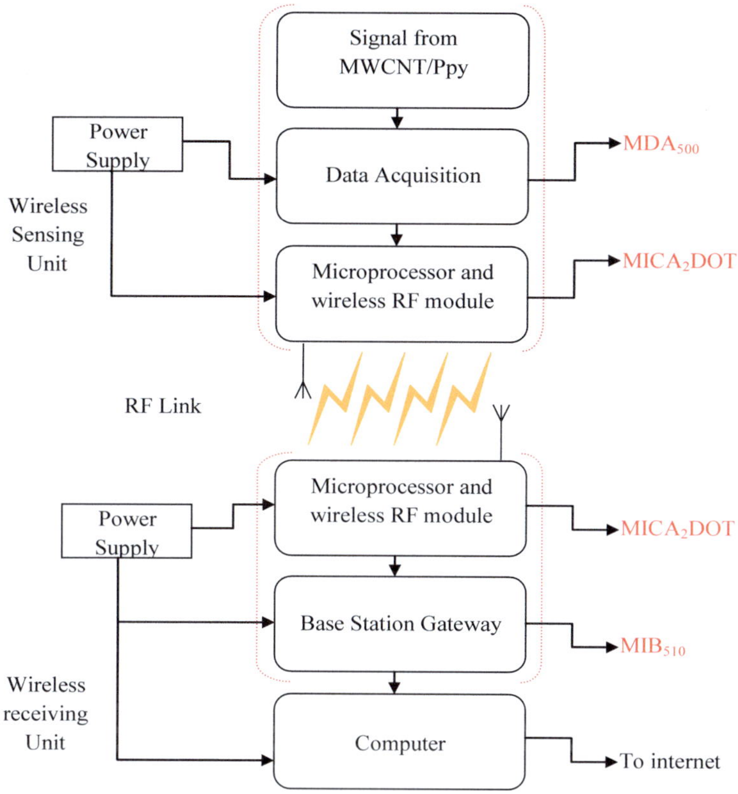

Fig. 9.20 Wireless gas sensor

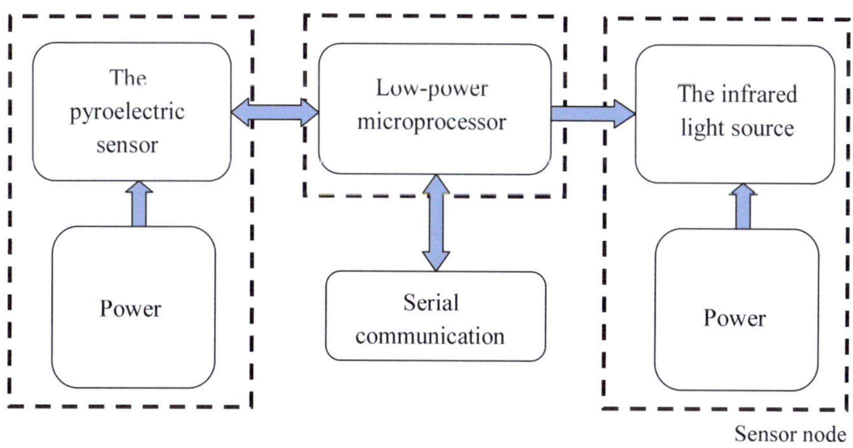

Fig. 9.21 Sensor node with circuit schematics

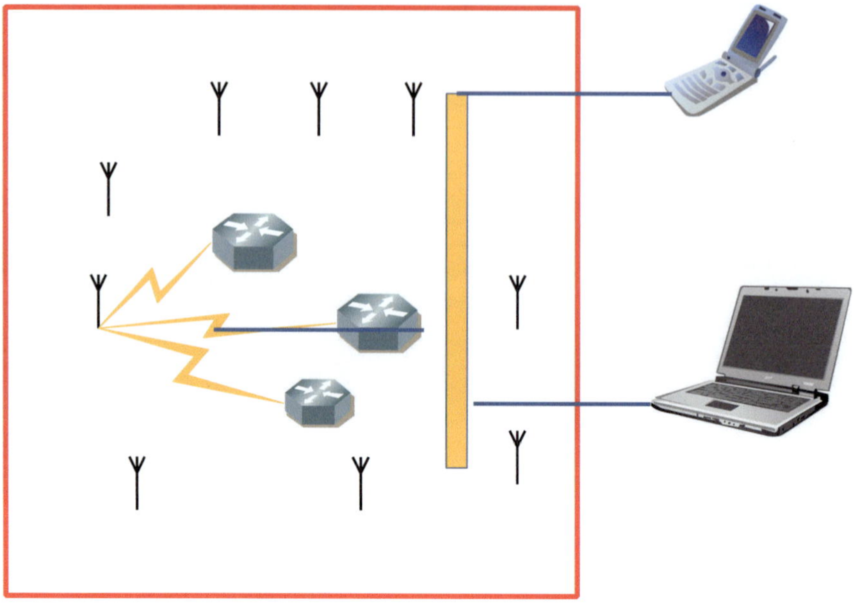

Fig. 9.22 Network configuration

A pyroelectric sensor has high accuracy and it has a thermal response time of 140 microseconds. The sensor works on the Lambert–Beer Law (Shabanov and Gastellu-Etchegorry 2018). Also, a dual pyroelectric sensor can reduce external interference:

$$I = I_0 e^{-\alpha(\lambda)CL} \tag{9.16}$$

where I is the intensity of light striking the detector (I, W/cm^2), I_0 is the measured intensity of an empty sample chamber (I_0, W/cm^2), α is the absorption coefficient (cm^2/mol), c is the CO concentration (mole/cm^3), and L is the absorption path length (cm).

Zigbee gives us a greater accuracy, and the price is little bit high. Wi-Fi promises to deliver high data rates over a large area to a large number of users and network forms to transfer data directly to cell phone and computer. Wi-Fi is the most economical option to transfer data in the present scenario. As shown in Fig. 9.22, network system consists of a personal computer, access point, and station. The communicating sensor node and the access point are connected through Wi-Fi transceiver module. Data collected from sensor nodes to the PC or smartphone is also called a router. The system network has a Network Configuration, Network Architecture, and Network Implementation.

The base station is based on million instructions/sec. program construction, running at 360 MHz and featuring two wired ports of Ethernet and one port of

Table 9.6 Parameters of the base station

System message	Parameters	Wi-Fi	Parameters
Processor	MIPS program (32 bit, 360 MHz)	Transmission distance	Max 400 m (open areas, 3 dB antenna)
RAM/Flash	8 M/1 M	Frequency range	2.412–2.484 GHz

RS232 (RS232 is a serial interface). The device is equipped with programmable FLASH of 1 MB, 8 GB random access memory, in an open environment. Up to 400 m transmission range is obtained. It has 32 TCP connections that can collect information from 32 base stations. The base station has the following parameters as shown in Table 9.6. The RS232 and the sensor node are connected through the base station and data is received by routers. To control the sensor, switch instructions are given by routers to sensor node through the base station. The energy consumption is optimized by the radio parameters such as antenna and signal modulation are as follows (Bor-Yaliniz et al. 2016).

(a) Access point

The access point is the midway between computers and base station. The routers collects data from each base station, transmit it to the computer, and sends these instructions from the computer to the base station. The router runs on Linux OS and 32-bit ARM7tdmi RISC processor. The device is also equipped with 8 MB of programmable FLASH, 32 MB of random access memory, and outlying function modules, comprises of timers, programmable I/O ports, Universal Asynchronous Receiver/Transmitter (UART), and an interrupt controller (Giust et al. 2015)

(b) Network Architecture

As shown in Fig. 9.23, three routers are placed in an industrial plant. Each station has the service set identifier (SSID) and mac address. The signal strength and mac address of the station are received by a router, the distance receive signal srength indicator (RSSI) from each base station is acquired (Gupta and Jha 2015). Given the distance of every station to three routers, the triangle positioning method is proposed, and its principle is coarsely displayed in Fig. 9.23

A plane Cartesian coordinate system for the industrial plant is formulated. When routers receive the same mac address, they will define the sensor node as p_1, and we will obtain the distance between p_1 with every router as d_1, d_2, d_3. If p_1: (x, y) sensor node coordinate, the following formulas are formulated as follows:

$$\sqrt{(x_1 - x)^2 + (y_1 - x)^2} = d_1 \rightarrow \{1\} \tag{9.17}$$

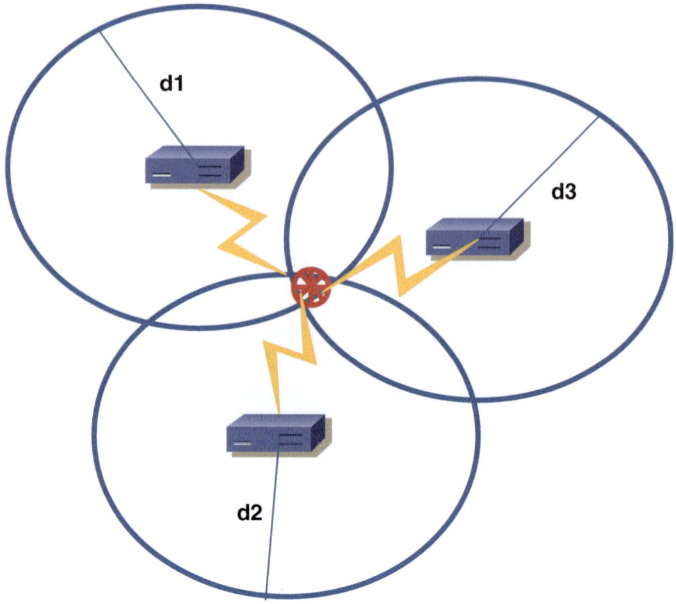

Fig. 9.23 Triangle positioning method

$$\sqrt{(x_2 - x)^2 + (y_2 - x)^2} = d_2 \rightarrow \{2\} \tag{9.18}$$

$$\sqrt{(x_3 - x)^2 + (y_3 - x)^2} = d_3 \rightarrow \{3\} \tag{9.19}$$

we can get $(x, y) = (x_{12}, y_{12})$, similarly, we get $(x, y) = (x_{23}, y_{23})$, $(x, y) = (x_{13}, y_{13})$

$$\bar{x} = \frac{x_{12} + x_{23} + x_{32}}{3} \tag{9.20}$$

$$\bar{y} = \frac{y_{12} + y_{23} + y_{32}}{3} \tag{9.21}$$

Let p1: $(\bar{x}, \bar{y})$ be a sensor node coordinate and by using the same way, we can get the sensor node. We can get the address of the nodes on the smartphone and PC, and with the triangulation, we can locate the address of every sensor node.

(c) Network Implementation

Sensor nodes obtain the concentration of CO all the time with the help of the sensor network. An open Wi-Fi module requires more power. And, it is useless to have the concentration of CO continuously. The three designed states of a Wi-Fi module are sleeping, wake up, and execution. Wi-fi module has a 5 min sleeping mode, and after 5 min, one router subjects an instruction to every station for waking

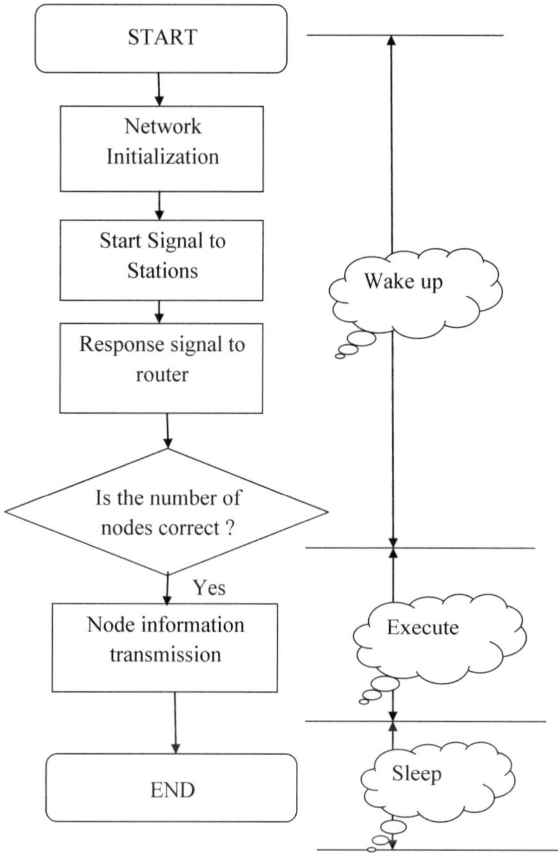

Fig. 9.24 Network and state of work

up. A response signal is returned by all the sensor nodes, and the position of the node in the coordinate system of PC or smartphone is located. So, sensor quality is determined and the newly added sensor nodes will be displayed in real time on a coordinate system. Data acquisition is for a minute and with that, CO concentration of each sensor node can be displayed.

9.3.4 RFID Based Gas Sensor for H_2S

To develop the wireless sensor on the RFID tag, one of the commercial 13.56 MHz RFID tag with a flexible substrate has been used with EM4034 standard IC. The antenna is mounted on the topside of the RFID tag with an adhesive paste on the other side to connect it to the body covering. In order to put the sensing film on the top of RFID tag, it was covered with adhesive tape except the antenna surface to shelter superfluous surface of the tag from chemical treatment. Then, flower-shaped

CuO-SWCNT (Asad and Sheikhi 2016b) materials, which was obtained on different temperatures, were dispersed in Dimethylformamide (DMF), and drop-casted on the topside antenna and heated to a required temperature (at 80 °C in this case) for ~2 min to obtain a good adherence to the tag. The adhesive tape was then removed from the top of the surface and RFID tag was provided with proper thermal treatment to provide better adhesion between functional materials and antenna of the tag. The fabricated wireless sensor was then attached to the body for further revelation in healthcare monitoring.

9.3.5 NFC-Based Gas Sensor for Biochemical with a Smartphone

Many electronic phones have been introduced into the marketplace over the past few years, and these are integrated with NFC modules. There is a huge demand for food safety, health diagnosis, and environmental monitoring. This paper is about the development of flexible NFC tag sensors for biochemical sensing at low-cost. The construction of NFC tag is as follows:

Electronic devices are incorporated into near-field communication (NFC) modules which could wirelessly power when the object is close enough (~3 cm). These NFC tags are based on three important steps. An unmodified NFC tag is shown in Fig. 9.25 as Step 1, and the power is wirelessly transmitted from the smartphone to the resonant circuits through inductive coupling, which energizes the IC chip on the parallel branch. So, the NFC tag is readable and can be marked as "On". In Fig. 9.25, Step 2 shows the conducting aluminum film which was removed from the IC chip to disconnect the IC chip from the resonate circuit so that it no longer gets the power supply to communicate with the smartphone. At this time the NFC tag becomes unreadable, that could be marked as "Off". In Step 3, two short wires connected the modified NFC tag in Step 2 with electrodes, such as gold electrodes and platinum electrodes to detect analytes in a solution like ions and bacteria, or carbonic interdigitated electrodes covered with monolayer graphene for gas detection in this study. There was an "On to Off" threshold value Rtof the NFC tag. If the impedance of the electrodes R sensor was smaller than Rt, the NFC tag sensor could be read by the smartphone so it was "On". Once R sensor exceeded Rt with the exposure of target analytes, the IC chip could not get enough power and the sensor was "Off". The NFC tag sensor is turned red "On" to "Off" when an analyte is detected (Xu et al. 2017b).

Figure 9.25 shows Commercial Near-Field Communication tags. These are simple resonant circuits which consist of inductance (L) on a flexible polyethylene terephthalate (PET) substrate, an integrated circuit (IC) chip (RIC), and capacitor (C).

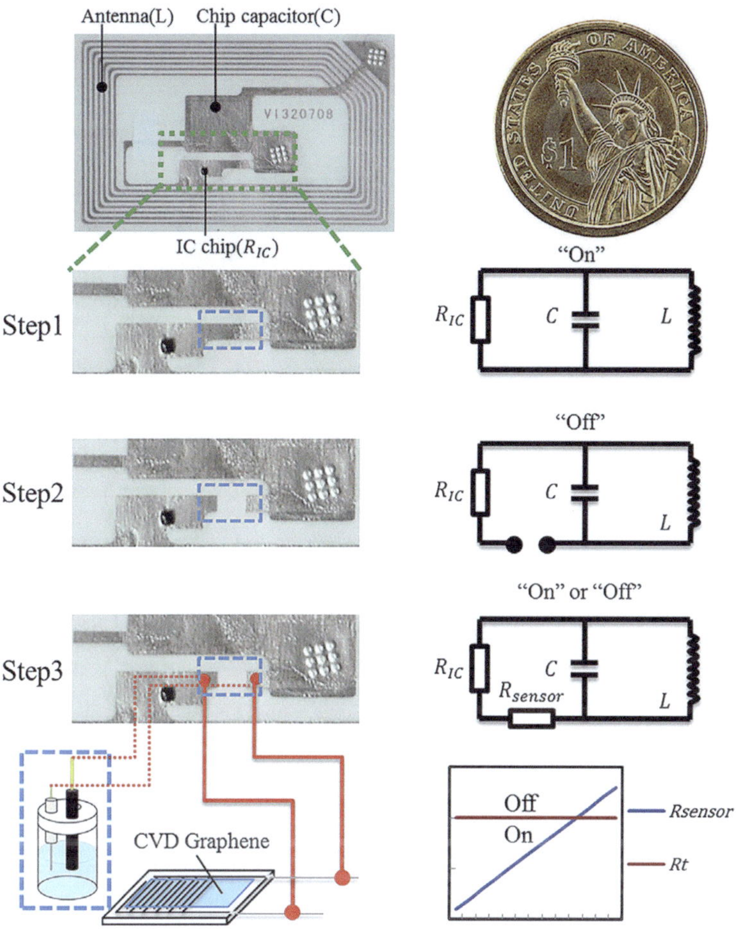

Fig. 9.25 Commercial near-field communication tags (reprinted with kind permission from Xu et al. (2017b))

This tag sensor cannot give a concentration of the detected analyte. So, rheostat (R_0) was added on the electrodes branch. It is turned off only when the sum of R_0 and R_{sensor} in series combination is calculated. The circuit shown below in Fig. 9.26 gives the total idea of NFC-based gas sensor.

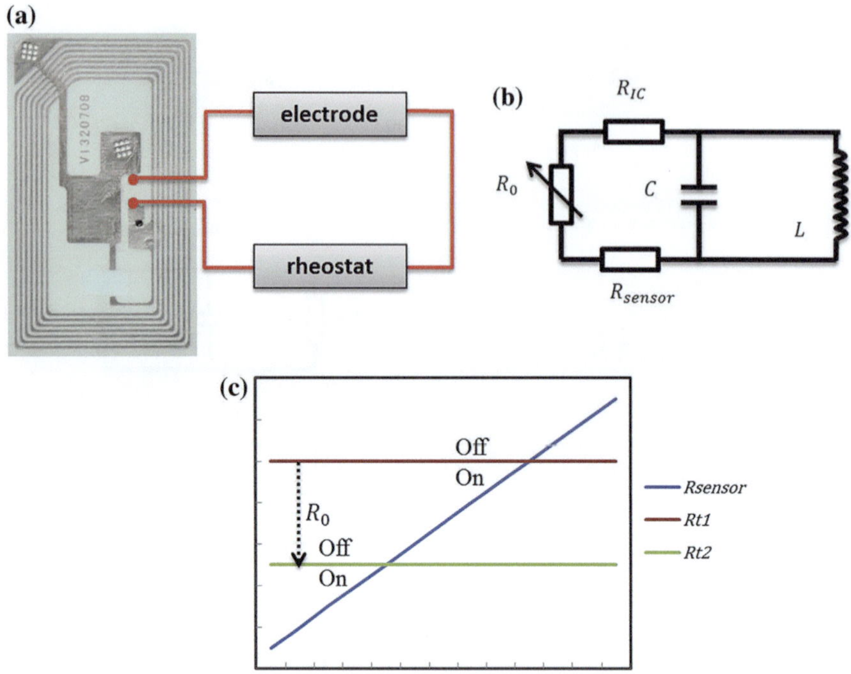

Fig. 9.26 NFC-based gas sensor (reprinted with kind permission from Xu et al. (2017b))

9.3.6 A Bluetooth Module-Based Alcohol Gas Sensor

A flexible and transparent nanostructure of indium oxide and platinum nanoparticles is used for real-time monitoring of alcohol vapors. The integration of the Bluetooth module to this gas sensor gives wireless processes of alcohol (C_2H_5OH) sensor using computers, smartphones, and electronic gadgets. The operating temperatures of this device range from −40 to 125 °C. The fabrication is on an ultra-thin (2 μm) polyimide (PI) substrate. This material has good mechanical properties in bending tests, which makes it a potential material for fabricating smart

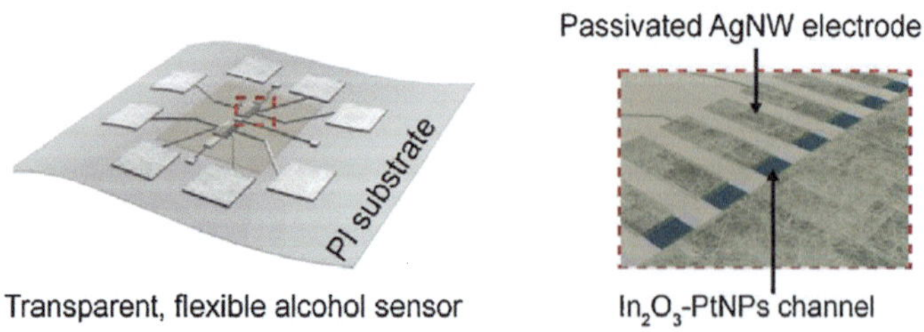

Fig. 9.27 Bluetooth module gas sensor (reprinted with kind permission from Kim et al. (2018))

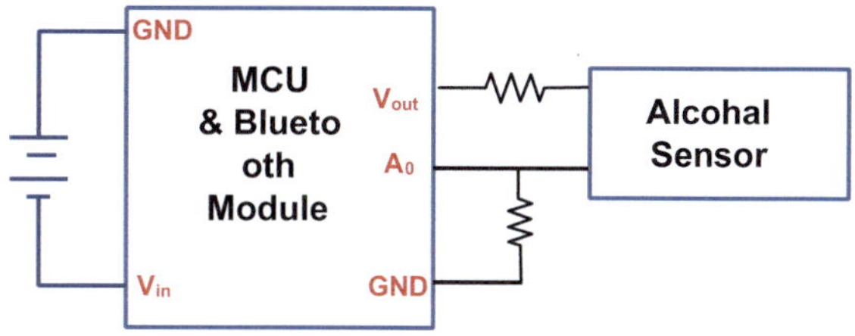

Fig. 9.28 Bluetooth module gas sensor

electronics. Then, a channel layer of In_2O_3-PtNP was made and networks of AgNWs randomly were patterned using photolithographically as electrodes. ZLater, except for the channel area, the SU-8 passivation layer was coated with openings. Finally, the supporting substrate is delaminated from fabricated devices (Kim et al. 2018). The final device has been shown in the Fig. 9.27.

The Bluetooth module is integrated with the gas sensor to obtain the results. The circuit of the Bluetooth module with the alcohol gas sensor is given Fig. 9.28.

The ethanol sensor included with the Bluetooth module is used for measuring electrical differences. The data is processed by an alcohol sensor component, and transmitted by Bluetooth module. An analog-to-digital converter is used to convert the analog input signal of the sensor to a digital output. A customized application is installed on the smartphone, and the transceiver facilitates transmission of the data wirelessly.

9.4 Advantages of Wireless Gas Sensors

Harmful gases can leak anytime. When this happens, there may be a loss of human life or any other living beings. Wireless gas detection systems like man-down GPS and alarm signals alert you the moment gas or other chemicals start to leak. These tools give us more time to estimate the situation and call the relevant emergency response units. If we install this remote wireless sensing data technology for our commercial or industrial units, we can monitor in real time (Kim et al. 2018).

This installation setup saves much time and is economical. It saves money because they require non-electrician-trained personnel. Wireless gas detection systems also have little maintenance costs compared to wired gas detection systems.

(a) Ease of Use and Implementation

Wireless gas detection systems are easy to use and implement. When setting them up, you simply need to mount the units and turn them on. No wiring technology is required from the monitor to the sensor or from the sensor to an external power source. A good wireless system provides 12 months of monitoring before the battery is replaced.

(b) Multiple Receivers

With wireless systems, you can monitor your commercial or industrial units remotely. The alarms monitor your structures without the need for your physical presence. You can use multiple receivers to monitor multiple units of any size.

(c) Safe and Dependable

Another benefit of a wireless gas detection system is that they are one of the lowest risk gas detection systems in the market. They are safe and dependable. There are no wires that need to be connected or cut. So there are low risks of damaging the system or injuring yourself. They are also dependable because you can be assured of their accuracy.

(d) Cost Effectiveness

While safety and efficiency are the major factors to consider when looking for a gas detection system, the cost is also a valid concern. Wireless gas detection systems do not break the bank like some other systems. There are several companies that sell affordable wireless systems that you may consider working with.

(e) Bottom line

If you intend to purchase Wireless gas detection systems, consider products that work dependently without interfering with other unit systems. Consider systems that are field adjustable for background gas check. Also, consider products designed to reject electromagnetic interference and other forms of interference that may lead to false.

9.5 Challenges in Setting up Wireless Sensor

Wireless gas sensor networks have many similarities with other systems but they have unique constraints. These constraints make designing of a WSN difficult. Here, we discuss different design constraints of a WSN. Constraints of wireless sensor networks are not limited to the following (Ojha et al. 2015).

(a) Energy

While designing any sensor network energy is associated with it. Generally, DC batteries or any AC source is energy to the sensor which gets depleted. Is the battery replaceable or renewed? As a primary design challenge is to reduce energy or increase efficiency. Wireless sensor networks have a very limited energy source (<500 mAh, 1.2 V) and replenishment may be limited or impossible. Commercially, battery-operated sensors used are based on one Li-AA cell or two

AA alkaline cells. Energy consumption has three main functional domains: sensing, communication and data processing, and processed or raw transmit. For example, the choices made at the physical layer of a sensor node affect the energy consumption of the entire device and design of higher level protocols. The main energy consumption of CMOS-based sensors is that their processors are primarily due to switching energy and leakage energy:

$$E_{\mathrm{CPU}} = E_{\mathrm{switch}} + E_{\mathrm{leakage}} = C_{\mathrm{total}} V_{2\mathrm{dd}} + V_{\mathrm{ddIleak}} \Delta t \tag{9.22}$$

Let C_{total} be the total capacitance switched by consumption, supply voltage be V_{dd}, I_{leak} is the leakage current, and Δt is the duration of the computation. It is believed that switching on the sensor dominates the energy consumption of processors but leakage of energy will also be responsible for energy consumption. There are few techniques to control the consumption by shutting down of idle components when idle and using Dynamic voltage scaling software techniques.

(b) Wireless Networking

The communications to wireless networks have different challenges, for example, RF signals fade while it propagates through a medium and while it passes through any opaque objects. The RF signals use inverse-square law

$$p_r \propto \frac{p_t}{d^2} \tag{9.23}$$

where the received power is proportional to the inverse square of the distance d from the source of the signal. If $p_r x$ is received power at distance x, and if the new distance is $y = 3x$, then the power decreases by $p_r y = p_r x/9$.

As increasing the distance between a base station and sensor node rapidly increases, then, the transmission power required is also increased. Therefore, it is always suggested that larger distances should be subdivided into multiple small distances leading to multi-hop communication. Therefore it is more energy efficient while transmitting devices. The multi-hop communications require that nodes in a network cooperate with one another to provide an efficient path and to function as relays. This challenge is further impaired in networks that duty cycles to conserve energy, i.e., many of the nodes are turned off when not in use. As a consequence, during these down times, the sensor cannot as a relay. Some networks depend on the wakeup demand strategies to ensure that nodes are active when needed. It involves using radio one with the low power to receive the wakeup response and other with high power to act as a relay.

(c) Decentralized Management

In a centralized solution network, one node, i.e., the base is allowed to take decision and decision is passed to other nodes. This type of management is optimal and not energy efficient. Whereas, if we use a decentralized (distributed) approach, each node can take a decision based on the limited local information. This type will not be optimal but it will be more energy efficient.

(d) Security

Numerous wireless sensor networks are brought together to bring sensitive information. The remote and unattended operation of sensor nodes may enhance the possibility of their disclosure to malevolent intrusions and attacks. Further, wireless communications provide a trouble-free mode for an adversary to eavesdrop on sensor transmissions. For example, one of the most challenging security threats is a denial-of-service attack, whose objective is to interrupt the correct operation of a sensor network. This can be done by using a variety of spasms, including a jamming attack, where wireless signals are used to prevent sensor communications.

There are various techniques and solutions for distributed systems that inhibit attacks or reduce the extent and damage of such cyber-attacks. As an outcome, sensor networks necessitate innovative solutions for key establishment and distribution, node authentication, and secrecy.

9.6 Different Commercially Available Wireless Modules

Numerous commercially available wireless communications nodes are including the Casira devices from Cambridge Silicon Radio, CSR, several Bluetooth kits, and including Lynx Technologies.

From the website Crossbow (http://www.xbow.com/), they produce three Mote processor radio module families—MICA [MPR300] (first generation), MICA2 [MPR400], and MICA2-DOT [MPR500] (second generation). Nodes have five sensors inbuilt in them. These are Acceleration/Seismic, Magnetic, Acoustic (Microphone), Temperature, and Light. These are especially suitable for surveillance networks for personnel and vehicles. We can install different sensors to the node depending on the requirement. These devices are small in size and consume less power which can be placed in any environment. Since each sensor node in a network can act as base stations, the network can configure itself and it has multi-hop routing ability. The frequency range is ISM band, either 915 MHz or 432 MHz, with a data rate of 40 KB/s. and a range of 30–100 ft. Each node can operate at a speed of 4 MHz and has a flash memory with 128 Kb, a low power microcontroller processor, and SRAM of 4 K bytes and EEPROM of 4 KB. It works on a tiny micro-threading distributed operating system (Tiny-OS), developed by UC Berkeley, with a Nested C (NES-C) source code which is similar to C language. A great deal of programming is required for installation of these devices. A workshop is offered for training (Haghi et al. 2017).

Microstrain's X-Link Measurement System (http://www.microstrain.com/) may be the easiest system to get up for running and to program. The frequency range used is 915 MHz, which lies in the US license-free ISM band. The sensor nodes are multi-hop, and it can be supported with eight sensors by a single wireless node. The onboard data storage is 2 MB. There are three types of sensor nodes—S-link (strain gage), G-link (accelerometer), and V-link (supports any sensors generating voltage differences). The sensor nodes have a preprogrammed EPROM, so a great deal of programming by the user is not needed. Sensor nodes use a 3.6 V lithium ion internal battery (9 V rechargeable external battery is supported). A single receiver (Base Station) addresses multiple nodes. Each node has a unique 16-bit address, so a maximum of 216 nodes can be addressed. The RF link between Base Station and nodes is bidirectional and the sensor nodes have a programmable data logging sample rate. The RF link has a 30 m range with a 19,200 bit per second rate. The bit per second rate on the serial RS-232 link between the Base Station and a terminal PC is 38,400. We can also use the Lab VIEW interface for support.

9.7 Conclusion

This whole article attempts to describe various wireless detection techniques especially for gas sensing, and their basics along with the coverage of various sensing strategies on the basis of research work is reported. It is not possible for the authors to cover each and every aspect of wireless sensing module and their miscellaneous applications in one article, therefore, this article may lack basic electronics involved in wireless modules, detailed analysis of covered sensing modules so and so forth.

In a nutshell, this article reviews various works performed related to wireless gas sensing systems along with a proper illustration of radar-based detection.

References

Abad E, Zampolli S, Marco S, Scorzoni A, Mazzolai B, Juarros A, Gómez D, Elmi I, Cardinali GC, Gómez JM, Palacio F (2007) Flexible tag microlab development: gas sensors integration in RFID flexible tags for food logistic. Sens Actuators B: Chem 127(1):2–7

Akyildiz IF, Su W, Sankarasubramaniam Y, Cayirci E (2002) Wireless sensor networks: a survey. Comput Netw 38(4):393–422

Alabaster C (2012) Pulse doppler radar. The Institution of Engineering and Technology

Asad M, Sheikhi MH (2016a) Highly sensitive wireless H_2S gas sensors at room temperature based on CuO-SWCNT hybrid nanomaterials. Sens Actuators B: Chem 231:474–483

Asad M, Sheikhi MH (2016b) Highly sensitive wireless H2S gas sensors at room temperature based on CuO-SWCNT hybrid nanomaterials. Sens Actuators B: Chem 1(231):474–483
Azzarelli JM, Mirica KA, Ravnsbæk JB, Swager TM (2014) Wireless gas detection with a smartphone viarf communication. Proc Natl Acad Sci 111(51):18162–18166
Barton DK, Cook CE, Hamilton PC (eds) (1991) Radar evaluation handbook. Artech House on Demand
Bhoir S, Goregaonkar S, Satre S (2017) IoT based gas detection system. Int J Eng Sci 10251
Blackman SS (1986) Multiple-target tracking with radar applications. Dedham, MA, Artech House, Inc., 463 pp
Bor-Yaliniz RI, El-Keyi A, Yanikomeroglu H (2016) Efficient 3-D placement of an aerial base station in next generation cellular networks. In: 2016 IEEE international conference on communications (ICC), 22 May 2016. IEEE, pp 1–5
Butler MA (1991) Fiber optic sensor for hydrogen concentrations near the explosive limit. J Electrochem Soc 138(9):L46–L47
Campbell JB, Wynne RH (2011) Introduction to remote sensing. Guilford Press
Capone S, Forleo A, Francioso L, Rella R, Siciliano P, Spadavecchia J, Presicce DS, Taurino AM (2003) Solid state gas sensors: state of the art and future activities. J Optoelectron Adv Mater 5:1335–1348
Chavali M, Lin TH, Wu RJ, Luk HN, Hung SL (2008) Active 433 MHz-W UHF RF-powered chip integrated with a nanocomposite m-MWCNT/polypyrrole sensor for wireless monitoring of volatile anesthetic agentsevoflurane. Sens Actuators A 141(1):109–119
Chen C, Tsow F, Campbell KD, Iglesias R, Forzani E, Tao N (2013) A wireless hybrid chemical sensor for detection of environmental volatile organic compounds. IEEE Sens J 13(5):1748–1755
Choi BI, Lee B, Song MK, inventors; NanoLambdaInc, assignee (2016) Nano-optic filter array based sensor. United States patent US 9,395,473, 19 Jul 2016
Curlander JC, McDonough RN (1991) Synthetic aperture radar. Wiley, New York, NY, USA
Curry GR (2012) Radar essentials: a concise handbook for radar design and performance. The Institution of Engineering and Technology
Dai J, Zhu L, Wang G, Xiang F, Qin Y, Wang M, Yang M (2017) Optical fiber grating hydrogen sensors: a review. Sensors 17(3):577
DeCuir J (2014) Introducing Bluetooth smart: Part 1: a look at both classic and new technologies. IEEE Consum Electron Mag 3(1):12–18
DeHennis A, Getzlaff S, Grice D, Mailand M (2016) An NFC-enabled CMOS IC for a wireless fully implantable glucose sensor. IEEE J Biomed Health Inform 20(1):18–28
Doni A, Murthy C, Kurian MZ (2018) Survey on multi sensor based air and water quality monitoring using IoT. Indian J Sci Res 17(2):147–153
Edde B (1993) Radar-principles, technology, applications. NASA STI/Recon technical report A, 93
Electrical apparatus for the detection of combustible gases in domestic premises. Test methods and performance requirements, EN 50194:2000 (2000)
Farina A (1986) Studer FA. Radar data processing. Volume 2-Advanced topics and applications. NASA STI/Recon Technical Report A. 1986;86
Ferrer-Vidal A, Rida A, Basat S, Yang L, Tentzeris MM (2006) Integration of sensors and RFID's on ultra-low-cost paper-based substrates for wireless sensor networks applications. In: 2006 2nd IEEE workshop on wireless mesh networks, WiMesh 2006. IEEE, pp 126–128
Formisano F, Massera E, De Vito S, Buonanno A, Di Francia G, Veneri PD (2015) Tinynose, an auxiliary smart gas sensor for RFID tag in vegetables ripening monitoring during refrigerated cargo transport. In: Sensors. Springer, Cham, pp 217–221
Geisheimer JL, Marshall WS, Greneker E (2001) A continuous-wave (CW) radar for gait analysis. In: 2001 conference record of the thirty-fifth Asilomar conference on signals, systems and computers, 4 Nov 2001, vol 1. IEEE, pp 834–838
Giust F, Cominardi L, Bernardos CJ (2015) Distributed mobility management for future 5G networks: overview and analysis of existing approaches. IEEE Commun Mag 53(1):142–149

Gupta A, Gangopadhyay S, Gangopadhyay K, Bhattacharya S (2016) Palladium-functionalized nanostructured platforms for enhanced hydrogen sensing. Nanomater Nanotechnol 1(6):40
Gupta A, Jha RK (2015) A survey of 5G network: architecture and emerging technologies. IEEE Access 3:1206–1232
Gupta A, Pandey SS, Bhattacharya S (2013) High aspect ZnO nanostructures based hydrogen sensing. In: AIP conference proceedings, 3 Jun 2013, vol 1536, no 1, pp 291–292
Gupta A, Pandey SS, Nayak M, Maity A, Majumder SB, Bhattacharya S (2014a) Hydrogen sensing based on nanoporous silica-embedded ultra dense ZnO nano bundles. RSC Adv 4 (15):7476–7482
Gupta A, Srivastava A, Mathai CJ, Gangopadhyay K, Gangopadhyay S, Bhattacharya S (2014b) Nano porous palladium sensor for sensitive and rapid detection of hydrogen. Sens Lett 12 (8):1279–1285
Haghi M, Thurow K, Stoll R (2017) Wearable devices in medical internet of things: scientific research and commercially available devices. Healthc Inform Res 23(1):4–15
Hansen RC (2009) Phased array antennas. Wiley
Headrick JM, Skolnik MI (1974) Over-the-horizon radar in the HF band. Proc IEEE 62(6):664–673
Ho CK, Itamura MT, Kelley M, Hughes RC (2001) Review of chemical sensors for in-situ monitoring of volatile contaminants. Sandia Report
IEEE Standard Letter Designations for Radar Frequency Bands, IEEE Std 521-2002, The Institute of Electrical and Electronic Engineers, New York (2003)
Jackson MC (1986) The geometry of bistatic radar systems. In: IEE proceedings F (communications, radar and signal processing), 1 Dec 1986, vol. 133, no 7. IET Digital Library, pp 604–612
Jeffrey T (2009) Phased-array radar design: application of radar fundamentals. The Institution of Engineering and Technology
Johnson DH (2006) Signal-to-noise ratio. Scholarpedia 1(12):2088
Karmaoui M, Leonardi SG, Tobaldi DM, Donato N, Pullar RC, Seabra MP, Labrincha JA, Neri G (2015) Novel nanosynthesis of In_2O_3 and its application as a resistive gas sensor for sevoflurane anesthetic. J Mater Chem B 3(3):399–407
Kim SY, Kim J, Cheong WH, Lee IJ, Lee H, Im HG, Kong H, Bae BS, Park JU (2017) Alcohol Gas sensors capable of wireless detection using In2O3/Pt nanoparticles and Ag nanowires. Sens Actuators B: Chem
Kim SY, Kim J, Cheong WH, Lee IJ, Lee H, Im HG, Kong H, Bae BS, Park JU (2018) Alcohol gas sensors capable of wireless detection using In_2O_3/Pt nanoparticles and Ag nanowires. Sens Actuators B: Chem 15(259):825–832
Krikorian KV, Rosen RA, inventors; Raytheon Co, assignee (2007) Technique for compensation of transmit leakage in radar receiver. United States patent US 7,202,812,. 10 Apr 2007
Krishna KL, Madhuri J, Anuradha K (2016) A ZigBee based energy efficient environmental monitoring alerting and controlling system. In: 2016 international conference on information communication and embedded systems (ICICES). IEEE, pp 1–7
Kurzekar R, Arora H, Shrestha R (2017) Embedded hardware prototype for gas detection and monitoring system in Android mobile platform. In: 2017 IEEE international symposium on nanoelectronic and information systems (iNIS). IEEE, pp 6–10
Lee JS, Oh J, Jun J, Jang J (2015) Wireless hydrogen smart sensor based on Pt/graphene-immobilized radio-frequency identification tag. ACS Nano 9(8):7783–7790
Li Z, Yang MH, Dai JX, Wang GP, Huang CJ, Tang JG, Hu WB, Song H, Huang PC (2015a) Optical fiber hydrogen sensor based on evaporated Pt/WO_3 film. Sens Actuators B: Chem 206:564–569
Li Z, Yang MH, Dai JX, Wang GP, Huang CJ, Tang JG, Hu WB, Song H, Huang PC (2015b) Optical fiber hydrogen sensor based on evaporated Pt/WO_3 film. Sens Actuators B: Chem 206:564–569

Liu Y, Chen YP, Song H, Zhang G (2012a) Modeling analysis and experimental study on the optical fiber hydrogen sensor based on Pd-Y alloy thin film. Rev Sci Instrum 83(7):075001

Liu X, Cheng S, Liu H, Hu S, Zhang D, Ning H (2012b) A survey on gas sensing technology. Sensors 12(7):9635–9665

Lopez-Ruiz N, López-Torres J, Rodríguez MÁC, de Vargas-Sansalvador IP, Martínez-Olmos A (2015) Wearable system for monitoring of oxygen concentration in breath based on optical sensor. IEEE Sens J 15(7):4039–4045

Lorwongtragool P, Sowade E, Watthanawisuth N, Baumann RR, Kerdcharoen T (2014) A novel wearable electronic nose for healthcare based on flexible printed chemical sensor array. Sensors 14(10):19700–19712

Luo S, Fu G, Chen H, Liu Z, Hong Q (2007) Gas-sensing properties and complex impedance analysis of Ce-added WO_3 nanoparticles to VOC gases. Solid-state Electron 51(6):913–919

Ma GM, Wu Z, Zhou HY, Jiang J, Chen WX, Zheng SS, Li CR, Li X, Wang ZB (2016) A wireless and passive online temperature monitoring system for GIS based on surface-acoustic-wave sensor. IEEE Trans Power Delivery 31(3):1270–1280

Meharry DE, Hedges SA, inventors; BAE Systems Information, Electronic Systems Integration Inc, assignee (2009) Duplexer for simultaneous transmit and receive radar systems. United States patent US 7,633,435, 15 Dec 2009

Mirzaei A, Leonardi SG, Neri G (2016) Detection of hazardous volatile organic compounds (VOCs) by metal oxide nanostructures-based gas sensors: a review. Ceram Int 42(14):15119–15141

Mitra SK, Kaiser JF (1993) Handbook for digital signal processing. Wiley

Moon SE, Choi NJ, Lee HK, Lee J, Yang WS (2013) Semiconductor-type MEMS gas sensor for real-time environmental monitoring applications. ETRI J 35(4):617–624

Nayak M, Singh D, Singh H, Kant R, Gupta A, Pandey SS, Mandal S, Ramanathan G, Bhattacharya S (2013) Integrated sorting, concentration and real time PCR based detection system for sensitive detection of microorganisms. Sci Report 20(3):3266

Oikonomou P, Botsialas A, Olziersky A, Kazas I, Stratakos I, Katsikas S, Dimas D, Mermikli K, Sotiropoulos G, Goustouridis D, Raptis I (2016) A wireless sensing system for monitoring the workplace environment of an industrial installation. Sens Actuators B: Chem 224:266–274

Ojha T, Misra S, Raghuwanshi NS (2015) Wireless sensor networks for agriculture: the state-of-the-art in practice and future challenges. Comput Electron Agric 1(118):66–84

Parks AN, Sample AP, Zhao Y, Smith JR (2013) A wireless sensing platform utilizing ambient RF energy. In: IEEE topical conference on power amplifiers for wireless and radio applications (PAWR 2013), 20 Jan 2013. IEEE, pp 160–162

Pati S (2015) Gas sensors: an introduction. Int J Comput Math Sci 4(9)

Probert Jones JR (1962) The radar equation in meteorology. Q J Roy Meteorol Soc 88(378):485–495

Pummakarnchana O, Tripathi N, Dutta J (2005) Air pollution monitoring and GIS modeling: a new use of nanotechnology based solid state gas sensors. Sci Technol Adv Mater 6(3–4):251–255

Quddious A, Yang S, Khan MM, Tahir FA, Shamim A, Salama KN, Cheema HM (2016) Disposable, paper-based, inkjet-printed humidity and H2S gas sensor for passive sensing applications. Sensors 16(12):2073

Rappaport TS (1996) Wireless communications: principles and practice. Prentice Hall PTR, New Jersey

Roseline RA, Devapriya M, Sumathi P (2013) Pollution monitoring using sensors and wireless sensor networks: a survey. Int J Appl Innov Eng Manag 2(7):119–124

Ruck GT, Barrick DE, Stuart WD, Krichbaum CK (1970) Radar cross section handbook. Volumes 1 & 2. Plenum Press, New York

Sandra PS, Sandeep CM, Nair V, Vindhuja MV, Nair SS, Raja MP (2017) WSN based industrial parameter monitoring using smartwatch. In: 2017 international conference on circuit, power and computing technologies (ICCPCT). IEEE, pp 1–6

Sarkar TK, Wang H, Park S, Adve R, Koh J, Kim K, Zhang Y, Wicks MC, Brown RD (2001) A deterministic least-squares approach to space-time adaptive processing (STAP). IEEE Trans Antennas Propag 49(1):91–103

Scheer JA, Holm WA (2010) Introduction and radar overview. In: Richards MA, Scheer JA, Holm WA (eds) Principles of modern radar. Scitech Publishing, Raleigh, NC

Sekiguchi K, Morinaga W, Sakamoto K, Tamura H, Yasui F, Mehrjouei M, Müller S, Möller D (2010) Degradation of VOC gases in liquid phase by photocatalysis at the bubble interface. Appl Catal B 97(1–2):190–197

Shabanov N, Gastellu-Etchegorry JP (2018) The stochastic Beer–Lambert–Bouguer law for discontinuous vegetation canopies. J Quant Spectrosc Radiat Trans 1(214):18–32

Sikarwar S, Yadav BC (2015) Opto-electronic humidity sensor: a review. Sens Actuators, A 233:54–70

Singh R, Singh E, Nalwa HS (2017) Inkjet printed nanomaterial based flexible radio frequency identification (RFID) tag sensors for the internet of nano things. RSC Adv 7(77):48597–48630

Skolnik MI. Radar handbook

Smith JR, Sample AP, Powledge PS, Roy S, Mamishev A (2006) A wirelessly-powered platform for sensing and computation. In: International conference on ubiquitous computing. Springer, Berlin, pp 495–506

Somov A, Baranov A, Spirjakin D (2014) A wireless sensor–actuator system for hazardous gases detection and control. Sens Actuators, A 210:157–164

Spirjakin D, Baranov A, Karelin A, Somov A (2015) Wireless multi-sensor gas platform for environmental monitoring. In: 2015 IEEE workshop on environmental, energy and structural monitoring systems (EESMS). IEEE, pp 232–237

Spirjakin D, Baranov A, Akbari S (2018) Wearable wireless sensor system with RF remote activation for gas monitoring applications. IEEE Sens J 18(7):2976–2982

Stetter JR, Penrose WR, Yao S (2003) Sensors, chemical sensors, electrochemical sensors, and ECS. J Electrochem Soc 150(2):S11–S16

Trim RM (1990) Mode S: an introduction and overview (secondary surveillance radar). Electron Commun Eng J 2(2):53–59

Trouillet A, Marin E, Veillas C (2006) Fibre gratings for hydrogen sensing. Meas Sci Technol 17 (5):1124

Vena A, Sydänheimo L, Tentzeris MM, Ukkonen L (2015) A fully inkjet-printed wireless and chipless sensor for CO2 and temperature detection. IEEE Sens J 15(1):89–99

Villatoro J, Diez A, Cruz JL, Andres MV (2001) Highly sensitive optical hydrogen sensor using circular Pd-coated singlemode tapered fibre. Electron Lett 37(16):1011–1012

Voles R, inventor; EMI Ltd, assignee (1989) Identification of friend or foe (IFF) systems. United States patent US 4,862,176, 29 Aug 1989

Vyas R, Lakafosis V, Lee H, Shaker G, Yang L, Orecchini G, Traille A, Tentzeris MM, Roselli L (2011) Inkjet printed, self powered, wireless sensors for environmental, gas, and authentication-based sensing. IEEE Sens J 11(12):3139–3152

Walker DR, Pasquero J, inventors; BlackBerry Ltd, assignee (2017) Mobile wireless communications device providing guide direction indicator for near field communication (NFC) initiation and related methods. United States patent US 9,706,036, 11 Jul 2017

Ward HR, inventor; Raytheon Co, assignee (1984) Radar processor. United States patent US 4,488,154, 11 Dec 1984

Wheeler A (2007) Commercial applications of wireless sensor networks using ZigBee. IEEE Commun Mag 45(4)

Wu T, Wu F, Redoute JM, Yuce MR (2017) An autonomous wireless body area network implementation towards IoT connected healthcare applications. IEEE Access 5:11413–11422

Xu G, Zhang Q, Lu Y, Liu L, Ji D, Li S, Liu Q (2017a) Passive and wireless near field communication tag sensors for biochemical sensing with smartphone. Sens Actuators B: Chem 246:748–755

Xu G, Zhang Q, Lu Y, Liu L, Ji D, Li S, Liu Q (2017b) Passive and wireless near field communication tag sensors for biochemical sensing with smartphone. Sens Actuators B: Chem 1(246):748–755

Yadav BC, Sikarwar S, Bhaduri A, Kumar P (2015) Synthesis, characterization and development of opto-electronic humidity sensor using copper oxide thin film. Synthesis 2(11)

Yang L, Zhang R, Staiculescu D, Wong CP, Tentzeris MM (2009) A novel conformal RFID-enabled module utilizing inkjet-printed antennas and carbon nanotubes for gas-detection applications. IEEE Antennas Wirel Propag Lett 8:653–656

Yang J, Zhou J, Lv Z, Wei W, Song H (2015a) A real-time monitoring system of industry carbon monoxide based on wireless sensor networks. Sensors 15(11):29535–29546

Yang J, Zhou J, Lv Z, Wei W, Song H (2015b) A real-time monitoring system of industry carbon monoxide based on wireless sensor networks. Sensors 15(11):29535–29546

Yu J, Huang MF, Jia Z, Wang T, Chang GK (2008) A novel scheme to generate single-sideband millimeter-wave signals by using low-frequency local oscillator signal. IEEE Photonics Technol Lett 20(7):478–480

Zhang YN, Peng H, Qian X, Zhang Y, An G, Zhao Y (2017) Recent advancements in optical fiber hydrogen sensors. Sens Actuators B: Chem 244:393–416

Zhu R, Desroches M, Yoon B, Swager TM (2017) Wireless oxygen sensors enabled by Fe (II)-polymer wrapped carbon nanotubes. ACS Sens 2(7):1044–1050

Chapter 10
Sensors in Assembly Shop in Automobile Manufacturing

Mohit Pandey, Shreyansh Tatiya, Shantanu Bhattacharya and Shailendra Singh

Abstract Assembly shop is considered as one of the most important parts of automobile manufacturing in which the frame of the vehicle is converted to the final product. The automobile manufacturers take good care of the overall assembly shop. The equipment, robots, and the developed process of the assembly shop cost more than the final product sometimes. Sensors are deployed to protect them from damage and assist them for the smooth processing. The assembly shop being autonomous maintains the high standard of the quality due to the proper integration of the sensor. This review focuses on the sensors used in the assembly shop, their type, and importance at each step of the shop.

Keywords Assembly shop · Sensors for assembly · Automobile manufacturing sensors · Sensors · Safety

---

M. Pandey (✉) · S. Tatiya · S. Bhattacharya
Microsystems Fabrication Laboratory, Department of Mechanical Engineering, Indian Institute of Technology Kanpur, Kanpur 208016, UP, India
e-mail: pmohit@iitk.ac.in

S. Tatiya
e-mail: tshrey@iitk.ac.in

S. Bhattacharya
e-mail: bhattacs@iitk.ac.in

M. Pandey · S. Tatiya · S. Bhattacharya · S. Singh
Design Programme, Indian Institute of Technology Kanpur, Kanpur 208016, UP, India

S. Singh
Maruti Suzuki India Limited, Gurugram 122015, Haryana, India

S. Bhattacharya et al. (eds.), *Sensors for Automotive and Aerospace Applications*, Energy, Environment, and Sustainability,
https://doi.org/10.1007/978-981-13-3290-6_10

10.1 Introduction

An assembly line is defined as a production process in which the manufacturing of a product is divided into various steps which are completed in a predefined sequence. In this process, the product travels from one workstation to other for different processes, and each workstation contributes to the final development of the product. The important point in this type of production system is the predefined sequence which on altering can fail the product. The assembly line is used for the mass production of the products and also results in the reduction of the labor cost and increment in the productivity (Groover 2002).

The introduction of the concept of assembly line by Henry Ford in the year 1908 drastically changed the perception of the manufacturing of goods. He introduced the concept of the assembly of interchangeable parts before which only a small number of very skilled artisans were able to produce the coaches and buggies. His idea not only decreases the need for the fully skilled person but also increases the production and the repairing of the products due to replaceable parts. The introduction and implementation of the assembly line driven by the conveyor belts reduce the production time for a complete vehicle (Domm et al. 2009). The complete vehicle was divided into 45 steps for the complete assembly (Arnold and Faurote 1915; Weber 2008). With the drastic change in the productivity of the vehicle manufacturing unit, the assembly line method was considered to be the best. And with the focus on further increasing the productivity, the functioning of various sub-manufacturing systems is fine-tuned to make the assembly line balance and optimized (Tahar and Adham 2011). Also, to increase the productivity and reduce the health risk, automation in the overall process was introduced. After which, the various parameters affecting the automation process were studied to optimize the overall process (Jahanzaib et al. 2013; Jimenez et al. 2013). The amount of automation and the need of human interventions were studied and hybrid system with the proper use of sensors was studied with the proper simulation of the relevant parameters like material handling, energy inputs, etc., for optimizing the different set of assembly line problems (Pfeiffer 2016; Seha et al. 2017). Thus, for the proper functioning of the assembly line with the hybrid system of human and automation, sensors play a very crucial role.

10.2 Application of Assembly Line in Automobile Industries

The current automobile industries consist of several components for the manufacturing of their products. Due to this, the assembly line plays a very important role in the mass scale production of automobile industries product. The car is manufactured by moving through various processes before coming on the road, and this production is carried out by the concept of the assembly line. This has drastically changed the manufacturing model of the automobile industries. Reducing the demand for skilled labor, increasing the efficiency, decreasing the cost, decreasing

Fig. 10.1 Work flow

the production time, etc., are some of the key points where the assembly line is benefitting the automobile industries (Groover and Kolchin 1997; IFM Electronics).

Automobile final assembly line consists of three major sections: (1) Body shop, (2) paint shop, and (3) assembly shop. In the body shop, the main components of the vehicles are prepared through various processes like casting, sheet metal work, pressing, welding, etc. Later, these main components are sent to the paint shop where all the components are painted according to the need of the specific components. After the paint shop, the components are sent to assembly shop for the final assembly of the vehicle. Figure 10.1 shows the workflow for the overall vehicle manufacturing.

The overall assembly shop is further divided into four different types of lines:

(1) Trim Line
(2) Chassis Line I
(3) Chassis Line II
(4) Final Line

In the final assembly shop, the role of sensors is very crucial. All of the subassemblies are integrated with other systems and finally assembled in the vehicle in this assembly. The number of rejects by the inspection after the assembly shop can be reduced by the proper implementation of sensors in the assembly line.

10.2.1 Trim Line

It is the first section of the assembly shop in which the vehicle arrives from the paint shop. In this assembly line, many subassemblies like brakes, console box, cabin lamp, door assemblies, etc., are prepared. The trim line starts with the disassembling of the door from the vehicle which is then transported to the door module assembly. Here, the functionality in the door is provided by integrating all the electrical equipment. This door assembly is protected, and the movement is restricted in the defined area. Meanwhile, the vehicle proceeds further for the installation of the dashboard. The dashboard is removed from the lot and is installed in the vehicle. After this, the vehicle is positioned for the windscreen assembly which is done with the help of suction grippers. After these assemblies, the vehicle is moved to the Chassis Line.

In this assembly line, most of the work is completely automated, and thus, the precision level required is very high. The sensors are integrated with the overall control system to ensure the smooth flow of the manufacturing and the proper assemblies. The sensors which are used in the trim line are as follows.

10.2.1.1 Code Reader

The main function of the code reader is to read the coded tags and to relate the different subassemblies of a vehicle. Many subassemblies are processed in the parallel lines and are integrated later. Thus, to avoid any confusion, 1D/2D tags are used. This sensor works on the principle of different absorption of light by different colors. The white color absorbs the least amount of light and thus reflects most of the incident light while black, on the other hand, absorbs most of the light and reflects very little of the incident light on it. When the tag, with some pattern with black and white color, is exposed to light, the reflected light will also have the intensity pattern and thus can be easily decoded. In this sensor, light source and receiver are on the same side and works when the tag is in the vicinity (Product Catalog by Docooler). Figure 10.2 shows a 1D/2D code reader.

10.2.1.2 Inductive Sensor

The main function of the inductive sensor is to detect and monitor the position of the various components. This sensor is used when the position of the component fixture is very important, or the speed monitoring is required. This sensor is based on the change in the inductance value of a PCB coil. AC current is passed in the coil and due to which some inductance value is recorded. When some other object comes in the vicinity of the coil, the inductance value of the coil changes which is deduced to the presence of some object in the proximity. Also, it can be used to find out the position and movement by computing the change in the inductance and its rate. Figure 10.3 shows an inductive sensor working (Product Catalog by Baumer).

10.2.1.3 Capacitive Sensor

These sensors are used in the trim with multiple applications like a functional check of the electric equipment, part detection, position detection, etc. Having capacitive sensors in the assembly line gives us edge for its precise position and part detection.

Fig. 10.2 1D/2D code reader

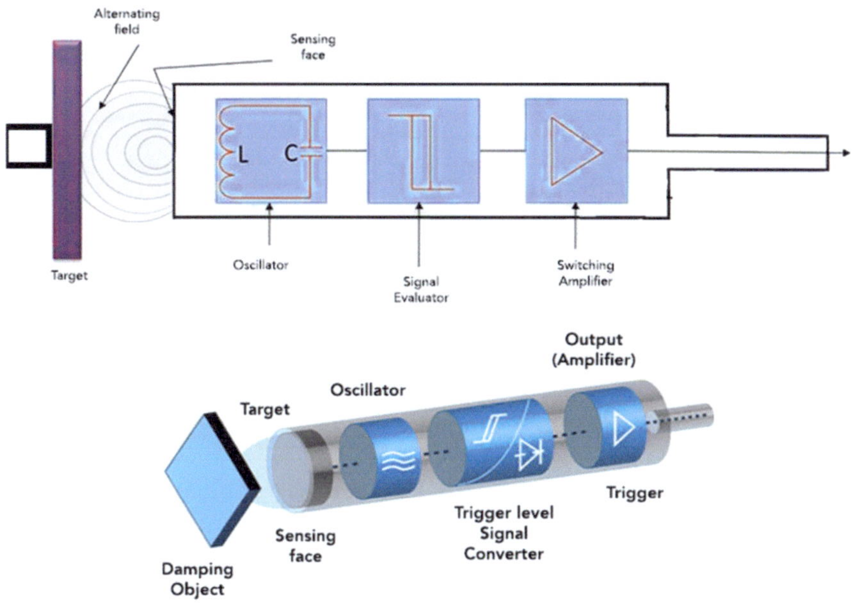

Fig. 10.3 Inductive sensor working

The working principle of all the capacitive sensors is the change in the capacitance value. It uses an alternative voltage which causes the electric field to change its direction. This change in direction also changes the charge that is collected due to the presence of an electric field. The capacitance value determines this amount of charge flow. Thus, bigger and closer the object is, greater will be the conductance and vice versa. Due to which it can be used for position as well as part detection. Figure 10.4 shows the working a capacitive sensor and its parts (Automation-insights blog).

10.2.1.4 Photoelectric Sensor

These sensors find their applications in the determination of exact position, for any approach check, for distance measurement, etc. It is based on the principle of reflection of light. It contains a source of light and a receiver to receive the reflected light. The position of the receiver depends upon the application of this sensor. Thus, based on applications, the three modes of operation of the photoelectric sensor are present (Keyence Corporation and automation-insights.blog).

(i) Reflective model—Both the light emitting and receiving elements are on the same side. In this case, the sensor senses the light reflected back from the target. Figure 10.5 shows the working of reflective model of photoelectric sensor.
(ii) Thru Beam Model—The light emitting and receiving elements are facing each other. In the presence of any object, the light received by the receiver is

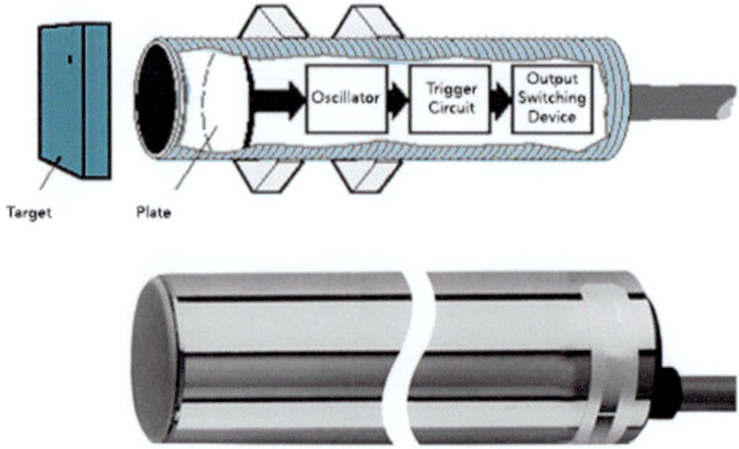

Fig. 10.4 Capacitive sensor and its parts (Automation-insights)

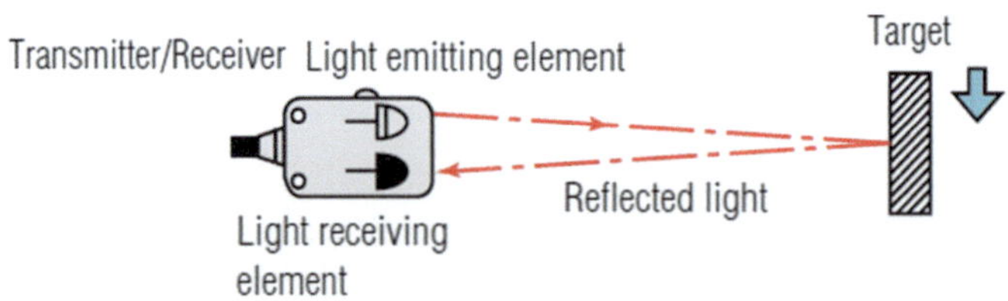

Fig. 10.5 Reflective model of photoelectric sensor (Reproduced from product catalog with permission from Keyence.)

interrupted. Figure 10.6 shows the working of thru beam model of photoelectric sensor.

(iii) Retroreflective Model—The light emitting and receiving elements are on the same side, and the light is reflected back by the reflector. There is no reflection when any object comes in between, and thus, the presence is detected. Figure 10.7 shows the working of retroreflective model of photoelectric sensor.

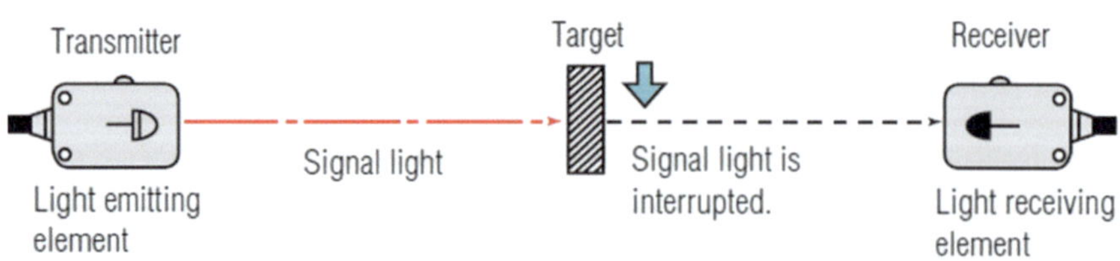

Fig. 10.6 Thru beam model of photoelectric sensor (Reproduced from product catalog with permission from Keyence.)

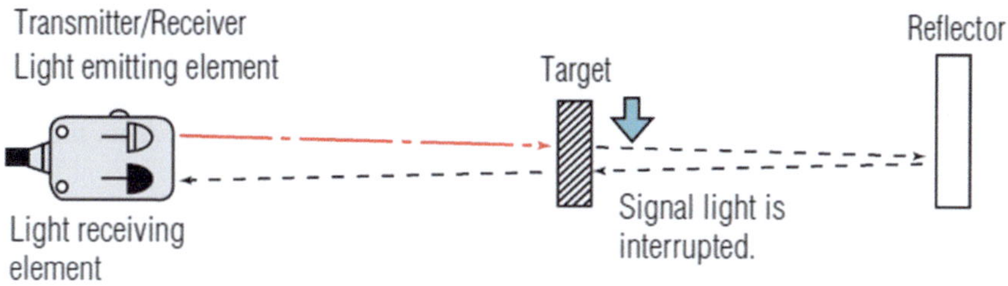

Fig. 10.7 Retroreflective model of photoelectric sensor (Reproduced from product catalog with permission from Keyence.)

10.2.1.5 Pressure Sensor

The pressure sensor finds its application for the monitoring of the suction pressure of the gripper. Many components are transferred from one place to other through gripper. Also, for the installation of the windscreen, the suction gripper is of great importance. The pressure sensing can be obtained by various principles such as a change in inductance, capacitance, etc., but generally, the piezoelectric effect is used. In the presence of pressure, the piezo-active material generates electricity which is proportional to the pressure. Thus, the pressure can be deduced by capturing electricity. Figure 10.8 shows below shows a pressure sensor in which the output is measured in electric signals (Keyence Corporation and IFM electronics).

10.2.2 Chassis Line I

After the trim line is completed the vehicle moved to the Chassis Line. In this

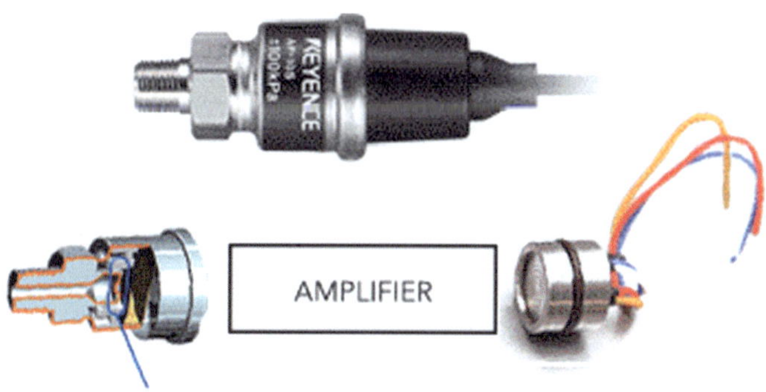

Fig. 10.8 Pressure sensor (Reproduced from product catalog with permission from Keyence.)

assembly line, the vehicle is integrated mainly with suspensions, silencer, fuel tank, etc. The quality assessment and monitoring of the suspension take place and only after passing, the suspension is assembled with proper precision. Meanwhile, for the exhaust system, the pipe for the desired length is cut by the automatic saw and processed to assemble with the silencer and exhaust in the vehicle. Also, the fuel tank assembly is integrated with the vehicle, after which, the vehicle is sent to the next section of the chassis line.

In this assembly line, suspension, fuel tank, exhaust, etc., are integrated with proper monitoring and feedback from various sensors. The sensors help in maintaining the smooth workflow and ensure the precision for the quality benchmark. Following sensors are used in the chassis line I.

10.2.2.1 Vision Sensor

This sensor is used in the chassis line for exactly locating the brake disc. Black and white sensor is used since there is not much color. The image captured by the camera is processed by the monochrome model to find the exact location. The light and dark color are separated, and the shape is formed in the processor. When the camera is fixed, the exact location of the component can be monitored for smooth functioning. The difference between the vision sensor and other sensor is the availability of multipoint inspection. Figure 10.9 shows a vision sensor (Keyence Corporation and IFM electronics).

Fig. 10.9 Vision sensor (Reproduced from product catalog with permission from Keyence.)

10.2.2.2 Capacitive Sensor

In the chassis line I, the use of the capacitive sensor is to monitor the release of the work step. The data is provided as the feed to the control system. The central control system then functions according to the state of the work step. Here, the change in the capacitance is observed when an object comes in contact.

10.2.2.3 Laser Sensor

The laser sensor consists of a laser source and a receiver that is used in a various application in the chassis line. It is used as the restricted curtain for some area in which if the laser is interrupted by some object, then the process is suddenly stopped to avoid any damage. Also, in the chassis line for the measurement of the length, the laser sensor finds its place very well. It helps the automatic saw to find the right point of working for having the exact length of the exhaust pipe. Figure 10.10 shows a laser sensor application in the safety curtain (Automation-insights.blog and IFM electronics).

10.2.3 Chassis Line II

After the integration of the suspension and fuel tank, the most important sub-assembly, the drive train is integrated with the vehicle. In this assembly line,

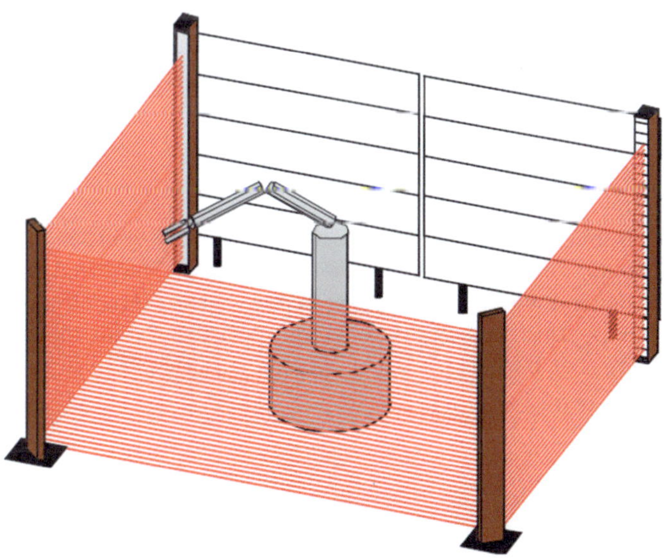

Fig. 10.10 Laser sensor application in safety curtain

engine, drive axil, disc, drum, and other responsible subassemblies for the motion of the vehicle are integrated with the vehicle. Due to this reason, the integration is also known as marriage. Apart from drivetrain, radiator, condenser, coolant pipes, blower connections, etc., are also integrated into this assembly line. First, the drivetrain is installed and after that radiator, condenser, etc., are connected. After this assembly line, the vehicle is moved to the last section of the assembly shop known and final line.

The drivetrain assembly and other subassembled components are fixed with a different type of fasteners with the vehicle and are completely monitored by the sensors. The various types of sensors involved in chassis line II are as follows.

10.2.3.1 Inductive Sensor

The inductive sensors in the chassis line II are deployed for the position detection and monitoring. When the drive train is integrated with the vehicle, the relative position plays a very crucial role in the assembly. The working principle remains the same, and thus, the change in inductance value is recorded for locating the components and/or fixtures for them.

10.2.3.2 RFID TAG Sensor

The full form of RFID is radio frequency identification and hence, RFID monitoring is completely based on radio waves detection. As the name suggests, it is used in the chassis line II for the identification of different components/subassemblies and their vehicles. The working of this sensor is based on the detection of radio waves and extracting information from them. The RFID tag contains a memory chip and antenna through which it sends the stored data of the memory to the reader. Figure 10.11 shows the functioning of the RFID tag.

10.2.3.3 Laser Sensor

The laser sensor in chassis line II finds its application in the collision protection and restricted curtain. The transporting autonomous arms are integrated with the laser sensor, and the data for any approaching component is sent as feedback to their motion planning. This decreases their velocity and thus avoids the collision and damage of any component.

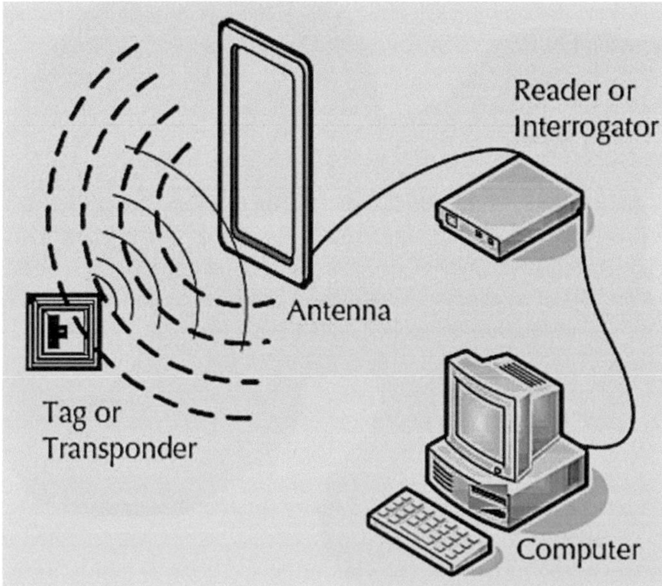

Fig. 10.11 RFID tag sensor working

10.2.4 Final Line

As the name suggests, it is the final section of the assembly shop after which the vehicle is ready to drive. Testing is included in the sequence after the final assembly of the vehicle. In this section wheel assembly, seats, door assembly, fluid filling, etc., are done. Once the vehicle is ready with integrated drivetrain in it, wheels are brought to the assembly station and are mounted with the vehicle. From the bunch of seats, specific seats related to the specific vehicles are detected through RFID and are brought to the station with suction grippers. The seats are assembled in the vehicle after which the door assembly is integrated into the vehicle. The vehicle is moved to the door station in which specific doors for the vehicle are detected and integrated. After the doors are integrated, the vehicle is moved for filling of various fluids like brake oil, coolant, etc. The bonnet is checked for its open or closed state with the help of sensors, and if it is open, then only the vehicle is sent to the fluid filling station where the automatic machines fill the required amount of fluids in different parts. After this, left out accessories are installed in the vehicle and vehicle is sent to the final testing section.

As in this section, all the left out subassemblies are integrated into the vehicle and the vehicle is tested for different weather and road conditions, and the sensors role becomes very important. Different types of the sensor with specific applications in the final assembly line are as follows.

10.2.4.1 Vision Sensor

The vision sensor in the final line is used for the pattern recognition of the wheel disc. The vision sensor does the mounting of the wheel to the wheel disc. Once the pattern of the disc and the wheel matches the set pattern, the automatic screwing arm fastens the wheel assembly. In this case, the image captured by the camera is processed by the monochrome model, and the whole pattern location is detected and monitored.

10.2.4.2 Inductive Sensors

In the final line, due to the integration of many small components to the vehicle, the part and its location detection become very crucial. The change in the inductance value is noted for the motion of vehicle during the door assembly integration and others. It is then processed for finding the parts and location of the parts.

10.2.4.3 Photoelectric Sensor

In this assembly line, the use of photoelectric sensor is present in various work. State recognition of the bonnet, marking the presence of the vehicle, etc., is done by photoelectric sensor. Although, the principle for detection remains the same, different intensities of light do it.

10.2.4.4 RFID Sensor

The RFID tag is used for the seat identification and is very helpful during the installation of the seat assembly in the vehicle. The principle and the way of application remain the same as mentioned in the chassis line II sections.

10.2.4.5 1D/2D Code Reader

The doors that were marked with 1D/2D tags are integrated with the vehicle in the final assembly line. Thus, the identification of the door is made by 1D/2D code reader. The working of the code reader is mentioned in the trim line section.

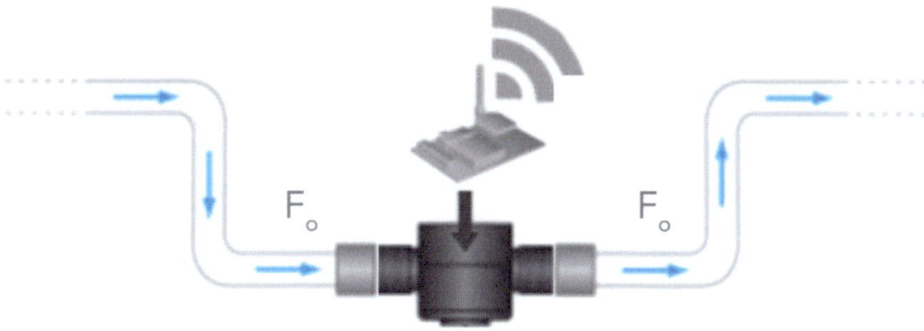

Fig. 10.12 Flow sensor

10.2.4.6 Pressure Sensor

The pressure sensor is used to monitor the suction pressure of the vacuum suction gripper. This gripper is used for the positioning and assembly of the seat in the vehicle. The functioning of the pressure sensor is already discussed in the trim line section.

10.2.4.7 Flow Sensor

This sensor is used to monitor the flow of the fluids during the filling of the coolant and oil fluid. This sensor gives the flow rate of the fluid flowing. We can calculate the amount of fluid flow with the help of the cross-sectional area and the data received from this sensor. Various methods are adopted to find the flow of any fluid. Four major types of flow sensors are available that are mechanical flow meters, vortex flow meter, ultrasonic flow meter, and magnetic flow meter. Mechanical flow sensor gives the flow by monitoring the speed of water flowing through the pipe that causes a piston or turbine to rotate. The vortex flow meter use vortices shed from the sensor immersed in the flow. The ultrasonic flow meter uses the sound wave to find the flow while the magnetic flow meter uses the magnetic field to measure the volumetric flow. Figure 10.12 shows a commercial flow sensor (Product catalog by Omega, product catalog by Zoro and IFM electronics).

10.3 Potential Failures in Assembly Shop Due to Sensor Failure

We have seen that the sensors play a very crucial role in the streamlining the workflow. Proper maintenance of all the sensors should be there otherwise there will be a completely different picture. The faulty 1D/2D code reader can wrongly identify the seats while inductive and capacitive sensors can give the wrong location of any component due to the malfunctioning of the sensors. This might lead to the halt of the whole conveyor belt and can stop the workflow. Any misleading photoelectric sensor reading can misguide about the exact position of the part. Also, it may not warn about any approach, and thus there might be collision which can damage the vehicle while in the production phase. The faulty pressure sensor could decrease the gripping power of the vacuum suction gripper which might drop the component in between and can break the fragile item and damage it. Without the proper functioning of the vision sensor, we would not be able to locate the brake disc, and laser sensor malfunctioning may create an entry for the restriction zone too. This can seriously damage the automation equipment and vehicles and is very dangerous for human life too. Thus, it is very clear that the use of sensors is inevitable for the smooth production assembly line.

10.4 Conclusion

The assembly shop is a very important process in the vehicle manufacturing. Proper vehicle manufacturing can achieve by the use of sensors properly. The assembly line is described with its origin. After this, the usage of the assembly line in automobile industries is explained. The different sections of the assembly shop are discussed with the sensor applied. The requirement of the sensor is a must and in the absence of which there will be many problems. Thus, the automobile industries improved drastically due to the increase in the use of the sensors and without the sensors, the productivity will be low and problems will be high.

References

Arnold H, Faurote F (1915) Ford methods and the ford shops. Adamant Media Corporation, Elibron Classics Series

Domm (2009) Michigan yesterday and today

Groover MP (2002) Automation, production systems and CIM

Groover MP, Kolchin MG (1997) Case study: automobile final assembly plant

Introductory guide to sensor basics by Keyence Corporation. Keyence Corporation

Jimenez et al (2013) Automation and parameters optimization in production line: a case of study. Int J Advanc Manufact Technol

Jahanzaib M et al (2013) Performance analysis of process parameters effecting the automated assembly system. Life Sci J

Processes in the Automotive Industry, IFM electronics

Product Catalog by Baumer, Baumer Technology

Product Catalog by Deal Extreme, Deal Extreme

Product Catalog by Docooler, Docooler

Product Catalog by Omega, Omega

Product Catalog by Zoro, Zoro

Pfeiffer S (2016) Robots, industry 4.0 and humans, or why assembly work is more than routine work. Soc J MDPI

Seha S et al (2017) Design and simulation of integration system between automated material handling system and manufacturing layout in the automotive assembly line. ICMER, IOP
Tahar RB, Adham AAJ (2011) An efficient assembly line balancing in automobile manufacturing. In: IEEE International summer conference of Asia pacific business innovation and technology management
Weber A (2008) How the model T was assembled. Assembly Magazine
https://automation-insights.blog

Chapter 11
Leakage Monitoring in Inflatable Space Antennas: A Perspective to Sensitive Detection of Helium and Nitrogen Gases

Pankaj Singh Chauhan, Geeta Bhatt and Shantanu Bhattacharya

Abstract Inflatable space structures have become an important part of space explorations due to their lightweight, simpler design, low cost, and fewer parts. These structures include antennas, solar arrays, solar concentrators, reflectors, etc. These structures are made of flexible polymers which can be folded and easily carried with spacecraft due to their small volume and weight. Structures, when reaching their destination, are inflated through internal pressurization to achieve desired structural integrity. In space, these structures are subjected to very harsh environment such as high radiation levels, structural vibrations, and micrometeoroid bombardments. The polymeric material used to fabricate these structures is susceptible to degradation under these harsh conditions. These structures are prone to lose their structural integrity over long-term degradation of the material. The most common problem associated with inflated space antennas is leakage of inflated gas. Hence, the health monitoring of these structures becomes crucial to avoid structural failure due to leakages which may cause loss of information, accuracy, and money. Gas sensors are used to detect leakages in these structures. A mixture of helium (He) and nitrogen (N_2) is used as inflating gas in space antennas. Helium is the lightest gas after hydrogen and has chemically inert, nonflammable nature which makes it an ideal inflating gas. However, the detection of He leakages is very difficult because of its nonreactive behavior with chemical species. Metal oxide based semiconducting (MOSs) materials are widely used sensing element for detection of various gases. Although it is very difficult to find out He gas leakages, vanadium pentoxide (V_2O_5) can detect even a small concentration of He through resistive changes. In this chapter, we will discuss the requirement of leakage

P. S. Chauhan (✉) · G. Bhatt · S. Bhattacharya
Microsystems Fabrication Laboratory, Department of Mechanical Engineering, Indian Institute of Technology Kanpur, Kanpur 208016, India
e-mail: pankajs@iitk.ac.in

G. Bhatt
e-mail: geetabht@iitk.ac.in

S. Bhattacharya
e-mail: hattacs@iitk.ac.in

S. Bhattacharya et al. (eds.), *Sensors for Automotive and Aerospace Applications*, Energy, Environment, and Sustainability,
https://doi.org/10.1007/978-981-13-3290-6_11

monitoring system for inflatable space antennas and He gas sensing properties of V_2O_5 semiconducting material.

Keywords Inflatable space structures · Gas sensors · Helium detection
Metal oxide semiconductors

11.1 Introduction

The large structures which are manufactured on Earth keeping gravity pull in mind come out to be heavyweight massive structures for a desired strength, but when we try to look after the application aspect of these structures in space, they fail miserably as the conditions there are very different, so the structures restrict the applicability rather than supporting and meeting the specified purpose (Thomas 1992). Hence keeping this factor in mind and also from the standpoint of launching capabilities of available launch vehicles, the concept of inflatable's was coined by NASA. The very first device of such kind, Echo I (passive communication inflatable satellite) was launched by US, which could be traced in its performance aspects. But after this, Echo II (another inflatable satellite), PAGEOS (Earth metric measurement satellite) and explorer satellites were launched, which were proved very successful. Keeping in mind the performance aspect, i.e., they should be able to keep their shapes even after exposing the pressure for elongated time, the initial inflatable based devices (Echo II and explorer satellites) were fabricated with aluminium and plastic films which were proved effective. Hence as the research on inflatable structures proved itself to be potentially relevant, the refinement aspects continued.

The inflatable structures technology has attained a lot of attention since last decade due to their unbeatable potential in aircraft applications (Cassapakis and Thomas 1995). Their advantages go way long to being lighter, occupying smaller volumes while packaging and being lesser expensive than their competitive mechanical parts. The typical inflatables offer approximately 50% reduction in the weight as compared to competitive mechanical parts for the comparable strength. Generally, flexible polymers are deployed to fabricate these inflatable structures and the employed fabrication material is expected to hold its structural integrity under internal pressurization for elongated time. NASA has also performed Inflatable Antenna Experiment (IAE) to access the application aspects of large inflatable antennas (Freeland et al. 1996). A 14 m-diameter reflector antenna was successfully deployed abroad the Space Shuttle, STS-77 and it was dedicatedly concluded that large antennas are of enormous use in the field of Earth remote sensing, space-based radar systems, and mobile communications. The inflatables have come up as very dedicated solution in these cases to perform the sensing operation due to the inherent advantages associated with inflatables.

In the progression, microsystems based health monitoring and control system (HMCS) has been fabricated for space inflatables (Tung et al. 2001). This HMCS

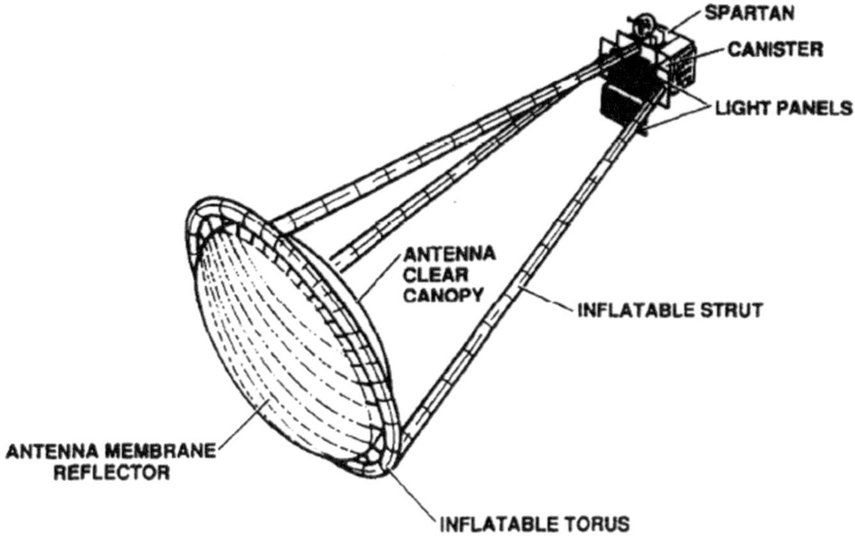

Fig. 11.1 Schematic of inflatable antenna with different components (Reprinted with permission from Freeland et al. (1995). Copyright (1995) Acta Astronautica.)

sensor (3-D multichip) is fabricated with flexible polymeric material, with an assembly of sensor layer at top, polyimide substrate in middle and actuator at the bottom, which can be directly fixed to the outer periphery of the inflatables for monitoring health. An interconnected architecture of MEMS-based sensors is used to formulate sensor layer which provides atmospheric conditions near the inflatables and also the structural vibration of the inflatable, while the electroactive polymers are used for formulating actuator layer to provide 2-D shape control capability to the sensor.

Figure 11.1 shows the schematic of inflatable antenna with nomenclature of different components. The sensor is attached at an important location in truss attached to the antenna, which helps monitor local environmental and structural conditions through the complete assembly. When sensor layer senses the strain or vibration through the correspondingly attached sensors, the actuator layer will take the corresponding preliminary actions to remove it, like it will deform surface contour of inflatables to remove the present wrinkles and correspondingly will reduce the generated vibrations. While exploring from low earth orbit (LEO) to geostationary earth orbit (GEO) or next, the conventional sized communication systems can just not be implemented in the existing form. In the respective field, Cube satellite (CubeSat) technology, which is an alternate to the conventional satellites for variety of applications like space research, astronomy, weather forecasting, and communications has also found application of inflatables relevant to it. As the CubeSat technology itself aims at lightweight, small and concise satellite fabrication in an affordable manner, inflatables play an important role to this also. Additionally, the antennas which are generally employed for CubeSat are patch or dipole antennas with limited gain, the inflatable antennas being able to be packaged efficiently, they provide large dimension of dish as well as correspondingly higher gain after being deployed. An S-band antenna based on inflatable antenna concept was designed and developed by MIT for CubeSat (Babuscia et al. 2014). A reflector incorporated antenna was programmed to operate at 2.45 GHz, having a gain of 10 dB for increasing data communication rate.

Hence working in space needs various types of sensors to sense the environmental conditions which can enhance the safety as well as operational characteristics of the device (Hunter et al. 1998). Additionally, for the inflatable structures, the generated pressure is of utmost importance for operation also and hence it becomes very important to sense the pressure of gases (or the stress generated through this) used for the operations. The basic example which quotes the need for sensor is in 1990 when space shuttle leakage delayed the fleet and it was delayed until the leakage source was identified. This detection of low concentration hydrogen at lower temperature was done through mass spectrometer connected through an array of sampling tubes. But the process was time consuming as well as it was not that efficient to measure the exact location of leakage, so the need of pinpoint hydrogen sensors to sense in lesser time emerged. And while designing the sensor also, there should be some conditions which should be fulfilled like it should be efficient enough to sense the low concentrations of the chemical species and the corresponding pressure (through the lower exposure limits) as well as the sensor should be able to survive in 100% gas exposure as well as should maintain its calibration. The sensors are also required to sense the emitted gases from aircraft engines, which demands for sensors to be able to operate in high-temperature conditions. For enhancing the efficiency of fire safety devices also these days, the sensors are being incorporated which can sense the chemical constituents of fire to help get a reliable alarm. Hence, these issues gave rise to the need for developing sensors with higher efficiency (so that a lesser number of sensors can give the desired results as well as lesser operational power is required) in lesser cost for different applications. The most common gases which are required to be sensed in various perspectives are hydrogen (Hunter 1992), oxygen (Takeuchi 1988), nitrogen oxides (Gregory 2007), hydrocarbons (Grosshandler 1995), and helium (Chauhan and Bhattacharya 2018). Silicon has always been a very significant sensor fabrication platform to sensor variety of gases, using microelectromechanicalsystem (MEMS) based fabrication techniques. As silicon substrate based sensor can work up to 600 °C, for high temperature applications, silicon carbide (SiC) based platforms are devised for application in aircrafts and automobiles (Maboudian et al. 2013) and several other modifications are also proposed. The method of detection of these chemical species, mainly gases, includes detection of their concentration through looking at response signal at different concentrations.

So this chapter aims at the detailed discussion of the application of inflatables in aircrafts, mainly in antennas; requirement of gas sensors and their monitoring and other detection aspects. It also touches on various aspects of detection like substrate requirements, measurement conditions, measurement aspects (like concentration/

pressure), etc. The main emphasis is on the detection of helium and nitrogen gas mixture, which are generally used in the aircrafts.

11.2 Sensor System for Inflatable Space Antenna

After a successful deployment of inflatable space antenna, its health monitoring is also an important task to ensure proper functioning. The health monitoring system monitors structural stresses, shape and geometry, internal pressure, and gas leakages. To monitor and control these parameters, various types of sensors and actuators are added with the inflatable structure. Lightweight and low-volume restrictions in space applications encourage to utilize microelectromechanical (MEMS) sensors which are lightweight and compact (Wang and Qian 2014). MEMS sensors are fabricated by using photolithography technique with chemical etching on silicon or any lightweight metal.

11.2.1 Design Considerations

The important issue that one has to consider before designing a sensor system for inflatable space structures is the extreme environmental condition at space. The near earth orbit where the sensor is expected to operate is 300–2000 km above the surface of the earth. The radiation level in these orbitals is very high with temperature ranging from −50 to 100 °C and pressure ranging from 10^{-6} to 10^{-10} mbar (Grossman and Gouzman 2003). The presence of atomic oxygen is also a cause of concern for the sensor designer. Hence, there are two important concerns for a sensor operating under these conditions: (1) the erosion of polymeric material due to the impact of atomic oxygen species, and (2) high radiation levels which can damage the MEMS sensor and its electronic circuit.

11.2.2 Sensor Layer Substrate

Polyimide films are ideal material for space applications due to a wide range of thermal stability and chemically stable nature. The processing and thin film deposition of polyimide is also simple. Kapton, a polyimide material is commonly used as flexible polymeric material for inflatable space structures. To avoid environmental erosion is space, the thickness of polyimide can be increased. The other method to avoid erosion is by passivating the exposed layer with thin film coating of metals such as Mg alloy and SiO_2 (Liu et al. 2017; Qi et al. 2017). Similarly to avoid radiation damage to MEMS electronics, shielding of aluminum is provided (Urban et al. 2017).

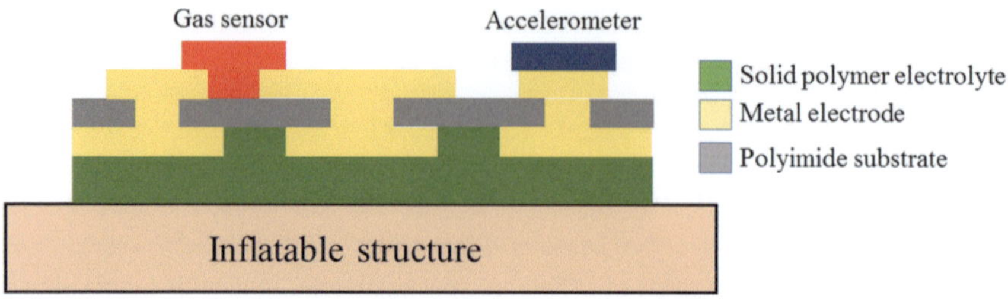

Fig. 11.2 Schematic diagram of MEMS-based sensing system for inflatable structures

Figure 11.2 shows the schematic diagram of MEMS-based monitoring system consisting of gas sensor and accelerometer for monitoring gas leakages and structural vibration of the inflatable structure. The monitoring system consists of three layers; the first one is sensor layer at the top, the second one is polyimide substrate layer in the middle, and the third one is actuator layer of solid polymer electrolyte.

Metallic coating is also required over polyimide film for electrode connections. The metal coating of copper (Cu) can be done by sputtering up to a thickness of 1 μm. The other method is by laminating the polyimide film with Cu layer. Further, the patterning of Cu electrodes is done by using photolithography technique. The patterning is required for making electrical connections. Flexible polyimide film needs stiff support during photolithography process which is done by using Si wafer as a temporary support. A schematic flow chart describing various fabrication steps associated with sensor formation on flexible substrate is shown in Fig. 11.3.

11.3 Metal Oxide Semiconducting Materials for Gas Sensors

Oxides of transition and non-transition metals show semiconducting behaviour and have been proved to be efficient materials for gas sensing application (Huang and Wan 2009; Wang et al. 2010). These materials can be used as sensing element for reducing, oxidizing and combustible gases. Some examples of these metal oxide based semiconductor materials are: TiO_2, Fe_2O_3, Cr_2O_3, ZnO, V_2O_5, WO_3, CuO, Mn_2O_3, CeO_2, etc. (Chauhan and Rai 2017). Transition metals (Fe, Cr, V, etc.) are those materials which possess multiple oxidation states while non-transition metals (Zn, Sn, etc.) show only one oxidation state (Fortunato et al. 2012). These type of semiconducting materials show change in conductivity by undergoing chemical reaction with reducing/oxidizing gases (Chauhan and Bhattacharya 2017). Transition metals undergo a change in oxidation state which plays a critical role in gas sensing mechanism. Hence, transition metals oxide prove to be better material in comparison with non-transition metal oxides.

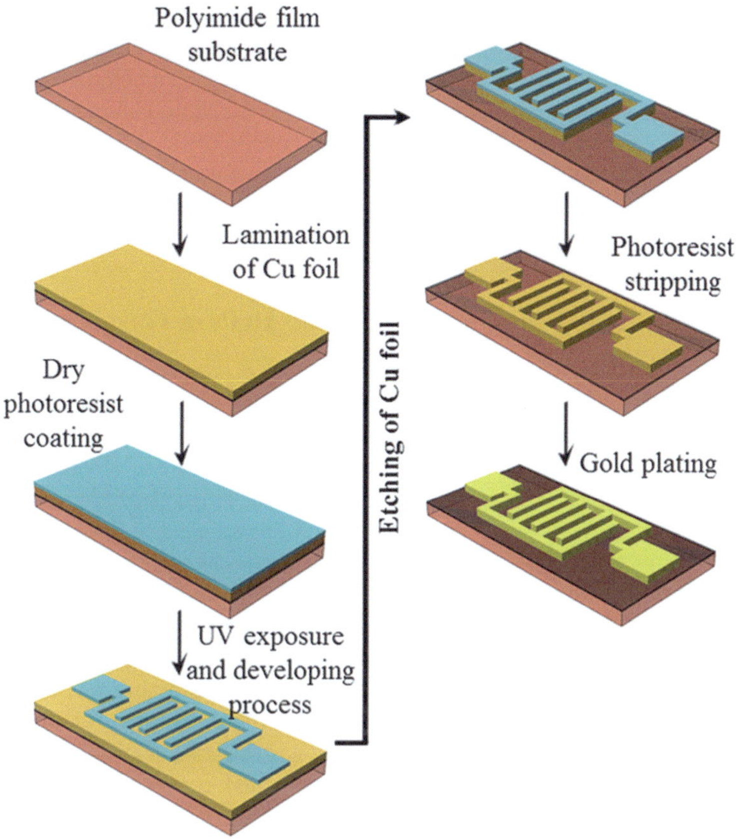

Fig. 11.3 Interdigitated gold electrodes formation process flow chart on polyimide film substrate. (Reprinted with permission from Haniffet al. 2015. Copyright (2015) Nature Scientific Reports.)

11.4 Considerations for Gas Sensor Development

Due to easy production, lower cost, and compact size, metal oxide gas sensors have been used predominantly as solid-state gas sensing devices. There are two distinguishing characteristics of metallic nanostructures. First, the Debye screening length is compatible with their lateral dimensions and second, long nanorods can be fabricated easily to be used as long semiconductor channels. As the size of the particle goes down to double thickness of the space-charge layer, there is a remarkable improvement in sensitivity and specificity of the sensor. This property is known as "size effect". The basic mechanism of sensing by metal oxide nanostructures relies on the change in conductivity due to the trapping of electrons by the adsorbed molecules. When oxygen gas is passed through the metallic nanostructures, the oxygen atoms get adsorbed onto their surface. These nanostructures also capture electrons from the inner of the metallic nanostructures. These entrapped negative-charged electrons form a depletion layer which results in reduced conductivity. When these oxygen doped nanostructures are exposed to the reducing gas, the entrapped electrons return back to the metallic nanostructures leading to a decrement in the potential barrier and an increase in the conductivity. The gas sensing metallic nanostructure devices mainly consist of three parts: the sensing film, a pair of electrodes, and a heater. Metallic nanostructures, when exposed to target gases, reflect a change in the resistance which is measured by the electrodes.

11.5 Vanadium Pentoxide (V_2O_5) for Helium Gas Detection

There are few reports of helium detection by using V_2O_5 nanostructures in the past. Yu et al. (2005) fabricated V_2O_5 nanowires by polycondensation of vanadic acid in water. The detection device was fabricated by making Ti/Au electrodes over SiO/Si substrate. The nanowires were deposited by well-known electrophoresis technique. V_2O_5 nanowires were well aligned across the electrodes as can be seen in Fig. 11.4.

The helium gas was detected by placing the device in a vacuum chamber and measuring the current, flowing through electrodes and nanowires after filling the chamber with helium gas. The current response of the device was measured as the function of pressure of He gas inside the chamber. The chamber was filled evacuated with helium gas, cyclically to obtain a continuous response and recovery curve shown in Fig. 11.5. The evacuation of He gas was done by using a leak valve. It can be observed from the response curve that the conductance of the V_2O_5 nanowires increases in presence of He gas.

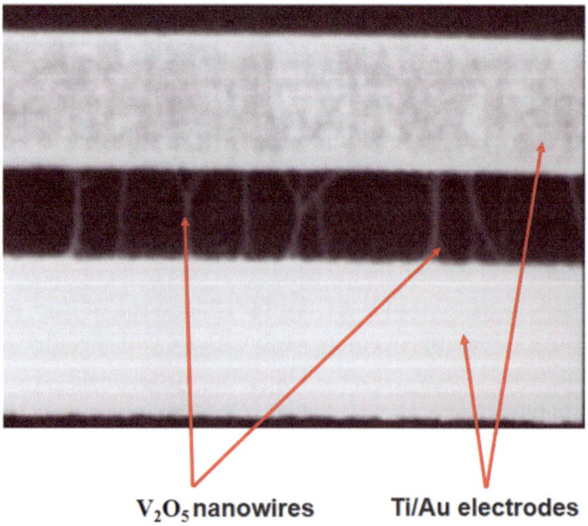

Fig. 11.4 SEM image of V_2O_5 nanowires deposited over metallic electrodes (Reprinted with permission from Yu et al. (2005). Copyright (2005) AIP publishing, Applied Physics Letters.)

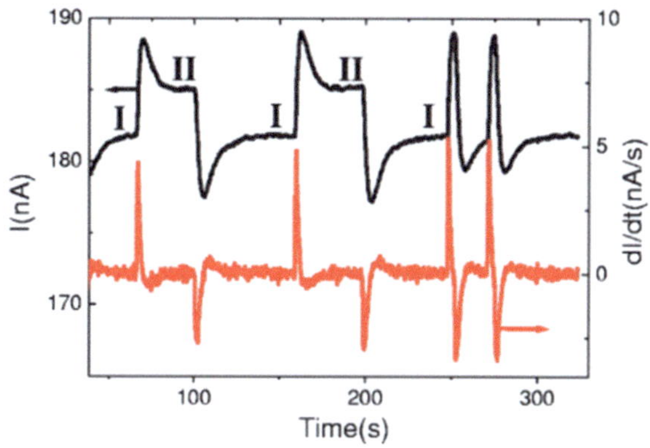

Fig. 11.5 Conductance/current responses to the repeated changing of the pressure from 1000 to 10^{-3} mbar. Where I is a stable state when the chamber pressure is 1000 mbar (filled with He), and II is stable when 10^{-3} mbar (He removed) (Reprinted with permission from Yu et al. (2005). Copyright (2005) AIP publishing, Applied Physics Letters.)

Chauhan and Bhattacharya reported the detection of helium gas by using hydrated layered structures of vanadium pentoxide ($V_2O_5 \cdot 1.6H_2O$) (Chauhan and Bhattacharya 2018). Quick response and recovery were achieved by using hydrated vanadium pentoxide nanowires and nanostars. The $V_2O_5 \cdot 1.6H_2O$ nanostructure was synthesized by oxidation of vanadyl sulfate $V_2O_5 \cdot 1.6H_2O$ precursor in presence of strong oxidizing agent $KBrO_3$ (Chauhan and Bhattacharya 2018). Further, the synthesized nanostructures were drop casted over interdigitated copper (Cu) electrodes. Cu electrodes were fabricated by using photolithography technique over a printed circuit board (PCB). In this method, the Cu coating of PCB board was selectively etched after masking to obtain the desired interdigitated shape. The

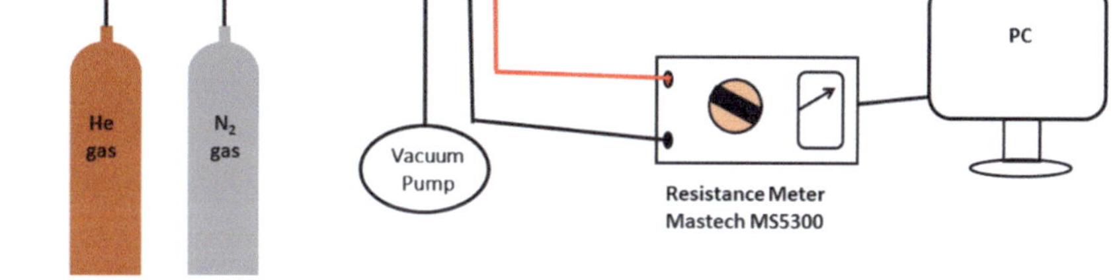

Fig. 11.6 Schematic diagram of the experimental setup (Reprinted with permission from Chauhan and Bhattacharya (2018). Copyright (2018) Elsevier, Materials Letters.)

test was conducted inside a vacuum chamber which was cyclically filled and evacuated with He gas by using mass flow controllers (MFCs) and vacuum pump, respectively. MFCs also help in maintaining the desired concentration level of He gas inside the test chamber. The entire setup is shown schematically in Fig. 11.6.

Figure 11.7 shows the morphological analysis of the synthesized nanostructures through finite element scanning electron microscope (FESEM), transmission electron microscope (TEM), and high-resolution transmission electron microscope (HRTEM) analysis. FESEM images evident the full-grown nanowire and nanostar type structures. The diameter of nanowires was about 30 nm, while nanostars have irregular dendritic type shape with rough surface which can be observed in TEM images. HRTEM images show fringes of the nanostructures with inter-planar distance of 0.19 nm, 0.22 nm, and 0.36 nm corresponding to (006), (005) and (110) plane, respectively. Dynamic response of the sensing material in the presence of different concentrations of He gas is shown in Fig. 11.8. The response of the sensor was calculated by percentage resistive change in presence of He gas. It can be observed that the response of the sensor increases with increase in gas concentration. The response and recovery time is very low (<10 s) for both shape. The response was calculated by using the following formula: Response (%) = (Resistance in the air—Resistance in He gas) * 100/Resistance in the air. Nitrogen (N_2) gas can be used as carrier gas in place of air.

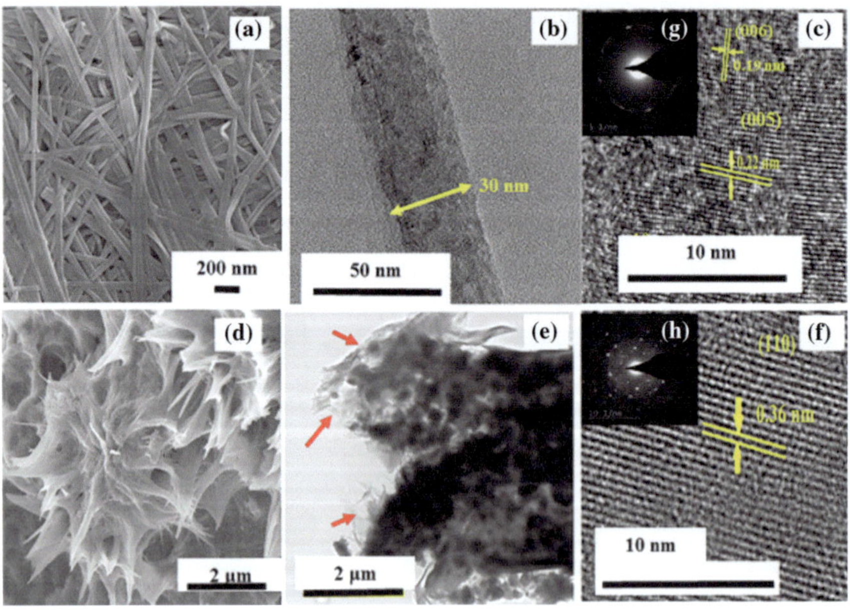

Fig. 11.7 **a** FESEM, **b** TEM, and **c** HRTEM image of $V_2O_5 \cdot 1.6H_2O$ nanowires. **d** FESEM, **e** TEM, and **f** HRTEM image of $V_2O_5 \cdot 1.6H_2O$ nanostars. The inset images show corresponding SAED pattern of nanostructures (Reprinted with permission from Chauhan and Bhattacharya (2018). Copyright (2018) Elsevier, Materials Letters.)

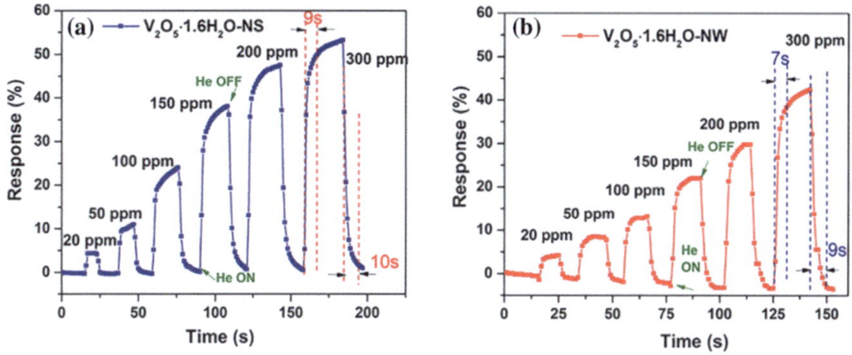

Fig. 11.8 Dynamic response of the sensor in presence of different He gas concentration (20–300 ppm) at room temperature for **a** $V_2O_5 \cdot 1.6H_2O$ nanostars, and **b** $V_2O_5 \cdot 1.6H_2O$ nanowires (Reprinted with permission from Chauhan and Bhattacharya (2018). Copyright (2018) Elsevier, Materials Letters.)

11.6 Sensing Mechanism

Helium gas molecules are very small in size with low atomic mass and can diffuse very fast through the porous structures. In abovementioned studies, He gas sensing mechanism was explained by polaron hopping model given by Mott (1968) and Muster et al. (2000). According to this model, the conductivity through vanadium oxide structures is governed by the hopping of polaron on various sites called hopping sites. The number of hopping sites depends upon defects created at the fabrication stage. In V_2O_5 nanostructures, vanadium ions are present in different oxidation states as defects and act as hopping sites for polaron. According to Mott, the conductivity of a material is inversely proportional to the hopping distance, i.e., smaller hopping distance facilitates faster charge transfer and higher conductivity.

Helium gas molecules get physically adsorb on the interlayers of V_2O_5 nanostructures and act as a hopping site for polaron. This phenomenon decreases the average hopping distance and increases in conductivity was observed. As the gas concentration was increased, number of He gas molecules get adsorb, which results in a large increase of conductance. As the conductivity increases in presence of He gas, the resistance decreases and the response was calculated as previously described manner.

11.7 Hydrogen Gas Detection Through V_2O_5

Hydrogen gas molecules are very small in size and diffuse rapidly through the nanostructures. Atmospheric oxygen gets adsorbed on the surface of nanostructures and captures free electrons, which creates a depletion region. This phenomenon causes the increase in resistance (Chauhan and Bhattacharya 2017). When the hydrogen gas is passed, it reacts with adsorbed oxygen species and generates water molecules. Hence, the captured electrons are again transferred back to the nanostructure surface which decreases the resistance. The variation in resistance is the indication of the presence of hydrogen gas.

11.8 High Precision Gas Sensor

The high precision in gas sensors can be obtained by using nondispersive infrared principle which provides better stability of the system. A second infrared source can be added which recalibrates the primary infrared source and redundancy check etc. This type of system was used by Frodl et al. for the detection of CO_2 gas (Frodl and Tille 2006). They found the accuracy more than ±5.5% for the whole temperature and pressure range. The resolution of gas detection was below 15 ppm and the response time achieved was less than 5 s. In this type of sensing system, the infrared source directs a wave of light towards the cuvette tube, which is detected by an IR light detector. When the light passes through the tube, any gas molecule with the same size as the wavelength of the IR light absorbs that wavelength while the other wavelengths are allowed to pass through. After that, an IR detector reads the amount of unabsorbed wavelength of light. The difference between the amount of light radiated from the IR source and the amount of light received back at the detector is measured. This difference is proportional to the number of gas molecules present in the tube.

11.9 Conclusion

Inflatable space structures are widely used in space explorations due to their compact design and lightweight structure. Extreme environmental conditions at near earth orbit may damage the structural integrity of these structures which are made of flexible polymeric material. A slight deviation from basic design may cause loss of data, poor performance, and damage to other systems. Hence, the health monitoring becomes an essential component to be integrated into the structure. Detection of gas leakages is one of the important parameters to be monitored. Various types of gas sensors are available to monitor the gas leakage. Space structures are inflated with helium gas due to its lightweight and chemically inert nature. Detecting the He ga leakage is very difficult by using conventional methods and materials. Vanadium pentoxide is such a material reported by researchers in the past to detect He gas molecules with small concentration through conductance change. Gas sensing properties of vanadium pentoxide films are attributed to the polaron charge transfer mechanism through hopping over the physisorbed He gas molecules.

References

Babuscia A, Van de Loo M, Wei QJ, et al (2014) Inflatable antenna for CubeSat: fabrication, deployment and results of experimental tests. In: IEEE aerospace conference. https://doi.org/10.1109/aero.2014.7024296

Cassapakis C, Thomas M (1995) Inflatable structures technology development overview. In: Space programs and technologies conference, p 3738

Chauhan PS, Bhattacharya S (2017) Vanadium pentoxide nanostructures for sensitive detection of hydrogen gas at room temperature. J Energy Environ Sustain 2:69–74

Chauhan PS, Bhattacharya S (2018) Highly sensitive $V_2O_5{\cdot}1\ 6H_2O$ nanostructures for sensing of helium gas at room temperature. Mater Lett 217:83–87. https://doi.org/10.1016/j.matlet.2018.01.056

Chauhan SP, Rai A (2017) Enhanced photocatalytic performance of vertically grown ZnO nanorods decorated with metals (Al, Ag, Au, and Au – Pd) for degradation of industrial dye. Mater Res Expr 4:055004

Fortunato E, Barquinha P, Martins R (2012) Oxide semiconductor thin-film transistors: a review of recent advances. Adv Mater 24:2945–2986. https://doi.org/10.1002/adma.201103228

Freeland RE, Bilyeu GD, Veal GR (1995) Validation of a unique concept for a low-cost, lightweight space-deployable antenna structure. Acta Astronaut 35:565–572. https://doi.org/10.1016/0094-5765(95)00018-U

Freeland R, Bard S, Veal G, et al (1996) Inflatable antenna technology with preliminary shuttle experiment results and potential applications. NASA Tech Paper, pp 2–7. https://doi.org/10.1109/131.9480

Frodl R, Tille T (2006) High precision NDIR CO_2 gas sensor for automotive applications. IEEE Sens J 6:1697–1705. https://doi.org/10.1109/JSEN.2006.884440

Gregory HL (2007) Measurement of trace atmospheric species by chemical ionization mass spectrometry: speciation of reactive nitrogen and future directions. Mass Spectr Rev 26:166–184. https://doi.org/10.1002/mas

Grosshandler WL (1995) A review of measurements and candidate signatures for early fire detection. Fire Res

Grossman E, Gouzman I (2003) Space environment effects on polymers in low earth orbit. Nucl Instruments Methods Phys Res Sect B Beam Interact with Mater Atoms 208:48–57. https://doi.org/10.1016/S0168-583X(03)00640-2

Huang J, Wan Q (2009) Gas sensors based on semiconducting metal oxide one-dimensional nanostructures. Sensors 9:9903–9924. https://doi.org/10.3390/s91209903

Hunter GW (1992) A survey and analysis of commercially available hydrogen sensors. NASA Tech, Memo, p 105878

Hunter GW, Chen LY, Neudeck PG et al (1998) Chemical gas sensors for aeronautic and space applications II. NASA Technical Paper

Liu K, Mu H, Shu M et al (2017) Improved adhesion between SnO_2/SiO_2coating and polyimide film and its applications to atomic oxygen protection. Colloids Surf A Physicochem Eng Asp 529:356–362. https://doi.org/10.1016/j.colsurfa.2017.06.024

Maboudian R, Carraro C, Senesky DG, Roper CS (2013) Advances in silicon carbide science and technology at the micro- and nanoscales. J Vac Sci Technol A Vacuum, Surfaces, Film. https://doi.org/10.1116/1.4807902

Mohammad Haniff MAS, Muhammad Hafiz S, Wahid KAA et al (2015) Piezoresistive effects in controllable defective HFTCVD graphene-based flexible pressure sensor. Sci Rep 5:14751. https://doi.org/10.1038/srep14751

Mott NF (1968) Conduction in glasses by—Mott. J Non Cryst Solids 1:1–17

Muster BJ, Kim GT, Krstic V et al (2000) Electrical transport through individual vanadium pentoxide nanowires. Adv Mater 12:420–424. https://doi.org/10.1002/(SICI)1521-4095(200003)12:6%3c420:AID-ADMA420%3e3.0.CO;2-7

Qi H, Qian Y, Xu J, Li M (2017) Studies on atomic oxygen erosion resistance of deposited Mg-alloy coating on Kapton. Corros Sci 124:56–62. https://doi.org/10.1016/j.corsci.2017.05.002

Takeuchi T (1988) Oxygen sensors. Sens Actuators. https://doi.org/10.1016/0250-6874(88)80058-1

Thomas M (1992) Inflatable space structures. IEEE Potentials 11:29–32. https://doi.org/10.1109/45.207143

Tung S, Witherspoon SR, Roe LA et al (2001) A MEMS-based flexible sensor and actuator system for space inflatable structures. Smart Mater Struct 10:1230–1239. https://doi.org/10.1088/0964-1726/10/6/312

Urban M, Nentvich O, Stehlikova V et al (2017) VZLUSAT-1: Nanosatellite with miniature lobster eye X-ray telescope and qualification of the radiation shielding composite for space application. Acta Astronaut 140:96–104. https://doi.org/10.1016/j.actaastro.2017.08.004

Wang JX, Qian XM (2014) Application and development of MEMS in the field of aerospace. Appl Mech Mater 643:72–76. https://doi.org/10.4028/www.scientific.net/AMM.643.72

Wang C, Yin L, Zhang L et al (2010) Metal oxide gas sensors: sensitivity and influencing factors. Sensors 10:2088–2106. https://doi.org/10.3390/s100302088

Yu HY, Kang BH, Pi UH et al (2005) V_2O_5 nanowire-based nanoelectronic devices for helium detection. Appl Phys Lett 86:1–3. https://doi.org/10.1063/1.1954894

Chapter 12
MEMS Sensors for Automotive Applications: A Review

Geeta Bhatt, Kapil Manoharan, Pankaj Singh Chauhan and Shantanu Bhattacharya

Abstract The automotive industry has always been an area, where the possibility to explore more and to achieve next-gen technology remains a challenge. As the market is growing, the possibilities are being explored on a wider scale. And these growing possibilities are increasingly demanding the research for the more accurate functioning of the automotive through the use of various sensors and the integration aspects associated with them. While enhancing the quality/performance of the vehicle, the need to reduce the size of sensors always remains a challenge. Over the period of time, the technology has advanced and these challenges are taken care through the use of MEMS (Microelectro mechanical systems) integrated systems. MEMS helps in devising microscale sensors with higher accuracy in small size and low cost. In the automotive, there exists a huge need for these installing these sensors and utilizing them to refine the performance characteristics of the vehicles. This chapter aims at discussing important MEMS-based sensors and their applications in the automotive-like tire pressure and monitoring systems, engine management system, vehicle stability/passenger safety system, and emission control.

Keywords MEMS · Engine management · Tire pressure and monitoring Vehicle stability · Passenger safety

G. Bhatt (✉) · K. Manoharan · P. S. Chauhan · S. Bhattacharya
Microsystems Fabrication Laboratory, Department of Mechanical Engineering, Indian Institute of Technology Kanpur, Kanpur 208016, UP, India
e-mail: geetabht@iitk.ac.in

K. Manoharan
e-mail: kapman@iitk.ac.in

P. S. Chauhan
e-mail: pankajs@iitk.ac.in

S. Bhattacharya
e-mail: bhattacs@iitk.ac.in

S. Bhattacharya
Design Programme, Indian Institute of Technology Kanpur, Kanpur 208016, UP, India

S. Bhattacharya et al. (eds.), *Sensors for Automotive and Aerospace Applications*, Energy, Environment, and Sustainability,
https://doi.org/10.1007/978-981-13-3290-6_12

12.1 Introduction

Automotive is the most fascinating and emerging area since centuries. The growing interest in the area brings up more responsibilities toward making the vehicles more refined with every new design/model that comes up in the market. This is a huge challenge from the manufacturer's point of view as the competition is growing exponentially and the customers are all open to the market for opting any new vehicle. These growing responsibilities are taken care by improving the performance, safety, and comfort aspects. While discussing the application of MEMS in automotive, MEMS allows for reduction of the size of the sensor in lower cost. It is accomplished by integrating several microsized components of semiconductor electronics/Integrated chip technology. The dedicated purposes are sought by including minor modifications or add-on parts to the general circuitry. Pressure sensing has been the first successful sensor device to be incorporated into industrial/automotive applications, which consists of one or more pressure-sensitive membranes to sense the pressure. These membranes can be fabricated through either bulk micromachining or surface micromachining and are used to measure the pressure of the range 10 kPa (vacuum pressure)–180 MPa. Depending upon the pressure conditions of the environment, various types of pressure sensors are used, like silicon diaphragm based piezoresistive micro-machined sensors (Schuster and Czarnocki 1997) (lesser flexibility sensors) for lower pressure and capacitive touch mode micro-machined silicon diaphragm sensors (Baney et al. 1997), which is more flexible (the pressure measurement is done through observing capacitive change while diaphragm moves to the base plate under the pressure) for higher pressure. And for other harsh automotive applications like power steering and brakes/suspension, capacitive ceramic module sensors (Sabetti et al. 1988) are used, while for diesel engine fuel pressure, piezoresistive polysilicon-on-steel sensors (Dunbar and Sager 2000) are used. MEMS-based pressure sensors are used for various applications in automotive, like manifold/barometric air pressure, oil pressure, in-cylinder pressure, fuel vapor pressure, exhaustive gas filters pressure, and side airbags pressure sensing with excellent reliability (Shaw et al. 2008) using mainly capacitive/piezoresistive detection schemes, where the measurement is done through a Wheatstone bridge principle. Figure 12.1 shows the basic architecture of the pressure sensor (Yulong et al. 2003), which comprises of membrane, SOI (Silicon on Insulator) strain gauge, and transmission rod in the complete housing.

Apart from pressure sensors, the other significant types of automotive sensors are speed sensors, position sensors, temperature sensors, and mass/air flow sensors. Speed sensors (Fleming 2008) measure crankshaft speed and angles for spark and fuel injection timing control. These sensors also control input/out transmission shaft speed for the gear shift and also detect misfire in the engine (which causes a modulation in crankshaft speed). Additionally, these sensors also provide input to automotive stability system and braking system. Magnetic MEMS can also be used for sensing speed (Treutler 2001). Figure 12.2 shows the principle of working of a wheel speed and incremental position sensor. A periodically varying magnetic field

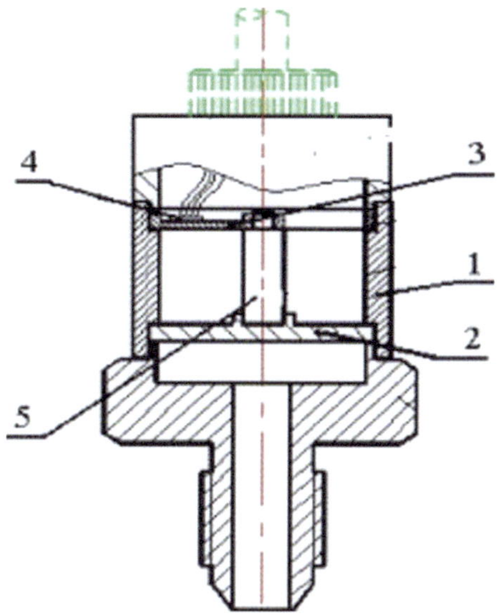

Fig. 12.1 Basic mechanical structure of a pressure sensor: (1) housing; (2) membrane; (3) cantilever beam; (4) SOI strain gauge; (5) transmission rod (Reprinted with permission from (Yulong et al. 2003) Copyright (2003) Elsevier.)

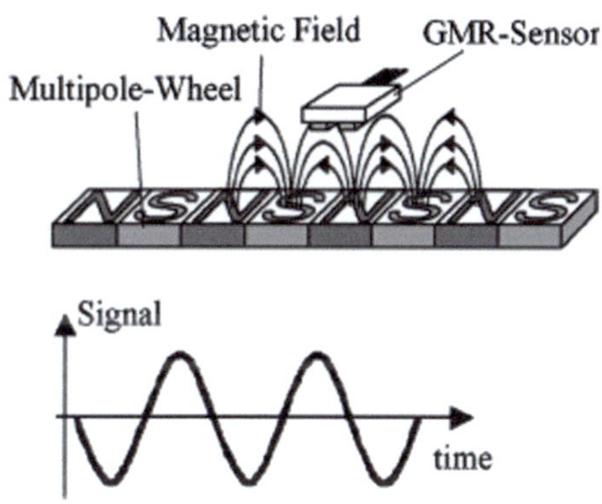

Fig. 12.2 Principle of wheel speed/incremental position sensor (Reprinted with permission from (Treutler 2001) Copyright (2001) Elsevier.)

is created through magnetic multipole wheel and the GMR (Giant magnetoresistance) sensor is used to detect the generated field, which is further converted to the corresponding wheel speed.

Various sensors which are used for speed sensing are variable reluctance sensors, which are also called inductive-type sensors (Pawlak et al. 1991) (electromagnetic devices), based on generating output voltage signal due to time-varying magnetic flux fluctuation through rotation of parts; Wieland effect sensors (Dlugos 1998) which consist of magnetic alloy wire with radial gradient magnetization varying from core to the periphery of wire and depends on interaction of applied magnetic field, hall sensors (Woerth et al. 1999) which produce voltage corresponding to the changing magnetic flux with respect to rotating parts and magnetoresistor sensors (Dietmayer 1999) producing change in resistance corresponding to the magnetic flux density.

Angular and linear position sensors measure the linear displacement/position of a specific component and the linear displacement up to 1 μm has been measured through MEMS-based sensors (Fleming 2001). The position sensors have their application in measuring steering wheel angle, chassis height link bar angle, float arm angle (to measure fuel level), and most importantly engine throttle plate angle (Fleming 2008). For the measurement of position, various sensors are used. Among them are potentiometric type of sensors (Frodl and Schoedlbauer 1998), which depend on the property that the resistance of the film varies proportionately with the length; hall voltage sensors (Woerth et al. 1999), which rely on the principle that the hall voltage varies as a function of sine of angle between flux density and bias current in the magnetic field; AMR anisotropic magnetoresistive sensors (Dietmayer 1999) which rely on the principle that the resistance of the sensor changes with respect to the rotating external magnetic field and optical encoder sensor (Madni and Wells 2000) which senses steering wheel angle.

Temperature sensors are the other significant categories of the automotive MEMS sensors, which are used in almost all the sensing devices to express the exact characteristics. Silicon-based ICs (Lacanette 1997) are used for temperature measurement up to 150 °C through resistance-based bulk charge conduction or voltage difference in PN junction; ceramic oxide thermistors (Carter 1996), which exhibit negative or positive temperature coefficient (temperature change is recorded as a function of resistance) are used for measuring temperature of the range −55 to 1000 °C and RTD (Resistive temperature detector)-based sensors (Kato et al. 1996), which possess platinum sensing film inside an alumina-ceramic structure are used for very high temperature sensing above 1000 °C.

Mass/air flow sensors are mounted in the bypass channel of the air intake to measure mass air flow in the automotive engine excluding the pulsating reversal of the air (Arai et al. 1990). If pulsating reversal of the air has to be included for sensing, upstream, and downstream thermal flow detection sensors are required to be installed (Konzelmann et al. 1995). Apart from these sensors, there are many more sensors which are incorporated in the automotive to accomplish different functions like exhaust gas sensors, engine knock sensors, linear acceleration sensors, angular rate, etc.

Depending upon the requirement to sense particular characteristics of the automotive, different sensors out of the above-mentioned sensors are mounted individually, or in combination. Out of the very broad spectrum of the automotive sensing, we are discussing here some major sensing-based systems which are important for the vehicles and these are automotive tire pressure monitoring sensors (TPMS), pressure and flow sensors for engine management, vehicle stability/ passenger safety sensors, and emission control sensors. This chapter briefly discusses the basic MEMS fabrication techniques and integration of MEMS schemes with the above-stated sensors-based systems. The basic requirements of the specific sensor system, its working aspects, and correspondingly integrated sensors scheme is discussed in the chapter.

12.2 MEMS Fabrication Techniques

There are various techniques for fabrication of MEMS-based features like photolithography, replication and molding process, microscale replication by double inversion (MRDI), soft lithography, LASER-based fabrication, and inkjet printing (Sundriyal and Bhattacharya 2017) out of which photolithography is the most prominent technique.

Photolithography is a MEMS fabrication technique, which aims at fabricating micron-sized features on the substrate through UV exposure (Bhatt et al. 2016). There are three main steps in the lithography, spin coating of the photoresist (it can be positive tone photoresist like S1813 or negative tone photoresist like SU8), UV exposure through a printed mask and developing. This process is additionally accompanied by prebaking (baking before the UV exposure) or/and post-baking (baking after the UV exposure) process before being developed. Figure 12.3 illustrates the photolithography process, which shows that in case of positive tone photoresist, the portion which is exposed to the UV light dissolves after the development process (developing chemicals are specific to the photoresist solution) while in case of negative tone photoresist, the portion which is not exposed to UV light melts away on development.

The photolithography process is further categorized as contact photolithography (where mask and photoresist coated substrate are in perfect contact), proximity photolithography (mask and substrate have a gap of 20–30 μm in between) and projection photolithography (the gap is of higher order). If we have to fabricate

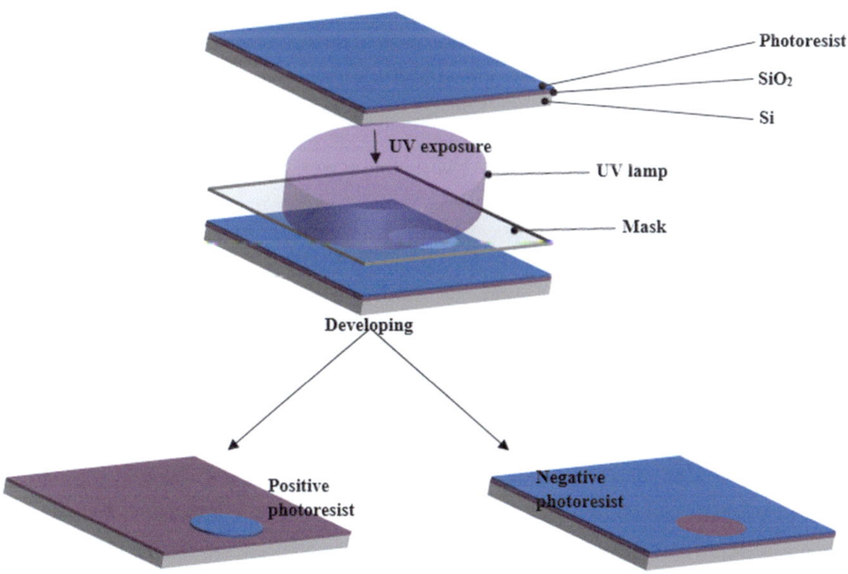

Fig. 12.3 Photolithography process

channels, then after the photolithography process is done, it is replicated with the PDMS (polydimethylsiloxane) (a soft polymer material) to get the inverted shape (channel) from the fabricated extruded feature, this process is called as replication and molding when the features fabricated through photolithography act as mold and the PDMS-based inversion is the making of replica (inverted one) of the mold. Several other features can also be developed using this replication and molding process. The researches have also reported micro-cantilever (Basu et al. 2016) and micro-valve fabrication through replication and molding using very thin wires as mold, instead of a patterned structure (Singh et al. 2014). The etching of the lithographed sample can also be done to get the micro-features on the substrate (Bhatt et al. 2017).

Further, if we need mass production of some features, going for lithography for each part would be expensive enough to handle. So to cut down the cost of expensive processes, other techniques are opted. One of them is MRDI, it is a low-cost process to fabricate intricate structures which are once machined/ fabricated through nanolithography (Chou 1996). It deals with making a master mold through stated processes and replicating it through PDMS and again replicating it to get back the shape of the master mold. This is the next step to the replication and molding process. The other micro, nanostructure fabrication process is soft lithography, it is a non-photolithographic process through replica molding/ self-assembly. The process which falls in this category is micro-contact printing (Whitesides et al. 2001), where PDMS stamp/PDMS slab with some already made features is used to write molecules on the substrate.

The other non-lithography micro-fabrication processes are a LASER (Light Amplification by Stimulated Emission of Radiation)-based machining, which promote fabrication by ablating through strong LASER radiation, which is capable of making trenches and crevices on the substrate. These fabricated structures are well utilized by various researchers for microfluidics (Hellman et al. 2007) and manipulation of biological species (Quinto-Su et al. 2008), which is overall termed as Lab-on-chip device fabrication. LASER-based fabrication/material removal is mainly suitable for substrates like polystyrene, polymethylmethacrylate, polyvinylchloride, polyimide, and many more polymers. LASER processing can also be used for mask making/mask writing (Kumar et al. 2013), which are used for photolithography. Hence, all these techniques being the basic MEMS fabrication techniques are utilized in direct or indirect form for fabricating various MEMS-based sensors for automotive. The specialized techniques are discussed in the specific sections.

12.3 Tire Pressure Monitoring Systems

Since the legal regulation of vehicles became a mandate in the USA (October 2005), the TPMS has emerged as a growing research aspect. There are two types of TPMS systems, one is direct TPMS and the other is indirect TPMS. The system which utilizes MEMS-based pressure sensors integrated with electronic circuits (called application-specific integrated circuits (ASIC)) to calculate pressure signal and transmitting the information through radio frequency (RF) are called direct TPMS. The TPMS system, which utilizes wheel speed sensor information to observe under-inflation of the tire is called indirect TPMS.

These days, a battery-based TPMS is commonly installed on the rim of a wheel, which comprises of a battery as a power source, a MEMS-based pressure sensor, low-power microcontroller for processing, and an ultrahigh frequency-based transmitter for data transmission. Temperature and pressure sensors perform the sensing operation through the microcontroller-based data, which controls the frequency of data collection and transmits through the transmitter to the electronic central unit (ECU). Figure 12.4 shows the various components of a battery-based TPMS (Löhndorf et al. 2007) and it represents how pressure sensor, temperature sensor, and voltage sensor (acceleration sensor is optional) are linked to the high-frequency transmitter and low-frequency receiver through analog-to-digital converter and microcontroller. Hence, the TPMS sensor comprises the diaphragm-type capacitive pressure sensor, MEMS acceleration sensor, temperature sensor, voltage sensor, and transmitter on a single substrate. The MEMS has made possible to achieve the overall size of the sensor of the order of 7.7 × 7.5 × 4.2 mm (Shaw 2002).

The various functional aspects of each component is the acceleration sensor detects the gravity-inspired minima–maxima output signals as a function of starting of wheel rotation to identify left and right tire location (Stewart et al. 2006) and simultaneously also activates power saver mode for battery when tire rotation stops. The pressure sensors measure the tire pressure and temperature sensor monitors temperature by using pressure from the pressure sensor through gas law for ideal gases and also helps in providing protection of electronic shutdown by detecting the over-temperature condition. The voltage sensor provides details of TPMS battery life. RF signals received from the front and rear wheels vary in terms of their

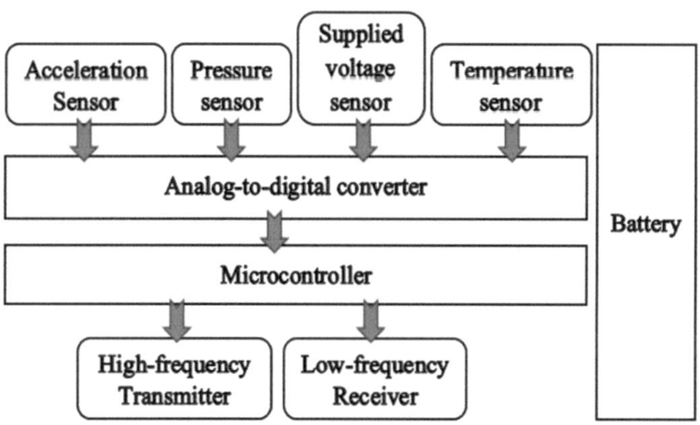

Fig. 12.4 Various components of a battery-based TPMS (Kraft and White 2013)

strength (as the distance between rear wheels is more than front wheels, the RF signals from rear wheels are of lower signal strength) and help to identify the rear/front wheel. The battery used in TPMS sensors is integrated lithium battery. Several localization techniques are also developed to determine the exact position of the tire, which further help in enhancing the efficiency of the TPMS systems (Lickfelt et al. 2013).

12.4 Pressure and Flow Sensors for Engine Management

MEMS-inspired sensor-based engine management was initiated in 1979 to decrease tailpipe emission (Eddy and Sparks 1998). This was done by utilizing a microprocessor-based engine control unit to control fixed air to fuel ratio (stoichiometric), which, in turn, can effectively reduce the amount of unburned fuel. This type of monitoring was accomplished through the incorporation of feedback sensors to sense air input and exhaust output. For this purpose, direct mass flow measurement and indirect air flow measurements were carried out using hot-wire mass air flow sensor and manifold absolute pressure sensor and electroceramic sensor at the outlet. Manifold absolute pressure estimates air density along with temperature sensor to get a good estimate of mass air flow into the engine. Silicon-based barometric absolute pressure sensors are used to calculate direct mass airflow for altitude compensation and closed-loop exhaust gas recirculation systems (Fleming 2001).

Pressure sensors are again used for the measurement of in-cylinder pressure in the engine combustion control system. The pressure-based feedback signal helps in regulating turbocharger operation, cylinder fuel injection timing, and exhaust gas recirculation. The application of the control system can be observed as reducing NOx emission by reducing combustion temperature via the control system. The piezoelectric ceramic sensor is also being used into diesel engine glow plug to monitor cylinder pressure in case of higher pressure. The increasing cylinder pressure generates pressure on the glow plug and heating plug acting as the pressure transfer pin applies pressure to the piezoelectric ceramic pressure sensor (Heinzelmann et al. 2006). Additionally, the mass/air flow sensors are utilized to sense the mass air flow in the automotive engine. This can be done in two ways, either by including the pulsating pressure reversal of the air or by excluding it. Measuring of mass air flow in the engine, excluding the pulsating reversal of air is detected by mounting airflow sensors in the bypass channel of the air intake (Arai et al. 1990). While in case if the pulsating reversal of air is required to be taken in account, upstream and upstream and downstream thermal flow detection sensors are installed (Konzelmann et al. 1995) and a combination of air flow sensor along with the set of thermal flow sensors detect the mass air flow.

12.5 Vehicle Stability/Passenger Safety Sensors

One of the most important features for selling a vehicle by a manufacturer is the importance given to passenger safety. Safety of the occupants in a crash event is now being effectively implemented with many recent advances in a number of technologies. Recent trends in various countries have shown a reduction in a number of deaths caused due to road accidents. It is a complex mixture of many factors and cannot be attributed to only improve the performance of vehicles or road conditions. One of the most important factors is an improvement in vehicle safety systems due to the developments in MEMS-based sensors. The safety system usually deploys the protection devices for frontal crash, side impact, and rollover. MEMS-based devices are used to sense, monitor, and detect vehicle behavior such as changes in accelerations and angular rates which are undesirable, positioning, and behavior of passenger and the surrounding conditions. Most of the vehicle manufacturers classify safety systems into the following domains:

- **Prevention Technologies**: These help the drivers to become better aware of the road conditions as well as minimize distractions and stresses for the drivers. For example, visibility is improved by radars and cameras.
- **Response Technologies**: These assist the driver when driving systems fail and things start to go awry. One such example is of anti-lock braking system (ABS) which improves the steerability and braking.
- **Protection Technologies**: These are designed to minimize the consequences due to accidents for the occupant passengers and other nearby road users. The examples include the restraint systems such as airbag systems which minimize the injury to passengers. The basic architecture of the protection system in vehicles is shown in Fig. 12.5.
- **Rescue Technologies**: These are adapted to help emergency services reach passengers in case of accidents.

Some of the important sensing systems associated with these technologies are car crash sensing system, rollover detection, and vehicle dynamic control system and these are discussed below.

Fig. 12.5 Protection system in vehicles (Kraft and White 2013)

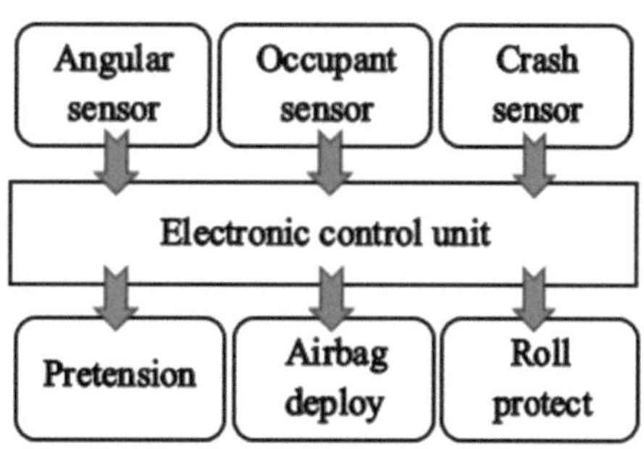

12.5.1 Crash Sensing Systems

Controlling the release of airbags in case of accidents is one of the widest application areas of MEMS sensors in the automotive industry. MEMS accelerometers are being used in the airbag control chip as acceleration sensors which eliminates the g-switches used previously. The g-switches are expensive which increases cost when distributed throughout the vehicle and also has lesser reliability. An accelerometer measures the acceleration and deceleration of the vehicle continuously. It is connected to a microcontroller which computes the acceleration or deceleration integral and checks with the predetermined threshold to show whether a large change has occurred or not. If the value is larger then the airbags get deployed. All g-switches can be replaced by a single MEMS module in the control terminal for airbags since these accelerometers can perform continuous (analog) measurements. Due to this, the reliability increases and the overall cost decreases. Another advantage is that these systems can perform robust self-testing with very high-reliability which enables the processor module to determine if the data received is reliable from the sensor or whether servicing is required. Higher G-range and very compact size are the need for this application of sensors (Weinberg 2002).

One of the key applications is for activating passenger side airbags, which have very low decision-making time due to the proximity of the side body to the passenger. Due to this, two to four MEMS accelerometers are deployed on the sides and connected to the main console for better results. To determine the impact of frontal crash front looking crash sensors can also be added just behind the front bumper, which compares the acceleration signatures with the main central console and help in modulating the rate of airbag inflation according to deceleration of the vehicle.

12.5.2 Rollover Detection

These sensors and systems are used to receive the roll angle and roll rate of the vehicle as to determine whether it is tipping or rolling over especially for vehicles with a higher center of gravity such as trucks, vans, and sports utility vehicles.

Rollover detecting systems use the gyroscope to read roll rates while an accelerometer reads the vertical acceleration (Z-axis) to determine whether rolling over will happen or not. A second accelerometer (in Y-direction) can also be added to measure lateral acceleration while moving or sliding sideways which when obstructed can also cause a rollover. This task cannot be performed by side crash detection accelerometers since the acceleration magnitude when sliding sideways is greater than the 100 g range, which is typically close to the noise floor used for impact detection on the sides. Dual-axis low G-range accelerometers are used for the reading these accelerations. The system fires the airbags from the sides if rollover happens (Weinberg 2002). Although the resolution of systems required is not very high, they should have excellent vibration and external shock resistance since the system becomes useless if it provides unreliable output just after a shock or crash event.

12.5.3 Vehicle Dynamic Control

Vehicle dynamic control (VDC) are systems which help in controlling automobiles in the event of skidding. These systems are also called as Electronic Stability Control (ESC), Electronic Stability Program (ESP), Dynamic Stability Control (DSC), Vehicle Stability Assist (VSA), or Vehicle Stability Control (VSC). The schematic of such a system of vehicle control is shown in Fig. 12.6.

The VDC system consists of a very low-g accelerometer, which determines the sliding, gyroscope which is used to compare the yaw or turn rate and the sensors for wheel speed at each of the wheels to determine the speed of the vehicle. If any lateral sliding or change in yaw rate is detected then torque reduction and single wheel braking are used to make the vehicle get back in line (Weinberg 2002). Introduction of MEMS-based gyros and accelerometers have made VDC practical as it has reduced overall cost and increased the life of the system for more than 10 years operation which is a need for any automobile nowadays. Figure 12.7 shows the architecture of an inertial sensor with negative spring and active control for enhancing the sensitivity of these inertial sensors (Handtmann et al. 2002).

A complete VDC system consists of the important sensors (Kraft and White 2013) like steering wheel angle sensor to helps in determining the direction in

Fig. 12.6 Architecture of stability control system

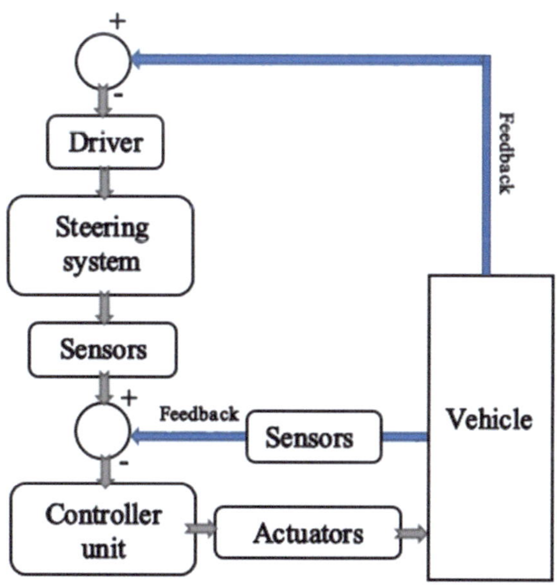

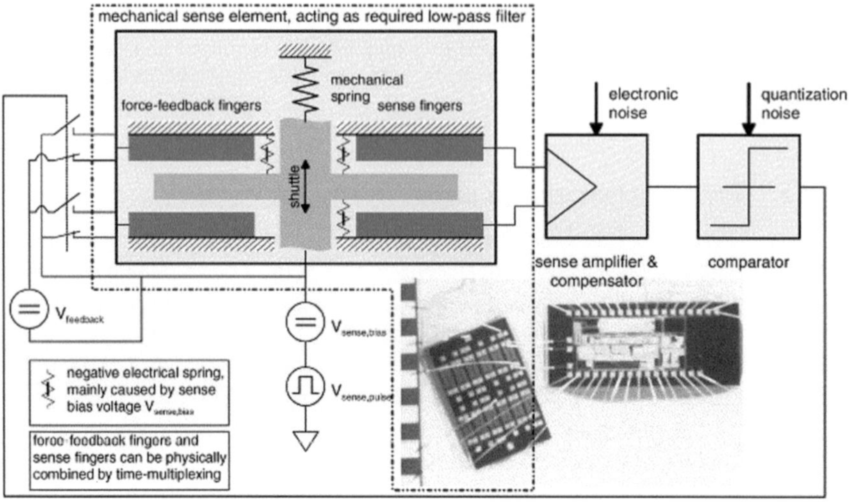

Fig. 12.7 Micromechanical Σ/Δ-architecture with negative springs and photograph of wire-bonded two-die system (Reprinted with permission from (Handtmann et al. 2002) Copyright (2002) Elsevier.)

which the vehicle is moving and is often based on Hall effect sensors; yaw rate sensors to measure the rate of rotation of car along the vertical axis and compares it with the data from a steering wheel angle sensor so as to determine if any corrective actions are required to avoid skidding; lateral acceleration sensor, and wheel speed sensor to measure wheel speed and their basic application is to assist in keeping control of vehicle by restricting speed.

12.6 Gas Sensors for Emitted Exhaust Gases

The major cause of environmental pollution is the generation of various harmful gases by rapidly growing automotive and industrial sector. The environment is getting polluted everyday by toxic gases such as: carbon dioxide (CO_2) (Kim et al. 2000), carbon monoxide (CO) (Li et al. 2009), nitrous oxide (NO_x) (Guo et al. 2008), various hydrocarbons (Herberger et al. 2010), explosive gases such as hydrogen (H_2) (Chauhan and Bhattacharya 2017) and other volatile compounds, which are released by industries and vehicles exhaust. The primary pollutants which are released into the atmosphere, either chemically transformed into secondary pollutants (by photochemical reaction or oxidation) or transported to different areas by the wind. Some of the pollutants get condensed and get deposited underground but most of the pollutants remain suspended into the air and affect human, animal, and aquatic life.

There are various metal–oxide-based semiconductors (MOSs), which are used as sensing materials and these are ZnO, V_2O_5, TiO_2, etc. (Chauhan and Rai 2017; Chauhan and Bhattacharya 2018). The main gases which are harmful to the environment and are emitted as the exhaust gases are discussed in detail.

12.6.1 Carbon Monoxide (CO)

Carbon monoxide (CO) is a colorless gas, which irreversibly binds to the iron center of hemoglobin and restricts the adsorption of oxygen. Hence, prolonged exposure to CO gas may cause a headache, nausea, dizziness, and sometimes death. The source of generation of CO gas is a by-product of poorly combusted fuels such as petrol, methane, oil, etc. Thermistor-type metal–oxide semiconductors are used for detection of CO gas, which detects the change in temperature due to the reaction of CO gas molecules with oxide surface. In the past, SnO_2 nanowires have been fabricated for the sensing of C_2H_5OH, CO, and H_2 gas (Chauhan and Bhattacharya 2017). Li et al. studied different phases of titanium oxide (TiO_2) films such as anatase and rutile. The conversion temperature from anatase to the rutile phase is around 600 °C which enables the improvement in the performance of the sensor (Li et al. 2009).

12.6.2 Carbon Dioxide (CO_2)

CO_2 is a most significant greenhouse gas, which absorbs infrared radiation and causes vibration of molecules. This phenomenon results in the passage of heat to the surrounding and increases the atmospheric temperature. Prolonged inhalation ion of CO_2 may cause heavy breathing, increased pulse rate, headache, dizziness, sweating, loss of consciousness, etc. The increase in the concentration of CO_2 in the atmosphere is caused by human activities such as combustion of carbon-containing fuels with oxygen (O_2). Hence, the detection of CO_2 gas for indoor and outdoor conditions becomes essential to monitor and control the concentration up to the safe limit. The sensitive detection of CO_2 was obtained by using the oxide of lanthanum as sensing material. Kim et al. synthesized different compositions of lanthanum oxide by using heat treatment at different temperatures and the CO_2 gas was detected for a wide range of gas concentrations (Kim et al. 2000).

12.6.3 Nitrous Oxides (NO_x)

Nitrous oxides (NO_x) comprises a group of gases such as nitric oxide (NO), nitrogen dioxide (NO_2), nitrous oxide (N_2O) etc. These gases are formed by the

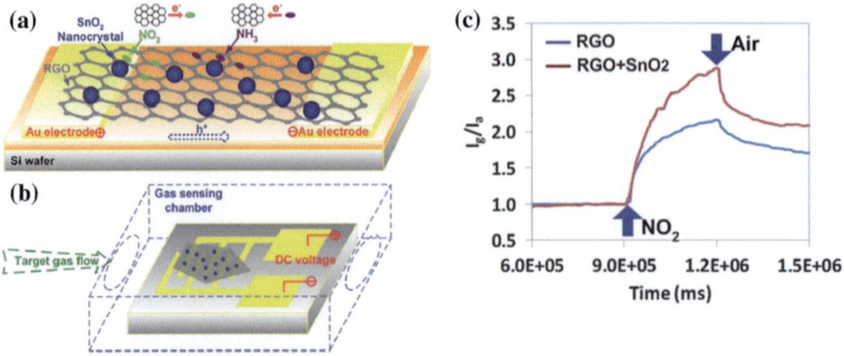

Fig. 12.8 **a** Schematic diagram representing graphene/tin oxide (SnO_2)-based NO_2 sensor, and **b** testing system respectively; **c** sensor response through direct current measurement (Reprinted with permission from (Mao et al. 2012) Copyright (2012) Royal Society of Chemistry.)

combustion of different fuels and a combination of nitrogen (N_2) and oxygen (O_2) gas molecules. Among these gases, nitric oxide (NO) is responsible for creating photochemical smog and forms a cloud after reacting with hydrocarbons and oxygen gas. Photochemical smog is irritating to the eyes and harmful to the plants. Similarly, NO_2 is harmful to the lungs and respiratory system after prolonged exposure. Hence, detection of these gases becomes essential, especially near industrial areas and vehicles exhaust system. Guo et al. have reported the synthesis of CdO nanowires through a hydrothermal method (Guo et al. 2008). Nitrogen oxide, one of the major air pollutants has been detected using the fabricated nanowires. The developed porous nanowires attribute to rapid response, higher sensitivity, high signal-to-noise ratio, and higher selectivity toward nitrogen oxide. Figure 12.8 represents a schematic diagram representing grapheme/tin oxide (SnO_2)-based NO_2 sensor, testing systems used in the sensing and corresponding sensor response through direct current measurement (Mao et al. 2012).

12.7 Conclusions

In conclusion, this chapter discusses various sensors/sensor-based systems which are employed in automotive. The important sensors which are generally used are pressure sensors, speed sensors, position sensors, temperature sensors, and mass air flow sensors. These sensors are used individually or in a combination with another sensor depending on the needed sensing system. The important sensing systems which are devised for the proper functioning of automotive are tire pressure monitoring systems, pressure and flow sensors for engine management, vehicle stability/passenger safety systems and sensing of emitted gases through automobiles. These systems are discussed along with their functioning and requirement aspects. Some basic MEMS fabrication techniques are also discussed which are generally used for fabricating MEMS-based devices.

References

Arai N, Sekine Y, Nishimura Y et al (1990) Advance design for a bypass type of hot-wire air flow sensor. In: SAE International congress & exposition

Baney W, Chilcott D, Huang X et al (1997) A comparison between micromachined piezoresistive and capacitive pressure sensors. In: SAE international truck and bus meeting

Basu AK, Sarkar H, Bhattacharya S (2016) Fabrication and resilience measurement of thin aluminium cantilevers using scanning probe microscopy. In: Foundations and frontiers in computer, communication and electrical engineering—proceedings of the 3rd international conference on foundations and frontiers in computer, communication and electrical engineering, C2E2, 2016

Bhatt G, Kant R, Mishra K et al (2017) Impact of surface roughness on dielectrophoretically assisted concentration of microorganisms over PCB based platforms. Biomed Microdev. https://doi.org/10.1007/s10544-017-0172-5

Bhatt G, Kumar S, Sundriyal P et al (2016) Microfluidics overview. In: Microfluidics for biologists: fundamentals and applications. Springer, pp 33–83

Carter J (1996) Thermistor control of engine coolant systems. Sens Mag 65–68

Chauhan PS, Bhattacharya S (2017) Vanadium pentoxide nanostructures for sensitive detection of hydrogen gas at room temperature. J Energy Environ Sustain 2:69–74

Chauhan PS, Bhattacharya S (2018) Highly sensitive V2O5·1.6H2O nanostructures for sensing of helium gas at room temperature. Mater Lett 217:83–87. https://doi.org/10.1016/j.matlet.2018.01.056

Chauhan PS, Rai A (2017) Enhanced photocatalytic performance of vertically grown ZnO nanorods decorated with metals (Al, Ag, Au, and Au–Pd) for degradation of industrial dye. Mater Res Express 4:055004

Chou SY (1996) Nanoimprint lithography. J Vac Sci Technol B Microelectron Nanom Struct 14:4129. https://doi.org/10.1116/1.588605

Dietmayer K (1999) New concepts for contactless angle measurement in automotive applications. In: SAE international congress & exposition

Dlugos D (1998) Wiegand effect sensors theory and applications. Sens Mag 32–34

Dunbar M, Sager K (2000) A novel, media-compatible pressure sensor for automotive applications. Sens Mag

Eddy DS, Sparks DR (1998) Application of MEMS technology in automotive sensors and actuators. Proc IEEE 86:1747–1755. https://doi.org/10.1109/5.704280

Fleming WJ (2008) New automotive sensors—a review. IEEE Sens J 8:1900–1921. https://doi.org/10.1109/JSEN.2008.2006452

Fleming WJ (2001) Overview of automotive sensors. IEEE Sens J 1:296–308. https://doi.org/10.1109/7361.983469

Frodl R, Schoedlbauer D (1998) Modern concepts for low cost, high performance position sensors. In: SAE international congress & exposition, p 980

Guo Z, Li M, Liu J (2008) Highly porous CdO nanowires: preparation based on hydroxy- and carbonate-containing cadmium compound precursor nanowires, gas sensing and optical properties. Nanotechnology 19:245611. https://doi.org/10.1088/0957-4484/19/24/245611

Handtmann M, Aigner R, Meckes A, Wachutka GKM (2002) Sensitivity enhancement of MEMS inertial sensors using negative springs and active control. Sens Actuators, A Phys 97–98:153–160. https://doi.org/10.1016/S0924-4247(01)00818-4

Heinzelmann S, Pechhold F, Marto A (2006) Pressure gauge glow plug
Hellman AN, Rau KR, Yoon HH et al (2007) Laser-induced mixing in microfluidic channels. Anal Chem 79:4484–4492. https://doi.org/10.1021/ac070081i
Herberger S, Herold M, Ulmer H et al (2010) Detection of human effluents by a MOS gas sensor in correlation to VOC quantification by GC/MS. Build Environ 45:2430–2439. https://doi.org/10.1016/j.buildenv.2010.05.005
Kato N, Ikoma N, Nishikawa S (1996) Exhaust gas temperature sensor for OBD-II catalyst monitoring. In: SAE international congress & exposition
Kim DH, Yoon JY, Park HC, Kim KH (2000) CO2-sensing characteristics of SnO2 thick film by coating lanthanum oxide. Sens Actuators, B Chem 62:61–66. https://doi.org/10.1016/S0925-4005(99)00305-6
Konzelmann U, Hecht H, Lembke M (1995) Breakthrough in reverse flow detection—a new mass air flow meter using micro silicon technology. In: SAE international congress & exposition
Kraft M, White NM (2013) MEMS for automotive and aerospace applications. Elsevier
Kumar A, Gupta A, Kant R et al (2013) Optimization of laser machining process for the preparation of photomasks, and its application to microsystems fabrication. J Micro/Nanolithog. MEMS, MOEMS 12:041203. https://doi.org/10.1117/1.JMM.12.4.041203
Lacanette K (1997) Using IC temperature sensors to protect electronic systems. Sens Mag 28–34
Li X, Ramasamy R, Dutta PK (2009) Study of the resistance behavior of anatase and rutile thick films towards carbon monoxide and oxygen at high temperatures and possibilities for sensing applications. Sens Actuators, B Chem 143:308–315. https://doi.org/10.1016/j.snb.2009.09.021
Lickfelt BK, Arai H, Strock N (2013) Localization of tire for TPMS and smart entry system 1–8
Löhndorf M, Kvisterøy T, Westby E, Halvorsen E (2007) Evaluation of energy harvesting concepts for tire pressure monitoring systems. Int Work Micro Nanotechnol Power Gener Energy Convers Appl 331–334
Madni A, Wells R (2000) An advanced steering wheel sensor. Sen Mag 28–40
Mao S, Cui S, Lu G et al (2012) Tuning gas-sensing properties of reduced graphene oxide using tin oxide nanocrystals. J Mater Chem 22:11009–11013. https://doi.org/10.1039/c2jm30378g
Pawlak A, Adams J, Shirai T (1991) Novel variable reluctance sensors. In: SAE international congress & exposition
Quinto-Su PA, Lai HH, Yoon HH et al (2008) Examination of laser microbeam cell lysis in a PDMS microfluidic channel using time-resolved imaging. Lab Chip 8:408–414. https://doi.org/10.1039/b715708h
Sabetti A, Bishop R, Charboneau T, Wiecek T (1988) Automotive pressure transducer for underhood applications. In: SAE international congress & exposition
Schuster J, Czarnocki W (1997) Automotive pressure sensors: evolution of a micromachined sensor application. In: SAE international truck and bus meeting
Shaw M, Ziglioli F, Combi C, Baldo L (2008) Package design of pressure sensors for high volume consumer applications. In: Proceedings—electronic components and technology conference. pp 834–840
Shaw ML (2002) Considerations to improve battery life in direct tire pressure monitoring. In: SAE international congress & exposition. pp 580–589
Singh RK, Kumar A, Kant R et al (2014) Design and fabrication of 3-dimensional helical structures in polydimethylsiloxane for flow control applications. Microsyst Technol 20:101–111. https://doi.org/10.1007/s00542-013-1738-7
Stewart WD, Boudaoud I, Molyneaux R et al (2006) Determination of wheel sensor position using a wireless solution
Sundriyal P, Bhattacharya S (2017) Inkjet-printed electrodes on A4 paper substrates for low-cost, disposable, and flexible asymmetric supercapacitors. ACS Appl Mater Interfaces. https://doi.org/10.1021/acsami.7b11262
Treutler CPO (2001) Magnetic sensors for automotive applications. Sens Actuators, Phys. https://doi.org/10.1016/s0924-4247(01)00621-5
Weinberg H (2002) MEMS sensors are driving the automotive industry. Sens Mag 1–9

Whitesides GM, Ostuni E, Takayama S et al (2001) Soft lithography in biology and biochemistry. Annu Rev Biomed Eng 3:335–373. https://doi.org/10.1146/annurev.bioeng.3.1.335

Woerth A, Winner H, Bosch R (1999) Apparatus for determining rotational position of a rotatable element without contacting it

Yulong Z, Libo Z, Zhuangde J (2003) A novel high temperature pressure sensor on the basis of SOI layers. Sens Actuators, Phys 108–111

Chapter 13
Sensors in the Joining and Welding Process in Automobile Manufacturing

Mohit Pandey, Shreyansh Tatiya, Shantanu Bhattacharya and Shailendra Singh

Abstract Welding and joining processes are one of the key processes in the automobile industries, which give the core strength to the joint and consequently to the vehicle. The complex design of the vehicle and small parts creates difficulty for the manual joining process and therefore the need for automation has increased. The manufacturing companies spend a lot of money on the equipment and processes to maintain the high quality of the vehicle. The sensors are integrated along with the equipment in the welding and joining processes to improve the quality and decrease the risk of damage associated with the process. This review starts with different types of welding methods and then focusses on the sensors and its types.

Keywords Welding and joining · Automobile manufacturing sensors
Sensors for welding and joining · Sensors and safety in the welding process

13.1 Introduction

Today, we are living in a world of hybrid machines in which our aim is to maximize the functionality without compromising the strength. To increase the functionality, many subsections are integrated together. This integration requires a large amount

---

M. Pandey (✉) · S. Tatiya · S. Bhattacharya
Micro Systems Fabrication Laboratory, Department of Mechanical Engineering,
Indian Institute of Technology Kanpur, Kanpur 208016, UP, India
e-mail: pmohit@iitk.ac.in

S. Tatiya
e-mail: tshrey@iitk.ac.in

S. Bhattacharya
e-mail: bhattacs@iitk.ac.in

M. Pandey · S. Tatiya · S. Bhattacharya · S. Singh
Design Programme, Indian Institute of Technology Kanpur, Kanpur 208016, UP, India

S. Singh
Maruti Suzuki India Limited, Gurugram 122015, Haryana, India

S. Bhattacharya et al. (eds.), *Sensors for Automotive and Aerospace Applications*, Energy, Environment, and Sustainability,
https://doi.org/10.1007/978-981-13-3290-6_13

of joints and therefore, the need for joining process comes into the picture. Welding is a commonly used process for joining metal parts in many industries. Some of the industries practice welding process via robotic and automated system, while some industries consider it as an art and are hugely dependent upon the skills and training of their labors. Although, due to complex shape and sizes of the different parts and safety of the labor in the harsh environment, many industries have shifted to hybrid operation in which some machineries will be assisting them for the joining and welding process. These machineries are equipped with many sensors and actuators that can endure harsh conditions and are suitable for complex designs. These sensors have a very high precision to withstand the six-sigma quality check for any industry.

Talking about the automobile industries, welding is one of the commonly applied processes for joining various small parts in the assembly. There are more than about 4500 weld points on an average steel vehicle, which truly reflects the need for highly precise sensors. Different parts have different functions and thus have different properties and requirement (Groover and Kolchin 1997; Automation-insights.blog).

13.2 Different Types of Welding

The automobile industries include the following types of welding methods.

13.2.1 Resistance Spot Welding

In this method, the joint is produced, at a spot between the two parts, by the heat generated due to the resistance of the workpiece. Two electrodes are present between them. The joints are overlapped and current is passed through the electrodes. Due to the high resistance of the workpiece, a high amount of heat is produced. Moreover, high pressure is applied through both the electrodes so that the metal fuses at the intersection and joint so formed becomes strong. As the name suggests, the weld is limited to the spot only and is not continuous (Devarasiddappa 2014) (Fig. 13.1).

13.2.2 Resistance Seam Welding

Unlike the spot welding, in this method, the joint is produced not in the form of the spot but along the length in any form of curve required. Due to the continuous weld, the joints so formed are leak proof. This welding may be made with the continuous or overlapping workpiece. Although instead of having rod-shaped electrodes, these

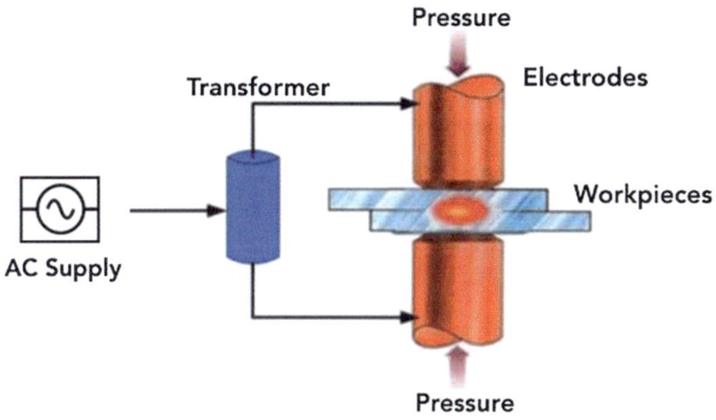

Fig. 13.1 Resistance spot welding

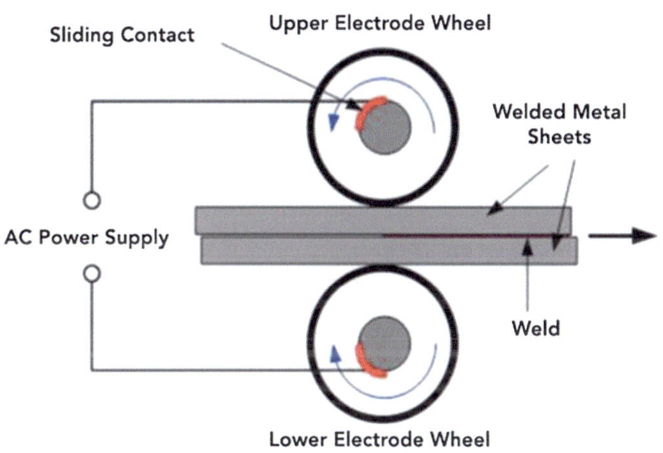

Fig. 13.2 Resistance seam welding

techniques have wheel shape electrodes which limit the use of it due to component shape and wheel access (Devarasiddappa 2014) (Fig. 13.2).

13.2.3 Laser Beam Welding

The advanced laser technology for the industrial welding process has gained the popularity because of having its distinct advantages like flexibility, economical, improved productivity, etc. This method uses the heat generated by a focused laser beam which strikes on the joint. In the automotive industries metal sheets, having a thickness ranging from 0.2 to 6 mm are welded easily (Jack and Jeter 1994; Devarasiddappa 2014) (Fig. 13.3).

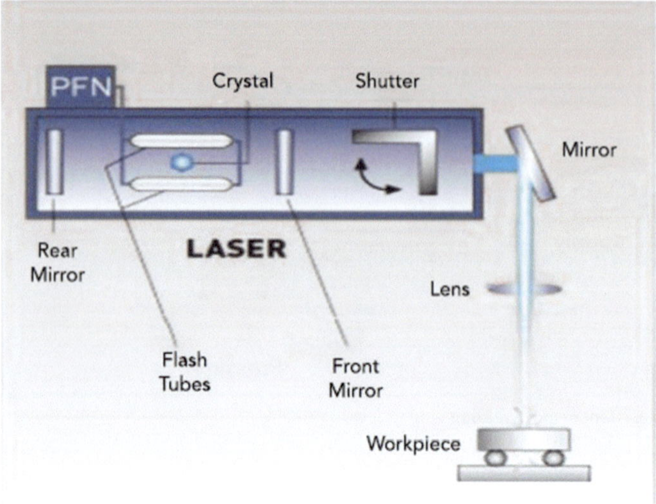

Fig. 13.3 Laser beam welding

13.2.4 Friction Welding

This method is a solid-state welding in which the joint is produced without any heating through electricity but by just the application of pressure, due to which there is no significant melting of any of the work parts. In this method, there is a sliding-motion-induced mechanically between the parts to be welded, which produces the required amount of heat. One part is rotated against the other part mechanically and due to the friction between the parts, heat is generated to weld. When the temperature exceeds the threshold level, the pressure is applied and the parts weld together (Dawes and Thomas 1996; Devarasiddappa 2014) (Fig. 13.4).

Fig. 13.4 Friction welding

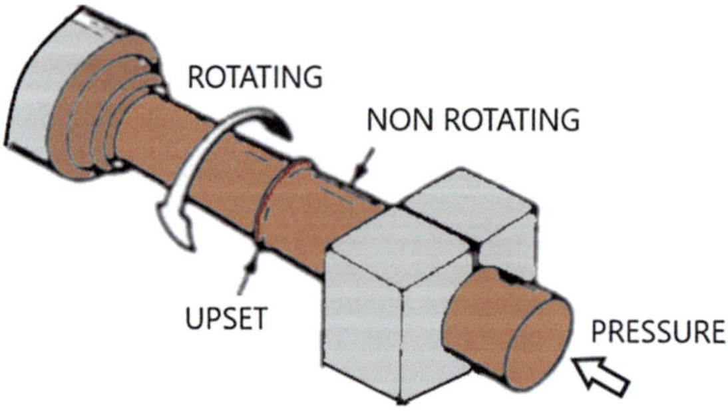

13.2.5 Magnetic Pulse Welding

In this method, dissimilar metals like aluminium and steel are attached without the use of additional metal at the weld joint. The welding process takes pre-shaped aluminium and steel tubular stock as input and subject them to high pressure with the help of precision-machined die cavities. A rapidly switching magnetic field causes one of the metallic component with sufficient velocity to impact the other stationary part. This impact at high velocity generates the required amount of force to create the magnetic pulse weld (Ben-Artzy et al. 2010; Durand 2001; Devarasiddappa 2014) (Fig. 13.5).

These different types of welding are used according to the required joint strength and functionality. But for the manufacturing to be consistent with the six sigma quality check, the welding process should be very robust. This robust process requires sensors which can sense the various parameters affecting the process, which will help us to improve and remains in the working region. Also, to smoothen the overall process and to ensure the safety of worker and material quality, various sensors are integrated into the process.

So, let us discuss these sensors in the order in which they are actually used. The joining and welding process for the automobile industry is divided into six sections. Various parts of the sections are assembled in that particular section. The order of the sections are as follows (Fig. 13.6).

The raw sheet metal is pressed and various components are formed which are sent for the welding and joining process. The welding process starts with the underbody of the frame. Starting with the front underbody and moving to rear

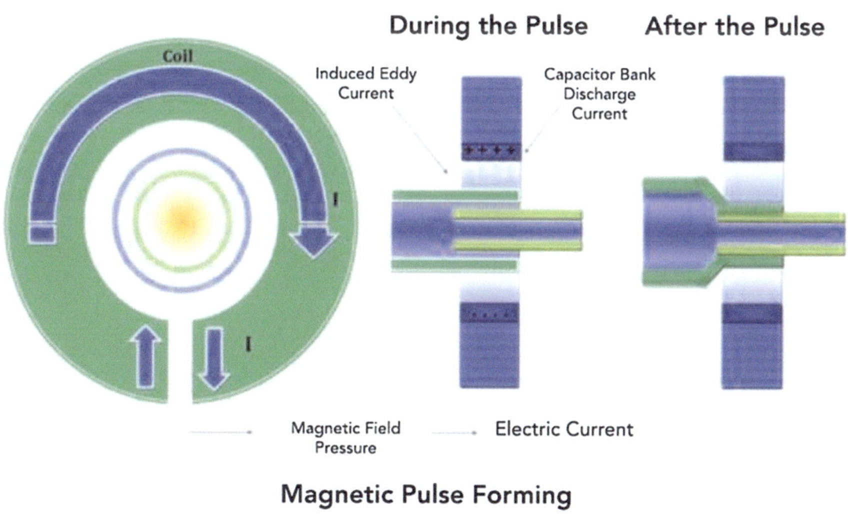

Fig. 13.5 Magnetic pulse welding

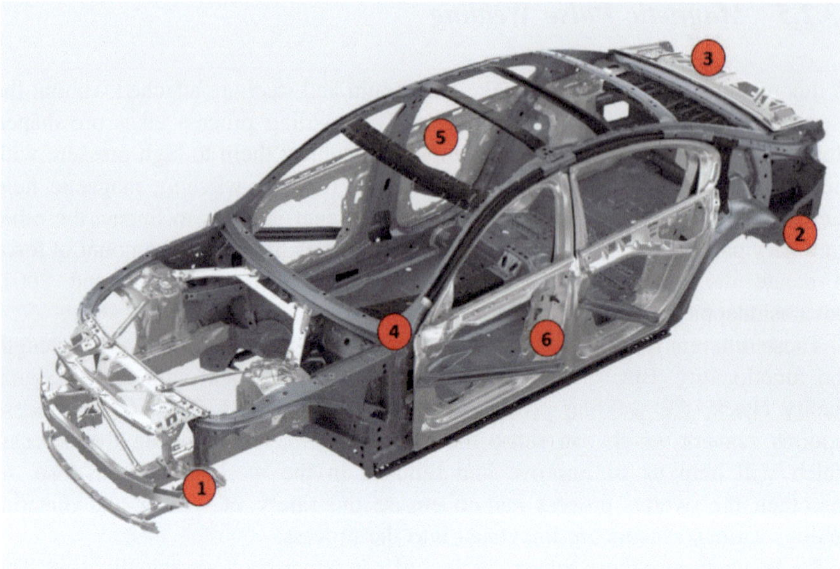

Fig. 13.6 Sequence for the welding process

underbody, the whole underbody structure welding is completed. After the completion of underbody welding, rear body and main body structures are welded. Once this is done, the roof is welded on the frame. The welding process ends with the welding of the remaining body components to form the whole structure. After this welding process, the vehicle is sent to the paint shop and hence all the components are welded before it.

The input resource for the welding process is the raw electricity, which is then transferred to the desired current and voltage combination according to the need of the weld via a transformer. This transformer is converting the 440 V input electricity into the desired output combination. This electricity is then fed to the two electrodes for the resistive spot welding. These electrodes, along with the different parts to be joined are pressed against each other with the flow of electricity. Due to this flow of electricity, the metal at the joint is heated and due to the pressure applied with the electrodes, one part is fused with the other to complete the welding process.

13.3 Different Types of Sensors Used in Welding and Joining Processes

The overall welding process is automated and thus the use of sensors and feedback system is very important. The different type of sensors that are used in the above processes for the welding are:

Fig. 13.7 Voltage sensor (Reproduced from product catalog with permission from Tifomy.)

13.3.1 Voltage Sensor

The main function of the voltage sensor is to detect and measure the AC and/or DC voltage levels. The transformer receives the current at 440 V, which is then converted to the desired value of the output. This output voltage is continuously monitored by the voltage sensor. The reading of the output voltage also works as the feedback for the transformer functioning. In the voltage sensors, the principle of measurement is the voltage divider. Mainly two types of sensors are available: Capacitive-type voltage sensor and resistive-type voltage sensors (Product catalog by Tifomy; IFM electronics) (Fig. 13.7).

13.3.2 Current Sensor

This device senses the amount of current flowing through the electrodes as the output of the transformer. The monitored value of current can be used as the feedback to the transformer as well as used for the simulation of the welding process. The simulation depends on the heat generated at the joint which is related to the amount of the current flowing. Thus, with the help of this sensor, we tune our transformer as well as our simulation for better results. The current sensors are divided into two types—direct and indirect sensors. The principle for the direct sensor is Ohms Law and Hall Effect. It has an integrated coil, which on placing around the current carrying conductor induces voltage which is proportional to the current in the system. Then by using various amplifiers such as shunt, it is easy to detect and measure. The indirect sensor works on the Ampere's and Faraday's Law.

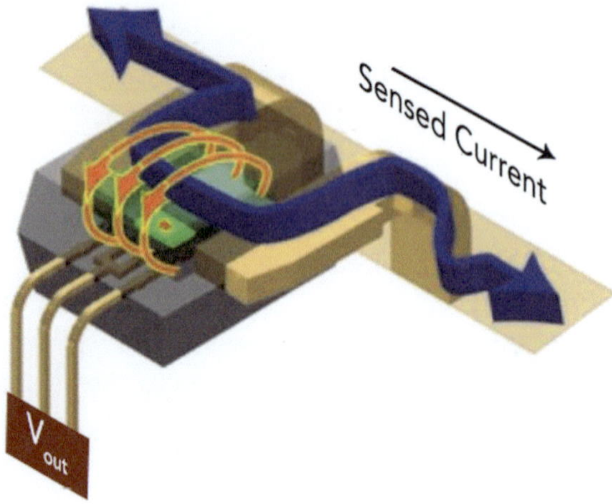

Fig. 13.8 Current sensor working

It measures the magnetic field around the current carrying conductor and consequently, detects and measures the current in the system. The indirect current sensor is used for high current applications, while for lower range direct type is used (Product Catalog by Allegro micro; an Introductory guide to sensor basics by Keyence) (Fig. 13.8).

13.3.3 Pressure Sensor

This sensor is very important for the welding process as only passing electricity will not join the two metal parts. The pressure fixes the parts to each other and for this reason, the right amount of pressure value is required for calculating the input force. Real-time monitoring of the force is done through this sensor, in which the amount of pressure exerted can be monitored. Also, for the vacuum suction grippers, the pressure monitoring is very important. The designing of a robust piezoelectric sensor has been a major challenge for the sensor designers. Generally, MEMS-based sensors are used for measuring the pressure. The principle behind the pressure sensor is the piezoelectric effect. When there is a force exerted on the piezoelectric active material, it generates electricity when is directly proportional to the amount of pressure exerted (Introductory guide to sensor basics by Keyence) (Fig. 13.9).

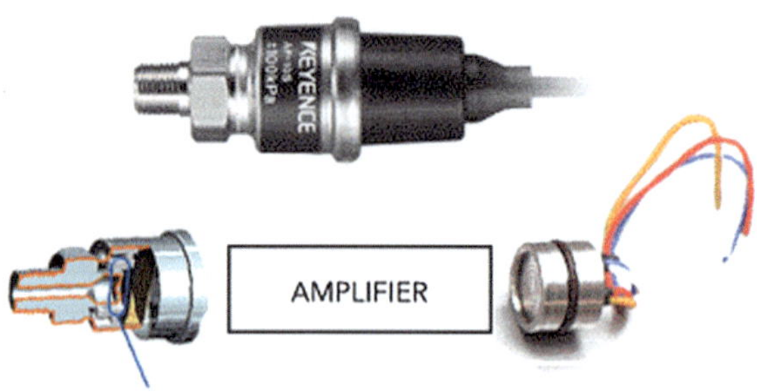

Fig. 13.9 Pressure sensor (Reproduced from product catalog with permission from Keyence.)

13.3.4 Thickness Sensor

The thickness sensor plays a crucial role in the joining process of automobile industries by detecting and monitoring the thickness of the two different parts. This thickness of the parts will ensure the smooth and efficient joint and will ensure the overall strength. This sensor incorporates the use of sound energy and use the property of reflection of sound from boundaries between dissimilar materials. From this device, we can measure the thickness of the sample when at least one face is exposed. Thickness Sensor precisely measures the time taken by the reflected wave from the far end of the workpiece, generated by a small probe called an ultrasonic transducer. The piezoelectric element is present in the transducer and the receiver. Thus, to and fro conversion of sound energy into electricity occurs and with the difference in the electrical signal we can monitor the thickness of the part (Introductory guide to sensor basics by Keyence; Product catalog by Omron) (Fig. 13.10).

13.3.5 Position Sensor

The position of the components is tracked by these sensors throughout the conveyer. The basic principle behind the functioning of this sensor is the change in the reflection of light. The change in the reflection of the light is detected and monitored by the photoelectric sensor. A light source radiates light to the target object via lens which on reflection is received back by the receiver. The distance between the source and receiving end determines the position of the object. Less the distance between source and receiver end, far is the target and vice versa. This sensor plays a very crucial role in the fully automatized manufacturing system in which the position of the part is very important for any application. The sensor can be calibrated to give absolute and/or relative position according to the need of the system (Product catalog by Omron; Product catalog by Omega) (Fig. 13.11).

Fig. 13.10 Thickness sensor (Reproduced from product catalog with permission from Keyence.)

Fig. 13.11 Position sensor (Reproduced from product catalog with permission from Omron.)

13.3.6 PICA–PICA Sensor

This sensor interfaces with the control system of the process and helps the operator for the next task. This sensor reduces the production time and increases production quality. The sequence of all the tasks are already feed in the control system and once any task is complete the light indicator tells the operator which of the next job is to be done and in which sequence. These types of sensor help the operator in reducing the error when multiple sequences are present and the next operation depends upon the previous operations. These sensor interfaces take the input from various operations and compute the next operation with the existing data. The accuracy of the sensor depends not only on the already existing sequence, but also on the various inputs from the different sensors and their integration by the control system. These sensors are very helpful in manually operated stock maintenance too (IFM Electronics; Introductory guide to sensor basics by Keyence) (Fig. 13.12).

13.3.7 1D/2D Code Reader

These types of sensors are used to mark the individual component/sub-assembly with some other component/sub-assembly and/or vehicle. This sensor works on the principle of different absorption intensity of light by a different color. The relevant information is coded in 1D or 2D pattern and is printed on the tag. This tag is then sensed by the code reader. It contains a light-emitting source and light-sensing source which detects and analyzes the reflected light from the code. Due to the difference in absorption of light, the same pattern is formed in the light intensity. This light intensity pattern is then decoded and the final information is extracted out of the signal (Product Catalog by Docooler) (Fig. 13.13).

Fig. 13.12 PICA–PICA sensor application

Fig. 13.13 1D/2D code reader

13.3.8 Capacitive and Inductive Sensor

The functioning of these types of sensors is found in the part identification and part positioning. These sensors are used, when the precise location of the fixture or component is needed for any job. Although the working of both the groups are completely different, it can be used for the same functioning. Inductive sensor has the conductor coil mounted on a PCB and the current is changed in the coil due to AC supply. When any component comes closer to this coil, the inductance changes

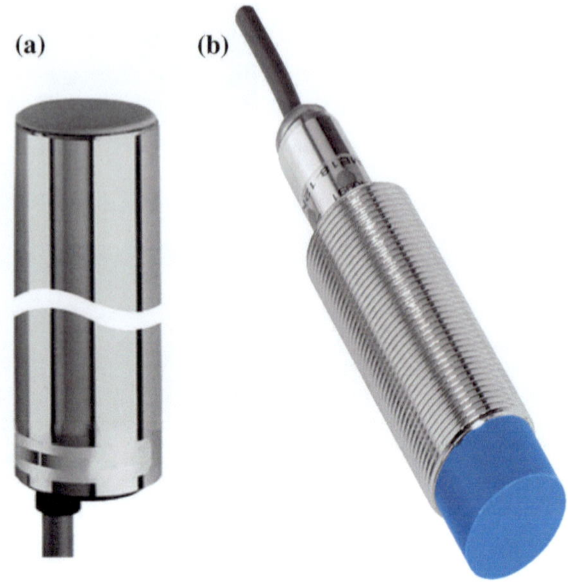

Fig. 13.14 **a** Capacitive sensor, **b** inductive sensor

which deduces the presence of the component. Through this change, the exact location of the component can also be found out. Similarly, the capacitive sensor monitors the change in the capacitance value. An alternative electric field caused due to the AC current results in a change in the charge collected due to the presence of electric field (Product Catalog by Baumer) (Fig. 13.14).

13.3.9 *Laser Sensor*

As the name suggests, this sensor consists of a laser source and a receiver, which can be used in various applications. The use of laser sensor as the restricted curtain is very common. The laser light emitted is constantly received by the receiver after reflection and thus, forming a curtain. Whenever any object crosses this section, the continuous receiver signal broke and shows the presence of entry/exit from the restricted area. These types of sensors also provide feedback to the central control system to modulate or terminate the ongoing process whenever there is the crossing of laser curtain. Another application of the laser sensor is the collision protection between the bots/components. Whenever there is a breaking of the continuous laser signal, the speed of the bot is modulated accordingly to prevent any collision. Sometimes laser sensors are also used for the measurement purpose (Product Catalog by Omron; Introductory guide to sensor basics by Keyence) (Fig. 13.15).

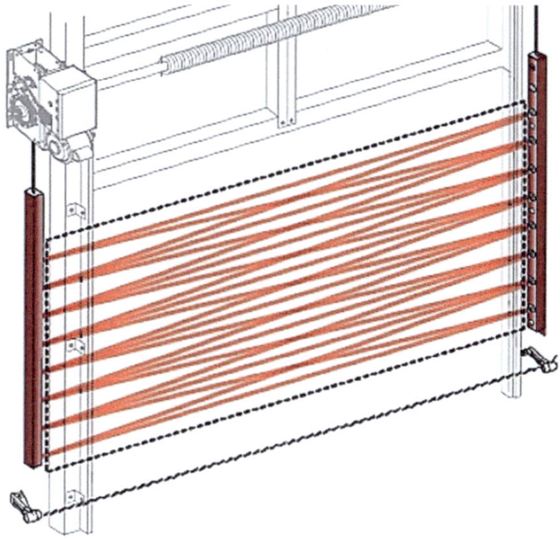

Fig. 13.15 Laser sensor

13.3.10 Safety Sensor

These types of sensors are used for the safety of the workers and the safety of the materials and ongoing process. These sensors are used in various ways according to the need of the process like for any unwanted fluctuation in the current, kill switches are designed which incorporate the current sensor with the switch while for any restricted or hazardous zone, the optical light sensor is used. The optical sensor uses a laser light source and a receiver, which receives the light travelling in a straight line (IFM Electronics). Different types of safety sensors incorporated in automobile manufacturing are as follows:

3.10.1 Optical Safety Sensor This sensor detects the presence of operators near any hazardous machine. This sensor includes laser light for the detection.

3.10.2 Safety Door Switch—This switch detects the mechanical guards (basically covers) are closed or not. These sensors may be mechanical or noncontact detection or actuated with manual key possessed by an operator.

3.10.3 Safety Limit Switch—This sensor checks that all the parameters are within the operational limit or not.

3.10.4 Emergency Stop Switch—These switches are integrated with different sensors in such a manner that whenever there is the malfunctioning of any operation then the operation will stop and thus, will prevent any situation compromising the safety and quality.

13.4 Potential Failures in Welding and Joining Due to Sensor Failure

Sensors are very helpful for the welding and joining process for the automobile industries. All the workflow is maintained smooth, error-free and robust due to the functioning of sensors but if the sensors are not functioning or giving improper readings then this can be very dangerous. Faulty voltage sensor is way too risky as the operating voltage range is high and can harm the costly equipment. Also, the welding process will not be proper due to which the strength of the overall vehicle is compromised. Any malfunctioning in the current sensor can cause serious damage to the components as the lower range woill not be able to melt the metal and higher one will harm the component and the chances of fire will be very high. The pressure sensor used for ensuring proper pressure while welding can cause the welding to be improper. Sheet metal processing and joining would be compromised and will be weak if the thickness is not the pre-defined thickness due to the monitoring of faulty thickness sensor. The components will not fit in the fixtures with the wrong data from the position sensor. A lot of confusion will be created from the malfunctioning of the PICA–PICA sensor. Seats and door will always be mismatched when the tag sensor will not work properly. Also, the restricted area will not be restricted anymore if the laser sensor does not respond. There will not be any safety for anyone. The workers, components, and sub-assemblies may be damaged and the plant may be shut down due to the malfunctioning of the safety sensors.

13.5 Environment and Health Issues

Welding being one of the important processes for the automobile industries comes with its own set of drawbacks. These drawbacks are directly or indirectly associated with the growth in the production of the industries, environmental problems and health issues of the workers associated with that atmosphere (Chang et al. 2015; Springer-Verlag 2003). During the joining process, a large amount of heat is produced which causes the melting of filler metal, base metal and its coating. This melting along with the gases formed releases the small, solid particles in the air which is commonly known as welding fume (Hariri 2012). This fume is a typical combination of many toxic gases like nitric oxide, carbon monoxide, nitrogen dioxide, etc. with many other oxides of metals which are very dangerous and have the potential hazards for inhalation (Antonini 2003; American Welding Society (Confined spaces, safety and health fact sheet no 11) 1995; American Welding Society (Fumes and Gas Fact sheet) 2005). The particle size may vary from 0.5 mm up to 2 mm, which on inhalation or dermal exposure is responsible for many health diseases like chronic respiratory disorder, damage to skin and eye, etc. (Golbabaei et al. 2012; Gonser and Hogan 2011; Fard and Fard 2016). Most of the fume particles have the initial size less than a micron in diameter but tend to increase due to agglomeration with an increase in time (Ashby 2002 and Mansouri et al. 2008). The toxic gases in the fumes can also cause asphyxiation, headache, drowsiness, pulmonary edema, etc. The concentration of the gases and PMs in the fumes easily crosses the permissible limit for a worker working in the workshop and thus, it is very dangerous and risky to work in such an environment (Sandström 1995). The vulnerability of the joining process for the workers and environment thus depends on the following factors (Blunt and Balchin 2000 and Hewitt 2001):

(1) Type of welding process
(2) Electrode material and presence of shielding gas
(3) Material being welded
(4) Presence of coat on the material
(5) Voltage and current type used
(6) Workbench design and type of ventilation of workshop.

To maintain the output quality, many input parameters are fixed according to the requirements and in spite of resulting in high pollution or wastage remains unchanged by many industries. In such a situation, workbench designs like downdraft worktable and ventilation of the workshop become very crucial (Kumar and Mistry 2015 and AWS, Confined Spaces, Safety and Health Fact Sheet). Powerful exhaust equipment can be installed which will work in accordance with the chemical sensor to check and maintain the concentration of the contaminants below the permissible limit.

13.6 Conclusion

The welding and joining processes play an important role in the automobile manufacturing and in which the sensors play an important role. A high-strength quality with the safety of the workers can only be achieved by automation and proper use of sensors. Different types of welding processes are described in the first section out of which resistance spot welding is used majorly. The sequence of the welding process is very important for the smooth working on the vehicle frame. The sensors required in the process and the importance of their functioning are discussed. Further integration of sensor data with the central control system can improve the automation of the welding process and without sensors there will always be a risk.

References

American Welding Society (2005) Fumes and gases, safety and health fact sheet no 1. Miami, FL
American Welding Society (AWS) (1995) Confined spaces, safety and health fact sheet no 11
Antonini JM (2003) Health effects of welding. Crit Rev Toxicol 33:61–103

Ashby HS (2002) Welding fumes in the workplace: preventing potential health problems through proactive controls, professional safety, pp 55–60, Apr 2002
Ben-Artzy A et al (2010) Wave formation mechanism in magnetic pulse welding. Int J Impact Eng
Blunt and Balchin (2000) Health and safety in welding and allied processes. Woodhead Publishing Limited, Cambridge
Chang YJ et al (2015) Environmental and social life cycle assessment of welding technologies. Elsevier
Dawes CJ, Thomas WM (1996) Friction stir process welds aluminum alloys. Weld J 75:41–45
Devarasiddappa D (2014) Automotive applications of welding technology—a study. Int J Modern Eng
Durand B (2001) Magnetic pulse welding process to decrease vehicle weight and increase fuel efficiency. Research and data for status report, 95-02-0055, pp 271–274
Fard MH, Fard MH (2016) Impacts of welding on environmental problems and health and solutions to overcome these problems. In: Proceedings of 41st IASTEM international conference
Golbabaei F et al (2012) Exposure to metal fumes among confined spaces welders. Ital J Occup Environ Hyg 3(4):196–202
Gonser M, Hogan T (eds) (2011) Arc welding health effects, fume formation mechanisms, and characterization methods. InTech
Groover MP, Kolchin MG (1997) Case study: automobile final assembly plant
Hariri A (2012) Preliminary measurement of welding fumes in automotive plants. Int J Environ Sci Dev 3(2):146–151
Hewitt PJ (2001) Strategies for risk assessment and control in welding: challenges for developing countries. Ann Occup Hyg 45(4):295–298
Introductory guide to sensor basics by Keyence Corporation
Jack AH, Jeter W (1994) Laser welding of P/M for automotive applications. In: Proceedings of SAE international congress and exposition, Detroit, Michigan, USA, 28 Feb–3 Mar 1994
Kumar P, Mistry J (2015) Impact of welding process on environment and health. Int J Adv Res Mech Eng Technol (IJARMET)
Mansouri et al (2008) Gravimetric and analytical evaluation of welding fume in an automobile part manufacturing factory. J Res Health Sci 8:1–8
Processes in the automotive industry, IFM electronics
Product catalog by Allegromicro, Allegromicro
Product catalog by Baumer, Baumer Technology
Product catalog by Docooler, Docooler
Product catalog by Omega, Omega
Product catalog by Omron Automation and Safety, Omron
Product catalog by Tifomy, Tifomy
Sandström T (1995) Respiratory effects of air pollutants: experimental studies in humans. Eur Respir J 8:976–995
Springer-Verlag (2003) The environmental management of fabrication by welding, welding in the world, vol 44, pp 42–51
https://automation-insights.blog

Chapter 14
Sensors Used in Automotive Paint Shops

Shreyansh Tatiya, Mohit Pandey, Shantanu Bhattacharya and Shailendra Singh

Abstract Paint job can be considered as the heart of the automobile manufacturing process from the customer's point of view. Automobile manufacturing companies spend a hefty amount on the process at times even more than the body itself. Paint process in an automobile is thus a rigorous and time-consuming process involving: pre-treatment, electrodeposition, sealing, priming, coating and then finally inspection. In a typical car manufacturing plant, more than 1000 cars are processed each day. This high quantum and superior quality standards have forced manufacturers to invest and maintain automated modern production line. Sensors and robots are thus employed at each step of the manufacturing process. This review focuses on the sensors used in the paint job, type, and their importance at each step.

Keywords Paint shop process · Automobile painting · Sensors in coating Paint process sensors

---

S. Tatiya (✉) · M. Pandey · S. Bhattacharya
Microsystems Fabrication Laboratory, Department of Mechanical Engineering, Indian Institute of Technology Kanpur, Kanpur 208016, UP, India
e-mail: tshrey@iitk.ac.in

M. Pandey
e-mail: pmohit@iitk.ac.in

S. Bhattacharya
e-mail: bhattacs@iitk.ac.in

S. Tatiya · M. Pandey · S. Bhattacharya · S. Singh
Design Programme, Indian Institute of Technology Kanpur, Kanpur 208016, UP, India

S. Singh
Maruti Suzuki India Limited, Gurugram 122015, Haryana, India

S. Bhattacharya et al. (eds.), *Sensors for Automotive and Aerospace Applications*, Energy, Environment, and Sustainability,
https://doi.org/10.1007/978-981-13-3290-6_14

14.1 Introduction

Paint coating protects the car's body and additionally enhances visual appeal. If the car is not aesthetically pleasing, the customer may not even get inside to check the technical specifications. Some even say the car's paint colour reflects a lot about the owner's personality. During the inception of automotive industry around a century back; cars were painted manually for several rounds with a varnish-like product. A car had to spend roughly 40 days from the first coat till the time car was ready to sell in the paint shop. But, the first breakthrough came in the year 1923 when E.I. DuPont De Nemours utilised spray guns in his nitrocellulose lacquer systems to speed up the process (Khanna 2008). Automation in the spray guns was only possible in the late 80s and it reduced the need of craftsmanship and even provided a uniform coat to both flat and curved surfaces (Liu et al. 2000).

A person often misinterprets the paint as just the outer colour of their car, but in reality, that is just the tip of the iceberg. Manufacturers have to think about so many other factors like corrosion, durability, finish, texture, UV resistance. Till date, 70% of the total energy cost of the manufacturing process is due to the painting operation and thus it contributes to approximately half the total production cost (Galitsky et al. 2003). Therefore, the vehicle is passed through stringent quality checks even before the start of the painting process.

14.2 Paint Shop Process

Modern paint job application in the automobile industry consists of five major steps. They are

i. *Pre-treatment*: The vehicle body is generally passed through high-pressure water jets that clean off foreign materials like dust, oil, and other debris from previous processes. The vehicle is then dipped in series of alkali filled dip tanks that degreases the metal work. A conditional spray pump filled with de-ionised water removes excess alkali prior to the phosphate bath (Smith 2002). To increase adhesion of paint to the metal the vehicle is submerged in a tank filled with phosphate liquid. Proper pre-treatment ensures a smooth and uniform paint finish.
ii. *Electrodeposition*: Body from the phosphate tank is moved to the electrodeposition chamber also called as E-coat. DC current is applied within the tank filled with electro-coat paint and the body is submerged. E-coat is followed by cleaning spray operation and drying that removes excess liquid from the surface.
iii. *Sealing*: After this, the body is moved to the masking/sealing stage that is considered to be the beginning of the finishing plant. Poly Vinyl Chloride (PVC) is generally used as a sealer between the seams and under the bonnet. Sealing eliminates water leaks and minimises chipping.

iv. *Priming*: Colour of the vehicle is decided at this stage and a layer of primer is applied either manually or through a robot between the surface and basecoat. Priming also imparts an even surface for subsequent layers of paint.
v. *Coatings*: Post-oven operation from the priming the vehicle is passed through the final stage of the paint job. A double coating of a basecoat followed by a clearcoat is applied usually using atomizing bells. Topcoats offer all the visible surface properties like colour, smoothness, gloss (Akafuah et al. 2016). Finally, the vehicle is dried in the oven and sent to the trim/assembly area where it undergoes inspection for defects.

14.3 Sensors Utilised for the Paint Job

A perfectly painted car surface is a hallmark of each car out of the production line. Achieving this level of perfection requires rigorous rounds of inspection and high precision sensors at each stage of the manufacturing. Paint job being the most energy and cost-intensive process of the automotive plant is also highly prone to defects. These defects can arise at any stage and even the most minute ones should be detected and repaired if possible.

Pre-treatment is a 13 stage long process that strengthens the vehicle for painting. It starts with a hot water rinse and an industrial water rinse having pressure and temperature sensor. Emulsified soap like liquid is sprayed onto the body in the process called as spray degreasing. It additionally has sensors to measure the volume of liquid flowing into the tank to avoid its overflow. Flowmeters are employed at various points to maintain liquid flow. Degreasing is not a stagnant process and the liquid is constantly circulated through the help of pressure pumps in the tank and there are sensors to maintain its pressure and temperature. Logarithmic sensor maps the temperature variation in a day and reports data with the help of a plotter every minute. Dip degreasing process involves submerging the whole body in a tank with a capacity of around 100 kl. It has the same circulation mechanism and sensors as spray type. Additionally, it has a mechanism for separation of oil from the top layer of the bath. The separated oil is carried to another tank that has a temperature sensor, the level sensor indicating the amount of oil and flow sensors to maintain a uniform flow. Surface coating stage has sensors to measure the physical and chemical properties like weight, normality, molarity of the chemicals used. It also has pressure sensors for maintaining an optimum pressure range. Impurities from the surface coating stage are removed in the industrial water spray which is followed by a de-ionised water rinse. It serves two functions. First, it removes unreacted chemicals from the surface and second, it prevents the transfer of chemicals from one stage to other. Quality of the rinse water is also vital for effective and contamination free cleaning. Therefore, apart from the temperature and pressure sensor DI rinse also has a sensor to measure the conductivity of the inlet water which is maintained at less than 2–3 μs. Additionally, pre-treatment also has industrial water spray (conductivity around 100 μs) at pre phosphating stage and a DI spray (conductivity less than 4 μs) at post phosphating stage. Iron phosphate systems are used for parts that form the interior of the vehicle. Zinc phosphate system, on the other hand, is applied to the parts that are exposed to harsh corrosive environments. It consists of a surface conditioner that is constantly supplied using metered pumps.

Electrodeposition is a process to produce metallic coatings in situ by an action of current on a conductive material that is dipped in aqueous dispersed pigments that need to be deposited (Basile et al. 2010). Coatings could be applied either via anodic or cathodic deposition. In the paint job application cathodic electrodeposition (CED) is favoured over anodic electrodeposition (AED). High performance and increased corrosion resistance are highly desired properties and that is provided by CED. It involves dipping the metal which acts as a cathode in the electro-coat paint. The metal surface being cathode/negatively charged attracts positively charged paint particles from the bath. As metal acts as a cathode, a negligible amount of soluble iron enters the paint bath. Resultant surface is thus durable and corrosion free. In the automotive industry, CED takes place in a tank of approximately 150,000 L. It has sensors to measure the flow of current, conductivity and level of the electro-coat paint. Vehicle passes through a series of ultrafiltration (UF) units that separate uncoated paint from the body. UF has a temperature, pressure, level, normality, molarity sensors. The concentration of the E-coat paint drops in subsequent UF units. Paint collected from the first three units is taken to a UF tank that has a centrifuge system fitted with sensors to measure rotational speed and to detect any imbalance. Separated paint is collected and rechanneled back to the CED chamber so that paint utilisation is more than 99% at all times. The vehicle is coated with an anti-corrosion ED coat in the final tank and all excess liquid is drained out. After ED coating the body is heated in an oven that has a plotter type sensor to report temperature changes every second. The oven used here is of "camelback" type that maintains a baking window of temperature and time in various zones so that there is no sharp temperature gradient. Convection type of heating requires maintaining the air flow and quality inside the oven.

Intermediate coat chamber comprises of a long platform with filtered air flowing from top and robots surrounding the conveyor. Electrostatic painting is carried out in the chamber with the help of the robots. Painting takes place from both the direction due to the presence of a magnetic field and an electrostatic wraparound effect is created. It sprays paint uniformly in all directions. Chamber comprises of sensors to measure current, magnetic field of electrostatic painting, temperature, airflow and quality, water flow sensors at the bottom and sensors to maintain relative humidity that helps in ensuring the viscosity of the paint. Topcoat painting process, on the other hand, has an additional clear coat for metallic paints. Sensors though are similar to that of an intermediate coat. Robots used in both the processes have sensors to measure the speed of the robot, current flow, proximity with the body, corrosion levels, etc. Finally, the vehicle comes to the manual inspection that does not have many sensors. One important sensor utilised at this stage is a lux/light intensity sensor that is used to match the desired and obtained the colour of the

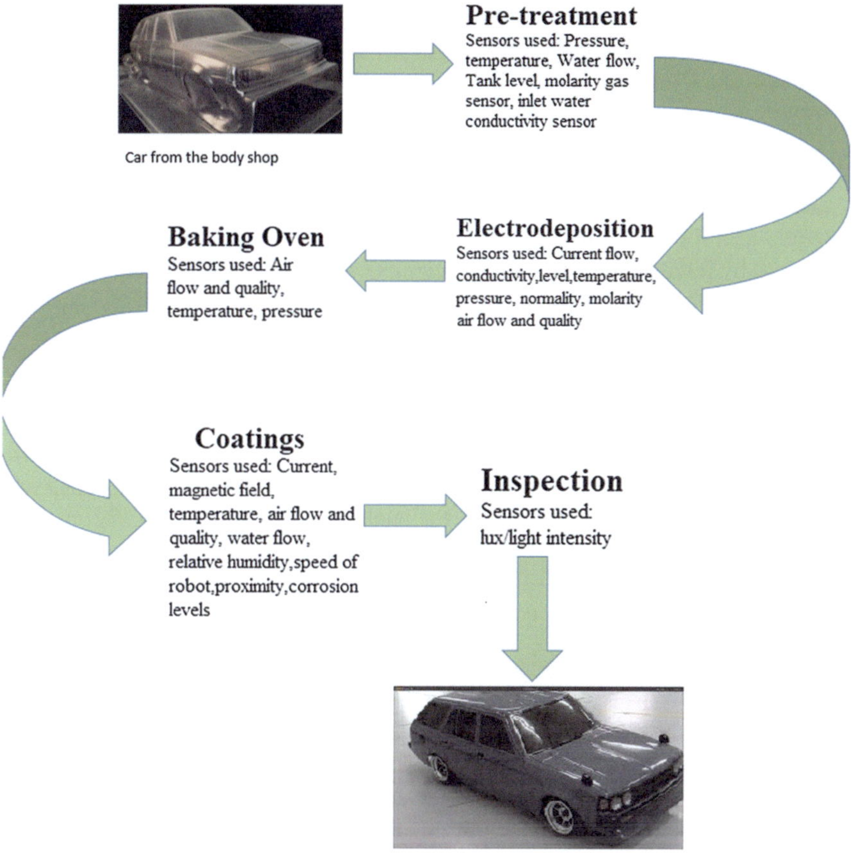

Fig. 14.1 Flow diagram indicating the major processes of the paint shop with associated sensors

vehicle. Figure 14.1 summarises all the major processes with sensors being used in the paint shop.

14.4 Major Classes of Sensors (Processes IFM)

- **Temperature sensor**—System's current state of operation is known through various types of sensors; temperature sensor provides one such key parameter to the electronic module (El-Awar et al. 1999). Comprehension and analysis of the output from the temperature sensor can help to optimise the efficiency and also control emissions from every process. Thermistor-based temperature sensor is preferred over conventional Bipolar Junction Transistor (BJT) type sensors. Temperature sensors work on the principle that change in temperature results in a linear change in resistance. In a posistor/PTC resistance increases with temperature, while in a thermistor/NTC resistance decreases with increase in temperature.

- **Hydraulic Pressure sensor**—Designing a reliable and robust piezoelectric sensor for the automotive industry has been a major challenge for sensor designers. Automobile manufacturing generally utilises MEMS-based sensors for calculation of manifold air pressure (MAP). Pressure sensors are used to ascertain the level of any fluid, detect any system leak and also in a spray-type equipment for paint transfer. Microprocessors detect deflection of the diaphragm caused by force applied to a sensing element of constant area and is converted to an electrical output. Pressure is calculated by:

$$\text{Pressure} = \text{Force}/\text{Area}$$

- **Tank level sensor**—Capacitive level instruments work on the principle of variation in capacitance and it is formed by sensors, tank wall, and dielectric material. Capacitance or storage capacity is directly proportional to plate area (A), their distance apart (d) and the dielectric constant (ε) of the material between the plates.

$$C = \varepsilon A/d$$

- **Position sensors**—Conveyer throughout the manufacturing process utilise photoelectric sensors to detect the position of the body on the skid. A photoelectric sensor utilises light as a medium and captures the change in reflection due to the position of the object. The sensor at the conveyor is based on background suppression and it works on triangulation. Light is transmitted through a lens in a straight line towards the target. The target reflects light back to the receivers at some angle. Sensor and target distance determine the angle in which light is reflected back to the receiver. The greater angle of reflection is obtained by closer objects and vice versa.
- **Flow sensors**—Measuring the flow of both liquids and gases with high precision, fast response time and low-pressure drop is essential for most of the processes of the paint job. Therefore, rather than the conventional mechanical flow metres an ultrasonic type flow metre is used in the painting process. Ultrasonic flow sensors use sound waves to detect the flow of liquid in a pipe. When no liquid is flowing in the pipe waves transmitted into the pipe and reflected from the liquid is same but during flow, due to Doppler effect, a change in reflected wave is observed. A signal from the transmitted and its reflected wave are processed by the transmitter to determine the flow rate.
- **Humidity sensor**—Hygrometer or humidity sensor measures the relative humidity of the surrounding air. Relative humidity is the ratio of actual moisture to the maximum moisture that can be held at a particular temperature. Hotter the air more is the moisture content it can hold.
- **Light sensor**—Robots used in the process of manufacturing have an inbuilt light sensor to measure the ambient light level of surroundings. A photoresistor

or a light-dependent resistor (LDR) is used for light detection. The resistance of an LDR changes with a change in light intensity as they are made from high-resistance semiconductor material like Cadmium sulphide that shows photoconductivity. These sensors are used for colour matching, inspection of the final painted car and also identification of defects.

- **Air Quality sensor**—Circulation of clean air in all the vents, exhausts and applications that require air drying is a must for a smooth and even finish of the paint layers. Air quality sensor comprises of an infrared LED source and a photodetector to measure the scattering of light from airborne particles. A low voltage pulse is generated every time a particle is detected. The concentration of the particulate matter in the air is directly proportional to low pulse occupancy percentage or the fraction of time the output signal is low.

14.5 Potential Failures in Paint Shop Due to Sensor Failure

A faulty or non-functional sensor if not addressed timely can cost huge loss in revenue or even bring the whole paint shop to halt. Incorrect reading on the most widely used sensors like temperature and pressure can leave spots/dirt from the body shop in the pre-treatment stage, improper mixing of chemicals in CED and of paint in the coating stage, it may even lead to ineffective drying in ovens. Faulty flowmeters and water level sensors can cause spillage and overflow from tanks, inadequate flow requirements through pipes, and other spray nozzles. Sensors determining the chemical and physical composition of liquids and gases directly affect the quality of paint layers thus must be monitored at all times. Sensors used in robots and conveyors are important for smooth movement of the automobile body, maintaining adequate distance and for deposition of uniform and thorough coatings. Malfunctioning in the relative humidity sensor can alter the viscosity of the paint at the base/clear coat stage. Finally, lux sensors play a very crucial role for high precision colour matching especially while repainting and during final inspections.

14.6 Conclusion

Usage of a plethora of sensors and automation in the paint shop for automobile manufacturing is inevitable. A perfect high-quality finish without any abrasions and a uniform protective layer against challenges like UV, dirt, and corrosion can only be ensured by automation. In the first section of the chapter, all the major paint shop processes for the automotive industry have been described briefly. The second section of the chapter deals with sensors and its application with all the subprocesses. Each sensor used is then described in detail with its components and working principle. Finally, the last section is dedicated to the effect of malfunctioning or complete loss of function of each sensor on the quality of painting and level of finish, gloss, or shine. Various sensors used are described in details in this chapter. The focus and importance of paint shop will not fade in the future. However, car companies may reduce the environmental burden specifically use of volatile organic compounds (VOC's) in the finishes they offer. Other than that, technological innovations to perform the work faster including faster drying, thinner but effective coats, better quality of adhesives will hallmark the future paint shop. Increase in visual appeal is evolving at a very rapid pace and technologies like colour-changing reflectants, high chromatic pigments, ground glass, use of matte clears, etc., will revolutionise the aesthetics of automobile manufacturing and sales.

References

Akafuah N, Poozesh S, Salaimeh A et al (2016) Evolution of the automotive body coating process—a review. Coatings 6:24. https://doi.org/10.3390/coatings6020024

Basile F, Benito P, Fornasari G, Monti M, Scavetta E, Tonelli D, Vaccari A (2010) A novel electrochemical route for the catalytic coating of metallic supports. Elsevier Masson SAS

El-Awar NY, Geer DJ, Krellner TJ, Straub PJ (1999) Automotive temperature sensing. 15857:1–8

Galitsky C, Worrell E, Berkeley L (2003) ENERGY STAR® Guides for energy efficiency opportunities, featuring the motor vehicle assembly industry, pp 118–129

Khanna AS (2008) High performance organic coatings

Liu Y, Sparer J, Woskie SR, Cullen MR, Chung JS, Holm CT, Redlich CA (2000) Qualitative assessment of isocyanate skin exposure in auto body shops: a pilot study. Am J Ind Med 37:265–274. https://doi.org/10.1002/(SICI)1097-0274(200003)37:3%3c265:AID-AJIM4%3e3.0.CO;2-O

Processes IFM Processes in the automotive industry

Smith R (2002) A guide to seals in the automotive industry, pre-treatment and paint plants

MIX
Papier aus verantwortungsvollen Quellen
Paper from responsible sources
FSC® C105338
www.fsc.org

If you have any concerns about our products,
you can contact us on
ProductSafety@springernature.com

In case Publisher is established outside the EU,
the EU authorized representative is:
Springer Nature Customer Service Center GmbH
Europaplatz 3, 69115 Heidelberg, Germany

Printed by Libri Plureos GmbH
in Hamburg, Germany